Ultrawideband Short-Pulse Radio Systems

For a listing of recent titles in the
Artech House *Antennas and Electromagnetics Analysis Library*,
turn to the back of this book.

Ultrawideband Short-Pulse Radio Systems

V. I. Koshelev
Yu. I. Buyanov
V. P. Belichenko

ARTECH
HOUSE

BOSTON | LONDON
artechhouse.com

Library of Congress Cataloging-in-Publication Data
A catalog record for this book is available from the U.S. Library of Congress

British Library Cataloguing in Publication Data
A catalog record for this book is available from the British Library.

ISBN-13: 978-1-63081-105-1

Cover design by John Gomes

© 2017 Artech House
685 Canton Street
Norwood, MA

10 9 8 7 6 5 4 3 2 1

Contents

Preface

The studies of ultrawideband (UWB) pulse radio systems date back to the pioneering experiments by Hertz (1887–1888). Since then the intensity of the research was closely related to new challenges faced the researchers. At the current stage, counted since the early 1990s of the last century, emphasis is given to the development of different types of high-power and low-power sources of UWB electromagnetic pulses and to their applications. Considerable work in this field is being performed at the Institute of High Current Electronics, SB, RAS in collaboration with specialists of the Radiophysics Department of Tomsk State University.

Over the past 20 years, the joint team, including the authors of this book, has gained a deeper insight into the physics of essentially nonstationary processes. A wealth of experience has been accumulated in research and development of transmitting and receiving antennas, antenna arrays, and high-power UWB radiation sources. Rather extensive studies have been related to the radar of objects and to the susceptibility of their electronic systems to short electromagnetic pulses. Therefore, the authors originally intended to restrict themselves to outlining and summarizing the original results obtained by the team.

However, following the logic of presentation of the results, they have come to realize that the current status of research on short-pulse UWB radio systems should be covered more broadly. Considerable attention is paid to the methods for solving boundary value problems related to the radiation and reception of UWB pulses, their propagation in media and channels, and their scattering by conductive and dielectric objects. In addition, the authors briefly review the application of UWB technologies in real-time location, Internet of Things, intelligent transport systems, MIMO radar and communications, in industry, in biomedical and biological areas, as well as for emergency braking of automobiles and for finding people trapped under rubble. An account is given of the development of dual-polarization radars intended for probing objects and media, with emphasis on those using the scanning of wave beams of orthogonally polarized radiation pulses.

We consider it our pleasant duty to express our sincere gratitude to the colleagues and coauthors of a number of results for the fruitful cooperation at different stages of the research work.

We also express our appreciations to T. K. Cherkashina for the intricate and arduous job of translating the book into English and to V. V. Plisko and E. V. Balzovsky for their help in preparing the manuscript.

CHAPTER 1

Introduction to Ultrawideband, Short-pulse Radio Systems

An ultrawideband (UWB) pulse radio system generally comprises not only a transmitter and a receiver incorporating a data processing system, but also channels (media) for the propagation of UWB radiation pulses, including scattering objects. The term ultrawideband electromagnetic pulse radiation applies to a radiation whose bandwidth is comparable to the center frequency.

To classify UWB radio systems, several defining approaches were proposed. We consider only two of them. The first definition [1] is oriented to the development of communication and radar systems. It is based on estimating the fractional bandwidth of a radiation pulse, which is defined by the equation

$$\eta = 2\frac{f_H - f_L}{f_H + f_L} \tag{1.1}$$

where f_H and f_L are, respectively, the higher and the lower boundary frequency of the pulse spectrum at a level of −10 dB. Here, the term UWB radiation refers to a radiation of fractional bandwidth $\eta \geq 0.2$ and frequency bandwidth ≥ 500 MHz. Subsequently, a radiation with the fractional bandwidth $\eta < 0.01$, will be treated as narrowband and a radiation with $0.01 \leq \eta < 0.2$ as wideband.

The second approach [2] to classifying UWB radiation is based on the use of ratio bandwidth $b = f_H/f_L$, which can be expressed in terms of the fractional bandwidth as follows:

$$b = \frac{(2 + \eta)}{(2 - \eta)} \tag{1.2}$$

In the context of this approach, four spectral bands are differentiated: (1) narrow band ($b < 1.01$), (2) moderate band ($1.01 \leq b \leq 3$), (3) ultramoderate band ($3 < b \leq 10$), and (4) hyperband ($b > 10$). This approach is oriented to the researchers who deal with the problem of the susceptibility of radioelectronic systems to intentional exposure to high-power electromagnetic pulses and to those engaged in studying electromagnetic compatibility (EMC). To narrowband radiation, there corresponds the first spectral band with $b < 1.01$, which is in accordance with the approach above ($\eta < 0.01$), and the other three bands are associated with ultrawideband radiation [3]. Ultrawideband signals are considered to fall within the fourth spectral band with $b > 10$ ($\eta > 1.63$) [4]. According to the first definition, for UWB radiation, we have $b \geq 1.22$. In the following, we will assess radiation pulses using the first approach to radiation classification.

At present, three basic types of UWB signals, which are associated with three types of radio systems, are differentiated:

1. frequency-modulated, phase-modulated, and noise-like (chaotic) signals;
2. trains of radio frequency pulses of different central frequency (time-frequency modulation), and
3. short electromagnetic pulses containing no radio frequency carrier.

In what follows, we will consider mainly the UWB radio systems in which the energy of short electric pulses is directly converted into electromagnetic radiation energy. These systems are intended primarily for studying nonlinear and nonstationary processes in various (including biological) objects and media exposed to UWB radiation. They can also be efficiently used in studying the susceptibility of radioelectronic systems exposed to electromagnetic pulses, as well as in developing up-to-date radars intended, in particular, for object recognition, and pulse communication systems.

By UWB antennas, we will mean radiators that ensure undistorted transmission of a signal. Physically, this corresponds to the requirement that the antenna should have a stable radiation phase center in the frequency range of a given UWB signal. This means that wide-range (frequency-independent) antennas [5], such as spiral and log-periodic ones, are not pertinent to ultrawideband antennas, as they distort the waveforms of radiated UWB pulses. In the subsequent consideration of radio systems, we will analyze in detail the UWB antennas radiating and receiving short electromagnetic pulses.

The major attention will be given to high-power UWB pulse radio systems. This is motivated that by an increase in energy of a short electromagnetic pulse, which is necessary for increasing the operational range of a radio system, is possible only due to an increase in the peak power of the pulse. The energy of a UWB radiation pulse is distributed in space according to the pattern of the radiator. For various applications, it is necessary to know the peak electric field strength of the pulse, E_p, in a far-field region, at a distance r away from the antenna in the main beam direction. In this connection, the key parameter used to characterize both high-power and low-power UWB radio systems is the effective potential of radiation defined as the product rE_p. By high-power radio systems, we mean systems capable of radiating pulses of peak power 100 MW and more.

1.1   History of the Development of Ultrawideband Radio Systems

The history of UWB radio systems [6–9] began with experiments by Hertz (1887–1888), who discovered a means of generating electromagnetic waves with the use of a transmitter consisting of a spark generator of damped electrical oscillations and an antenna (subsequently called a Hertzian electric dipole) and proved experimentally the existence of electromagnetic waves predicted theoretically by J. C. Maxwell (1865). To detect electromagnetic oscillations, Hertz initially used a loop antenna with a spark indicator, and in experiments with a mirror parabolic antenna radiating electromagnetic pulses, he also used an electric dipole. The receiver–transmitter system was tuned to a resonance frequency that was estimated beforehand by the formula for oscillations in an inductor-capacitor (LC) circuit.

In these experiments, broadband electromagnetic pulses (the bandwidth was determined by the number of damped oscillation periods) were radiated, and the receiving antenna cut out the narrowband radiation at the resonance frequency. This was dictated by the use of resonance receiving antennas and a spark indicator, that is, a spark breakdown in air between two spherical electrodes connected to the ends of a receiving loop or dipole, which was detected visually in darkness. There was yet no means for recording the waveforms of radiated electromagnetic pulses. An important result of these experiments was that the physical concept of the instantaneous activity of a force originated by I. Newton and others was changed by the concept of a short-range interaction proposed by M. Faraday (field lines) and developed in tandem by J. C. Maxwell (finite velocity of propagation of electromagnetic waves).

The next important stage in the development of UWB radio systems, using spark generators of damped electromagnetic oscillations similar to the Hertz generator, lasted from 1895 to 1913. It began with the invention of radio (A. S. Popov and, independently, G. Marconi) and ended with the development of a vacuum-tube oscillator (A. Meissner), which made feasible communication systems using undamped electromagnetic oscillations (narrowband radiation). An important goal at this stage was to increase the transmission range of radio systems. To attain beyond line-of-sight communication, diffraction of radio waves and excitation of electromagnetic oscillations at the resonance frequency of the antenna system were harnessed. This gave birth to large antenna facilities radiating in the frequency range 10–100 kHz. In 1901, Marconi realized long-range (3500 km) radio communication (between Great Britain and Canada) through the Atlantic Ocean. These experiments gave rise to the hypothesis (1902, A. Kennelly and, independently, O. Heaviside) that there is an ionized region in the Earth's atmosphere that reflects electromagnetic waves (ionosphere). This hypothesis was supported subsequently (1924–1925) by direct measurements of the altitudes of the ionospheric layers.

Signals were transmitted using the Morse alphabet (sequences of long and short damped oscillations). To increase the amplitude and decrease the duration of damped oscillations, that is, to increase the radiation power and bandwidth, it was necessary to seek new engineering solutions to ensure efficient operation of spark gaps and antennas. To decrease the duration of damped oscillations, it was proposed to shorten the time of deionization of the plasma in the discharge gap by an external action. Even the first investigations showed that to widen the transmission band of an antenna, it is necessary to increase its area or volume. The main designs of UWB antennas developed as applied to radio communication systems in that period and until the end of the twentieth century are given in the monograph by Schantz [10].

In the same period, a method for detection of metallic objects by using the radio waves reflected from them was proposed (1904, C. Hülsmeyer). The patented setup contained a spark transmitter, a directional antenna, and a receiver of the Branly–Lodge coherer type, identical to that used in the first experiments by A. S. Popov and G. Marconi. Subsequently, radar became one of the major, especially military, applications of radio pulses [11].

A great contribution to the development of high-power UWB radio systems was made by N. Tesla [12, 13]. His basic idea was to realize wireless transmission of energy and information over long distances. In developing this idea, Tesla observed that the detected signals from lightning discharges sometimes became more intense as the thunderstorm moved away. He supposed that a type of standing wave might occur in the resonator during an electrical discharge to the ground. Note that Popov, in his first experiments, also dealt with recording electric fields produced by lightning discharges (the Popov's first device was termed a storm indicator). To realize his idea, Tesla developed (in 1891) a resonant transformer, which is now termed a Tesla transformer, and obtained an output voltage of 1 MV even in the first experiments (1892) and 4 MV subsequently.

For energy transmission over long distances, Tesla intended to tune the transformer into resonance with the terrestrial globe and excite, using a high-voltage discharge to the ground, waves of frequency ~10 Hz propagating along the Earth surface in a cavity resonator by which he mentioned the space surrounding the Earth. Half a century later, Schumann [14, 15] suggested a feasibility of global resonance electromagnetic oscillations excited in the Earth–ionosphere cavity. He calculated the spectrum of fundamental frequencies of the Earth–ionosphere resonator and simultaneously suggested that lightning discharges are natural sources of oscillations of extremely low frequency. Two types of fundamental frequencies were revealed. The fundamental frequencies of the first type can be roughly estimated as follows:

$$f_n \approx 7.5n \text{ Hz}, \qquad n = 1, 2, 3 \tag{1.3}$$

and those of the second type as

$$F_n \approx 300 \, f_n \tag{1.4}$$

Resonances of the first and second types are termed, respectively, low-frequency (Schumann) resonances and high-frequency resonances. Note that the frequencies of Schumann resonances are in agreement with Tesla's estimates.

For the allocation of communication channels (selection of frequency bands), Tesla proposed (1893) to use resonance circuits in the transmitter and receiver, unaware of a similar proposition by W. Crookes (1892). Note that resonance circuits came into use in communication systems long after the first experiments by Popov and Marconi. In 1893, Tesla embarked on the development of wireless systems, having noted that his approach did not use Hertzian waves (now termed electromagnetic waves in free space).

In 1899–1900, Tesla created an experimental setup based on a resonant transformer that excited oscillations of amplitude 12 MV and frequency 100 kHz. The high-voltage end of the transformer winding was connected to a hollow metal ball of diameter 30 in located 142 ft above ground. Long-wave (presumably, kilometer wavelength) electromagnetic oscillations were detected during electrical discharges in the atmosphere. To produce short-wave oscillations, a setup was created that generated an output voltage of 400 kV. The discharge was initiated between the 30-in diameter ball and a grounded metal plate without current limitation. The generated electromagnetic waves (whose frequencies were in the range ~100 MHz, as can be suggested from the dimensions of the radiator) caused insulation breakdowns in the devices of the metering circuit. Furthermore, during an experiment with such a discharge, a short circuit occurred in the dynamo of the feed unit located at a significant distance from the laboratory. Most likely, these were first experiments with high-power electromagnetic pulses that caused damage to objects.

As mentioned earlier, once narrowband signals came into use in the 1920s of the last century, interest in UWB radio systems waned. The situation drastically changed in the 1940s and 1950s. This happened for two reasons. The first one was related to trials of atomic bombs in the USA (1945) and USSR (1949). The trials showed that the electromagnetic pulse (EMP) of a nuclear explosion (nuclear EMP) is a damaging factor. This pulse is ultrawideband. The second reason was related to that the researchers had realized that to improve the range resolution of radars (reduce $\Delta r \sim 1/\Delta f$) and to increase the capacity of a transmission channel, C, it is necessary to use radiation pulses having wide frequency bands Δf. According to Shannon's law [16, 17], we have

$$C = \Delta f \log_2\left(1 + \frac{P_S}{P_N}\right) \tag{1.5}$$

where P_S and P_N are the signal and the noise power, respectively. These two scientific areas developed simultaneously and independently, and only in the 1990s, when a demand was created for studying the action of high-power UWB pulses on large objects in free space and for developing high-power UWB radars, they began to approach one another.

We first discuss the development of UWB systems brought about by the need for testing various objects for susceptibility to nuclear EMP. Differentiation is conventionally made between the EMP of a high-altitude nuclear explosion and that of a surface nuclear explosion, and also between early-time and late-time EMPs. All this is due to the different physical processes involved in the formation of nuclear EMPs [18–20].

For simplicity, we shall consider in the main the early-time EMP of a high-altitude nuclear explosion. The physical mechanism of the conversion of nuclear energy into EMP is the following. Gamma photons interact with air atoms, knocking out fast Compton electrons and photoelectrons from the latter. The electrons moving in one direction with gamma photons (extraneous current) polarize the medium and give rise to a radial electric field. The fast electrons ionize the air, making it conducting. The polarization electric field induces a conduction current (inductive current) in the conducting air. The proportion between the extraneous current and the conduction current determines the resulting field strength. In reality, the extraneous and conduction current spatial distributions are asymmetric, and this gives rise to electromagnetic radiation that propagates from the EMP source over long distances.

The asymmetry of the effects of a high-altitude explosion is determined by locations of geoelectric and geomagnetic fields, vertical atmospheric irregularities, underlying terrain conductivity, and (possibly asymmetric) ionizing radiation flux distribution. For the early-time EMP of a high-altitude nuclear explosion, the following time and frequency characteristics are typical. At a peak field strength of 50 kV/m, the 10–90% rise time and full width at half maximum (FWHM) of the pulse are 2.5 and 25 ns, respectively. This is the high-frequency component of the EMP; the major part of its energy falls within the frequency range 0.1–100 MHz.

The frequency of late-time EMP is low; its duration is longer and the field strength is substantially lower. Because of the wide diversity of nuclear explosion conditions and nuclear component parameters, the EMP characteristics vary over wide limits; they can be found in the cited literature. We only note that the rise time of a super-EMP is 0.5 ns, and the EMP waveform is not necessarily monopolar and can have several lobes [21].

Once the USSR, the USA, and the Great Britain had signed the Limited Nuclear Test Ban Treaty (1963), intense work on developing nuclear EMP simulators began in many countries (the USA, the USSR, the Great Britain, France, China, Germany, Italy, Sweden, the Netherlands, Switzerland, and Israel). The simulators are classified as (1) simulators of a plane wave in free space (for testing rockets and airplanes in flight) and (2) simulators of a plane wave together with the plane wave reflected from the Earth surface (for testing objects on the ground).

Among numerous nuclear EMP simulators, three basic types can be distinguished as follows [22, 23]:

1. *Waveguide or transmission lines.* These are most widespread simulators intended for the simulation of uniform plane transverse electromagnetic (TEM) waves. The input of a line is connected to a generator of high-voltage pulses of a given waveform. The object is placed in the center part of the line, and its end carries a matched resistive load intended to absorb the passed pulse.

2. *Hybrid simulators.* The feature of hybrid simulators is that they combine devices of two types to simulate a high-frequency (early time) and a low-frequency (late time) electromagnetic field. Hybrid simulators can be used both with and without wave reflection from the ground.

3. *Electric dipoles.* Large electric dipoles with distributed resistive loads inhibiting oscillations in the dipole structure are generally used. The dipole is located at a distance from the test object, for example, a big ship. Antennas do not radiate at near-zero frequencies; therefore, these simulators have a low-frequency limitation. The characteristics of the now available simulators are given in the monograph by Balyuk et al. [20].

The development of nuclear EMP simulators gave impetus to advances in pulse power technology and in devising ultrawideband antennas radiating high-power electromagnetic pulses, and also sensors and oscilloscopes for detection and time-resolved recording of electromagnetic fields. In 1989, C. Baum, one of the leading developers of nuclear EMP simulators, proposed an antenna capable of radiating high-power UWB pulses of subnanosecond duration in free space, which was termed Impulse Radiating Antenna (IRA) [24, 25]. This was the natural result of the development of high-power UWB systems from nuclear EMP simulators to high-power radiation sources.

Since the 1990s, intense research on developing high-power UWB radiation sources has been carried out in many countries (the USA, Russia, China, the Great Britain, Germany, France, Ukraine, Israel, Korea Republic, and India). Ultrawideband radiation pulses of FWHM 0.1–1 ns with an effective potential of 0.1–5.3 MV at a pulse repetition frequency of up to 3 kHz have been obtained [3, 20, 26, 27]. The frequency content of the radiation falls in the range 0.1–10 GHz whose low-frequency edge adjoins the nuclear EMP spectrum. The radiation peak power reaches 1 GW.

In developing high-power UWB radiation sources, three basic lines can be distinguished [27]. The first one features the use of systems containing one high-voltage generator and one antenna. In these systems, to improve the radiation directivity and, hence, increase the effective potential rE_p, large-aperture antennas, such as IRA and TEM, are used. In experiments on the JOLT facility [28] using an IRA with an aperture diameter of 3 m and an input voltage 1 of MV, radiation pulses with an effective potential of 5.3 MV at a pulse repetition frequency of 200 Hz were obtained.

The second line is characterized by the use of antenna arrays with each antenna excited from an individual laser-controlled generator; that is, the number of generators equals the number of antennas in the array. The most impressive results were obtained in experiments with the GEM 2 facility [29] where UWB radiation pulses with an effective potential of 1.65 MV and a repetition frequency of 3 kHz were produced. On this facility, wave beam scanning within an angle of ±30° was realized. However, the facility was not put in practical use in view of its complexity (it incorporated as many as 144 laser-triggered semiconductor switches).

The third line features the use of multielement antenna systems excited from one generator through a power divider. In this case, using a UWB source with a 64-element array and an input voltage of 200 kV, radiation pulses with an effective potential $rE_p = 2.8$ MV at a repetition frequency of 100 Hz were produced [30]. In a subsequent experiment [31], the effective potential rE_p in a far-field region was increased to 4.3 MV. Each of these lines of research has merits and demerits.

The antennas and arrays of UWB radiation sources are excited with high-voltage pulses of different waveform: monopolar, bipolar, and double-exponential (with exponentially rising and exponentially falling output voltage of the generator, as in nuclear EMP simulators). A bipolar voltage pulse, owing to its spectrum, provides high radiation efficiency. In addition, a bipolar pulse having an amplitude of half that of a monopolar pulse produces nearly the same peak radiated field strength. This is the key factor for attaining the utmost parameters of UWB sources, as they are limited by the dielectric strength of the radiating systems.

Beyond the scope of the above research lines are UWB sources based on photo-emissive-cathode radiators excited by intense laser pulses [32, 33]. With these UWB sources, radiation pulses with an effective potential of 80–120 kV were produced. The effective potential can be increased by increasing the area of the photocathode.

Let us return to a discussion of the trends in developing UWB radio systems intended for radar. Here, three basic stages can be distinguished. The first stage is related to the changeover from simple harmonic signals to complex frequency- and phase-modulated and noise-like signals.

Let us briefly consider the systems using linear frequency-modulated (LFM) signals [34, 35]. The use of LFM pulses of duration τ_p and frequency bandwidth Δf made it possible to attain the compression of a signal in the receiver by a factor of $\tau_p \Delta f \gg 1$ and an increase in compressed signal amplitude by a factor of $\sqrt{\tau_p \Delta f}$. This served to improve the main lobe resolution of radar signals ($\Delta r = c/\Delta f$, where c is the velocity of light in free space) not reducing the limiting detection range. The main disadvantage of signals of this type is their limited capability of detecting small objects located near large ones, which is related to the presence of lateral lobes in the compressed signal.

Investigations on this line were carried out in many countries (the USA, the Great Britain, Germany, and the USSR). The basic problem here is to provide linear frequency modulation in an as wide as possible frequency band. In 1962–1963, Shirman et al. [36] performed a laboratory radar experiment in which a range resolution $\Delta r = 50$ cm was achieved using an LFM pulse of carrier frequency ~10 GHz and bandwidth 600–700 MHz. According to the above classification of UWB radiation ($\Delta f \geq 500$ MHz), it can be thought that this was a pioneering experiment on UWB radar. Note that the radars with LFM pulses of long duration (high energy) were originally designed, above all, for detecting distant objects in free space (atmosphere).

The second stage in the development of UWB radars was related to solving the problem of short-range subsurface probing of natural media (snow, ice, and ground) [37–39]. In this case, it was necessary to reduce the pulse duration to provide a reasonable space resolution and shift the spectrum toward the low frequencies to increase the depth of radiation penetration into the medium. To obtain short radiation pulses, shock excitation of the antennas was used. The number of damped oscillation periods depended on the Q factor of the antenna. Clearly, the lower the Q factor, the less the number of oscillation periods and the wider the relative frequency band of the radiation. Note that this was the second turn (since the pioneering work) in the study of the radiation of short damped oscillations by means of spark generators, but on a new engineering base.

The third, present-day stage in the evolution of UWB radars involves, first, the development of techniques and equipment for object recognition and radio imaging. This has demanded new methods for solving the problems of UWB pulse radiation, propagation, and scattering by objects and media. The complexity of the problems stimulated a more detailed exploration of transmitting and receiving systems. The main efforts were aimed at reducing the distortions in waveforms of both the radiated and the received UWB pulse, as the reflected pulse waveform just contains noncoordinate information on the object. The history of UWB radars is tracked in a series of monographs [39–49].

Of great importance in the development of UWB radio systems was the idea of pulse radio communication [50, 51]. At the heart of pulse radio is using coded short pulses for data transmission, which should ensure high information rates and stable communication in conditions of multibeam signal propagation. The spectral power density of UWB radiation can be lower than that of narrowband radiation. This can lead to a low probability of signal detection, which is of importance in developing safe (latent) communication systems. The evolution of UWB pulse radio communication began in the USA in the 1970s of the last century. However, extensive studies in this area date to the XXI century, which is related to that unlicensed use of the frequency range 3.1–10.6 GHz in pulse radio systems is permitted only if the spectral density is not above −41.3 dBm/MHz [1]. Another concept, being elaborated mainly in Russia, is the use of dynamic chaos in UWB communication systems [52].

1.2   Ultrawideband radar

Below we shall consider mainly UWB radars producing short UWB pulses of duration τ_p for which the condition $\tau_p \Delta f = 1$ is satisfied. Sometimes, pulses of this type are referred to as simple to distinguish them from variously modulated complex pulses. With a high degree of convention, UWB radar can be classified, depending on observation objects, as underground, terrestrial (overwater), and aerial (space) and, depending on distance, as short-range (less than 1 km) and long-range (~100 km) [53].

When an object is probed with UWB pulses, the reflected radiation consists of induced (early-time) radiation and natural resonance (late-time) radiation caused by the surface current. If the radial distance between the local scattering centers (point scatterers) of a complex object [54] is greater than the double spatial duration of the probe pulse $(2\tau_p c)$, the first portion of the signal contains a sequence of time-spaced pulses corresponding to these local scattering centers. Thus, the scattered radiation (its induced high-frequency component) contains information on the object geometry. The natural resonance radiation is characterized by the presence of resonant scattering frequencies, which depend on the object dimensions, shape, and material, and also bear information about the object. Note that the early-time radiation depends substantially on the aspect angle of the object, whereas the late-time radiation, by virtue of its resonant nature, varies only weakly with the object position relative to the probing direction.

For an exposed object of characteristic length L, the early-time component of the signal is formed in a time $2L/c$. In view of this, we may consider separately UWB radar of small-size ($L < \tau_p c/2$) and large-size ($L \gg \tau_p c$) objects. For efficient recognition of objects, the late-time and the early-time radiation component can be used in the first and in the second case, respectively.

The choice of a method for object recognition is dictated by the spectrum of the probe UWB pulse. To excite an object to natural resonances (late-time radiation), it is necessary that a substantial portion of the energy spectrum of the probe radiation pulse fall within the wavelength range corresponding to the characteristic dimensions of the object. In this case, a late-time component can be isolated from the signal.

If the condition $L \gg \tau_p c$ is fulfilled, the fraction of low-frequency energy in the probe pulse spectrum that is necessary for the excitation of resonant oscillations is small and the late-time component of the reflected pulse is difficult to distinguish. Note that the concepts of small-size and large-size objects are relative. The same object can be considered small-size when exposed to a long UWB pulse and large-size when exposed to a short UWB pulse.

It is obvious that different regions of radar observation require different UWB receiver-transmitter radio systems and different methods and rates of signal processing. In particular, to avoid ambiguity, the probe pulse repetition frequency should not be over 1 kHz for the distance to an object equal to 150 km, whereas for short-range radar, it can be ~100 kHz, allowing one to obtain averaged data for a short time interval. In probing underground objects, it is desirable to increase the low-frequency energy component of the probe radiation for increasing the radiation penetration depth. For short-range radar of objects in air and long-range radar in space, the duration of the UWB pulse can be less than 100 ps, whereas for long-range radar of objects in air, it is necessary to use pulses of duration 1 ns. This is dictated by the spreading of a picosecond UWB pulse during its propagation in the atmosphere [55] and by the loss of noncoordinate information about the object that is contained in the signal waveform.

For high-speed objects, it is necessary to use one probe pulse or a short sequence of pulses (pulse packet) and information processing for real-time recognition of objects, as the aspect angle of an object relative to the radar receiver-transmitter system can vary during the time interval between adjacent pulses or pulse packets. For immobile underground objects, accumulation of signals and time-shift information processing are usable.

The UWB radar has a number of advantages [56, 57] and all of them are due to the short pulse duration and, hence, the small volume occupied by the electromagnetic pulse in space. The most important of them are the high spatial resolution, promoting object recognition; the possibility of radar observation of low-flying objects on the background of the ground (water) underlying surface; the high probability of object detection, as, unlike narrowband radiation, the pattern of the UWB radiation reflected from an object does not contain deep interference nulls, and the low level of passive clutters (rainfall, fog, aerosols) due to the small radar cross section (RCS) of noise sources in a small pulse volume.

A challenge in UWB radar is numeral processing of signals, which demands analog-to-digital converters ensuring time-resolved digitization and processing of great bodies of digital information. An important problem of high-power UWB radars with spaced transmitting and receiving antenna systems is electromagnetic compatibility. Low-power (less than -41.3 dBm/MHz) UWB radars radiating in the frequency range 2.9–10.6 GHz [1], which are widely used in various fields, are free of this problem.

The signal that arrives at the receiving device in the process of UWB pulse probing can be presented as

$$y(t) = s(t) + n(t) + \gamma(t) \qquad (1.6)$$

Here $n(t)$ describes a normally distributed noise with a zero mean and root mean square deviation (RMSD) σ_N, and $\gamma(t)$ stands for the (active or passive) clutter. The reflected signal $s(t)$ is defined in terms of the convolution of the probe pulse $x(t)$ and the impulse response (IR) of the object, $h(t)$:

$$s(t) = \int x(\tau)h(t - \tau)\,d\tau \qquad (1.7)$$

An important parameter affecting the characteristics of a receiver is the signal/noise ratio (SNR), which, for UWB pulse probing, can be determined by the average power

$$q = \frac{\frac{1}{T_S}\int_0^{T_s} s^2(t)\, dt}{\sigma_N^2} \tag{1.8}$$

peak power

$$q = \frac{s_{max}^2}{\sigma_N^2} = \frac{\left(E_p l_e\right)^2}{\sigma_N^2} \tag{1.9}$$

and peak field strength

$$q_p = \frac{s_{max}}{\sigma_N} = \frac{\left(E_p l_e\right)}{\sigma_N} \tag{1.10}$$

Here $P_N = \sigma_N^2$ is the power of the noise, l_e is the effective length of the receiving antenna, and T_S is the duration of the signal. Note that the SNR estimated by (1.8) is less than that estimated by (1.9). When the decibel scale is used, the estimates determined by (1.9) and (1.10) coincide with each other. The time-dependent quantities may have dimensions of field strength or voltage.

For narrowband radar, the SNR is determined as follows:

$$q = \frac{2W_S}{N_0} \tag{1.11}$$

where W_S is the energy of the signal and N_0 is the spectral power density of the noise. Given the parameters of the probe pulse, the spectral power density of the noise can be estimated as $N_0 = \sigma_N^2/\Delta f = \sigma_N^2 \tau_p$. It follows that for identical conditions, the SNR estimated by (1.8) is lower than that estimated by (1.11) approximately by a factor of T_S/τ_p.

The major radar tasks are detection and recognition of objects. Let us consider in tandem some approaches used to solve these tasks in UWB radar.

1.2.1 Detection of Radar Objects

The UWB pulse probing of objects has the feature that the reflected signal differs from the probe pulse in waveform, and the reflected signal waveform is not known in advance. This gives no way of using either correlation processing of the reflected signal and the main (probe) pulse or matched filtration of signals widely used in narrowband radar.

For UWB radars, it is proposed to use suboptimal [58] and optimal [59] detectors of signals reflected from an object. In an optimal detector, the energies of signals received in two adjacent periods are estimated and the signals are subjected to inter-period correlation processing (IPCP). In a suboptimal detector, only IPCP is used.

The IPCP relies on knowing the period of repetition of the probe pulses. Two signals arrived at the input of the receiver in adjacent periods are subjected to cor-relation processing. If the peak value of the correlation function exceeds a specified threshold, the decision is made that a signal is present. As the readings of noise voltage in adjacent periods are uncorrelated, the correlation in the absence of a signal is zero.

A disadvantage of the method is its instability to clutter, which increases the correlation function, so that it may exceed a threshold level even in the absence of a signal. To enhance the stability, it was proposed to use a frequency filter correspond-ing to the amplitude spectrum of the probe pulse in both a suboptimal [60, 61] and an optimal (erroneously referred to as suboptimal) detector of UWB signals [62].

Let us discuss the results of a statistical examination of an optimal detector [62] on the assumption that the probed object is immobile. If the object has a high radial velocity, this will lead to loss of detection quality.

The voltage at the output of an optimal detector for two adjacent periods of time can be written as

$$u_j = 2\int y_f(t + jT)y_f\big(t + (j + 1)T\big)\,dt + \int \big[y_f(t + jT)\big]^2\,dt + \int \big[y_f\big(t + (j + 1)T\big)\big]^2\,dt \quad (1.12)$$

Here, the integration is performed within the signal observation window, j is the order number of the pulse repetition period, T is the pulse repetition period, and $y_f(t)$ is the received signal filtered by a filter K:

$$y_f(\omega) = Y(\omega)K(\omega) \qquad (1.13)$$

The first term in (1.12) represents an IPCP detector, and the second and third terms represent energy detectors.

As the voltage exceeds a specified threshold, $u_j > u_0$, the signal present decision is made. The threshold voltage u_0 is estimated using the statistical data obtained for a reference observation window (a selected window were a signal was absent). For the voltages associated with the reference observation window, the mean ζ and RMSD σ_u are calculated. In what follows, we shall assume that the voltage u is distributed according to a normal law:

$$W\big(u, \zeta, \sigma_u\big) = \frac{1}{\sqrt{2\pi}\sigma_u}\exp\left(-\frac{1}{2\sigma_u^2}(u - \zeta)^2\right) \qquad (1.14)$$

A numerical experiment has shown that the function (1.14) well approximates the distribution of the voltage at the input of an optimal detector and can be used to calculate the probability of false alarm

$$F = \int_{u_0}^{\infty} W(u)\, du \qquad (1.15)$$

The threshold voltage u_0 is dictated by the false alarm probability (1.15). For our consideration, we set $F = 10^{-6}$. Knowing the threshold voltage u_0, we can evaluate the probability of detection (POD) as

$$P = \int_{u_0}^{\infty} \tilde{W}(u)\, du \qquad (1.16)$$

Here, $\tilde{W}$ is the voltage distribution function for a signal + noise + clutter event. The probability (1.16) increases on decreasing the threshold voltage u_0, which, in turn, depends on the width of the distribution function (1.14) determined by σ_u. Note that pre-averaging of a signal $y(t)$ over N pulse repetition periods decreases the noise RMSD by a factor of $\sqrt{N}$. This results in a decrease in threshold voltage u_0 and, as a consequence, in an increase in the probability of detection (1.16).

The length of the signal observation window in this algorithm is limited by the computational power of the computer used for signal processing. To narrow the observation window, it was proposed [60] to perform signal pre-processing by which an averaged signal $\bar{y}(t)$ is replaced by its energy:

$$W(t) = \int_{0}^{t} \tilde{y}^2(\tau)\, d\tau \qquad (1.17)$$

The noise and clutter levels slowly varying in time ensure a gradual increase in energy $W(t)$. The slope angle of the energy curve, $dW(t)/dt$, changes stepwise at the starting and ending points of the signal $s(t)$. Thus, the algorithm of signal detection is applied only to those sections of the observation window where the slope angle of the energy curve changes stepwise.

In the numerical simulation, the UWB pulse of duration $\tau_p = 1$ ns presented in Figure 1.1 (curve 1) was used as a probe pulse. The probed object was a metal sphere, whose impulse response is well known [41]. The signal $s(t)$ reflected from a sphere of diameter $D = c\tau_p$, obtained by convolving the probe pulse $x(t)$ with the impulse response of the sphere, is also presented in Figure 1.1 (curve 2). Here, the second pulse in the reflected signal was caused by the creeping wave rounding the sphere.

In calculations, both kinds of UWB clutter, active and passive, were used. To simulate the active clutter $\gamma(t)$, pulsed signals were used whose duration was half or twice that of the probe pulse $x(t)$. The clutter and the signal did not overlap in the observation window. The use of an optimal detector for the suppression of this type of clutter is impossible when the clutter repetition frequency is multiple to the pulse repetition frequency and the frequency bands of the clutter and probe signals overlap.

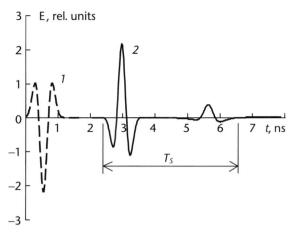

Figure 1.1   Probe pulse (1) and the signal reflected from a metal sphere (2). (With permission from EuMA.)

The passive clutter signal was simulated as the sum of the signals reflected from metal spheres that were much smaller than the probed object and randomly arranged around the object:

$$\gamma(t) = \sum_{m=1}^{M} a_m s_0 \left(t - \tau_m \right) \tag{1.18}$$

Here, a_m is the weighting coefficient distributed by a normal law with mean $\langle a \rangle$ and RMSD σ_a, τ_m is the random delay uniformly distributed within the observation window of the received signal, $s_0(t)$ is the signal reflected from a small sphere of diameter d, and M is the number of reflectors. For a sphere whose size is much less than the spatial length of the probe pulse, the signal reflected from the sphere can be presented with high accuracy as follows

$$s_0(t) \approx Ax(t) \tag{1.19}$$

where A is a constant depending on the sphere diameter.

We first discuss the results of an investigation of the effect of the noise level, determined by (1.8), on the probability of detection of a signal by various detectors with no filtration and UWB clatter (Figure 1.2). In the simulation, we used an energy detector (curve 1), an IPCP detector (curve 2), an optimal detector (curve 3), and a matched filter detector (curve 4). The latter is used in narrowband radar systems and demands knowledge of the reflected signal waveform, which is impossible in UWB radar. At the same time, the use of this type of detector allows one to evaluate the detection limits of a radar system.

The effect of a filter on POD in the absence of a UWB clutter was checked up with a pass band filter, whose bandwidth was specified by a certain level of the probe pulse spectrum amplitude, and a filter replicating the probe pulse waveform (optimal filter). The simulation results are presented in Figure 1.3. It can be seen that the presence of a filter (curves 1, 2) substantially reduces SNR compared to

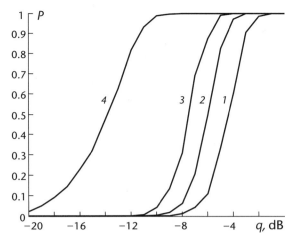

Figure 1.2 Probability of detection with an energy (1), an IPCP (2), an optimal (3), and a matched filter detector (4). (With permission from EuMA.)

the case of a detector with no filter (curve 3). The pass band filter (curve 2) with the bandwidth specified by the 0.5 level of the spectrum amplitude appeared to be somewhat less efficient than the optimal filter (curve 1). Comparison of the data presented in Figures 1.2 (curve 4) and 1.3 (curve 1) shows that an optimal detector equipped with a filter provides almost the same detection quality as a matched filter detector.

It was also investigated how a passive clutter of the form (1.18) affects the probability of detection. The number of scatterers, M, was set equal to 200. The object-to-scatterer size ratio (D/d) was set equal to 6 and 10. An optimal detector with an optimal filter was simulated. The simulation results are given in Figure

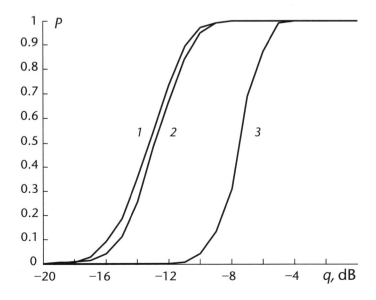

Figure 1.3 Probability of detection with an optimal (1) and a pass band filter (2) and with no filter (3). (With permission from EuMA.)

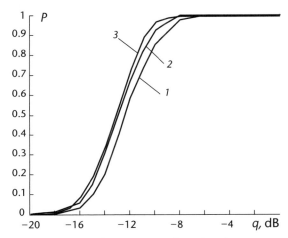

Figure 1.4 Probability of detection in the presence of passive clutter sources simulated by metal spheres with $D/d = 6$ (1) and 10 (2) and with no clutter source (3). (With permission from EuMA.)

1.4. Increasing D/d from 6 (curve 1) to 10 (curve 2) somewhat decreased SNR for a given POD. It can be seen that the results differ only slightly from those obtained with no clutter (curve 3), indicating that an optimal filter is highly efficient for passive clutter suppression.

When investigating the effect of active clutter on POD, the duration of the UWB clutter signal was varied along with the amplitude ratio of the reflected signal to the active clutter (in the range 0.5–2). It was observed that a decrease in duration and an increase in amplitude of the UWB clutter signal increased SNR to −3 dB at 0.9 POD.

The distinctive feature of ultrawideband radar is that the waveform of the pulse reflected from the object is different from that of the probe pulse. This difference is the more pronounced, the greater the ratio $L/\tau_p c$. In this connection, to estimate the limiting detection range of objects in free space for UWB probing, an energy approach was proposed [63]. According to this approach, the energy of the received signal is determined by the equation

$$W_r = \frac{W_t G A_e \sigma}{(4\pi)^2 r^4} \tag{1.20}$$

where W_t is the energy of the transmitted pulse, G is the gain of the transmitting antenna, A_e is the effective area of the receiving antenna, σ is the RCS of the object, and r is the distance from the receiver-transmitter system to the object. The maximum distance to the object (detection range) can be estimated by the formula

$$r_{\max} = \left(\frac{N W_t G A_e \sigma}{(4\pi)^2 W_{\min}} \right)^{1/4} \tag{1.21}$$

where $W_{\min}$ is the minimum received energy of the pulse reflected from the object, depending on the kind of noise and on the method of signal detection, and N is the number of pulses in the packet. The RCS of an object is an uncertain quantity.

There are several ways of estimating RCS based on the energy approach [64, 65]. The RCS of an object depends on the energy distribution of the probe pulse, $|S(\omega)|^2$, and, according to [64], it is defined as follows:

$$\sigma = \frac{\int_{-\infty}^{\infty} |S(\omega)|^2 \sigma(\omega) d\omega}{\int_{-\infty}^{\infty} |S(\omega)|^2 \, d\omega} \tag{1.22}$$

Generally, the RCS of an object depends on the polarization of the probe pulse field and on the aspect angle relative to the receiver-transmitter system. Therefore, to estimate r_{max}, it is necessary to use the RCS averaged over the observation angles of the object.

For rough estimation of the limiting range for loss-free detection of an object in space, we use a simplified approach based on the well-known equation applied in narrowband radar to estimate the RCS of an object located in the far-field region:

$$\sigma = 4\pi r^2 \frac{E_2^2}{E_1^2} \tag{1.23}$$

Here, E_1 is the amplitude of the field strength of the probe pulse in the region of the object, E_2 is the amplitude of the field strength of the signal reflected from the object in the region of the receiving antenna.

For UWB radar, the main measured characteristics are the waveform of the radiated pulse, $x(t)$, and its effective potential rE_p, which is a constant for the far-field region, and the waveform of the received signal, $s(t)$, and its peak field strength. After substitution of the corresponding quantities in (1.23), we obtain

$$r_{max} = \sqrt[4]{\frac{\sigma}{4\pi} \left(\frac{rE_p l_e}{q_p \sigma_N} \right)^2} \tag{1.24}$$

Here, q_p is the peak SNR (1.10) for a signal received by an antenna of effective length l_e. Note for a receiving antenna array with N_r elements, l_e should be multiplied by a factor of $\sqrt{N_r}$.

To estimate r_{max} [66], we use the parameters of an existing UWB radiation source producing pulses of duration 1 ns (with the central frequency of the radiation spectrum equal to 1 GHz) that is based on a transmitting array with the number of elements $N_t = 64$ and $rE_p = 2.8$ MV [30]. In this case, the peak radiation power is about 1 GW. Taking an active antenna array with $N_r = 16$ and the effective length of one element equal to 0.015 m [67] as a receiving antenna, for $\sigma = 1$ m^2, $\sigma_N = 0.5$ mV, and $q_p = 2$, we obtain $r_{max} \approx 7$ km.

Doubling the radiation effective potential (to 5.6 MV) by increasing the number of transmitting antennas in this UWB source four times ($N_t = 256$) and increasing the effective length of the receiving antenna array by increasing the number of elements to $N_r = 1024$ should increase the detection range to 33 km. In this version of a UWB radar, the aperture the transmitting antenna array should be 2.8 × 2.8 m and that of the receiving antenna array should be 1.7×1.7 m. Averaging the signal amplitude over N pulse repetition periods should increase the range, according to (1.21), by a factor of $\sqrt[4]{N}$. At a UWB pulse repetition frequency of 100 Hz, averaging the signal amplitude over a time interval of 1 s should result in the detection range $r_{max} \approx 100$ km. The above estimates indicate the ways in which UWB radio systems should be developed to enhance their detection range capabilities.

1.2.2 Recognition of Radar Objects

Radar recognition is a procedure of comparing the obtained information on the probed object with reference data and making a rule-based decision for classification of the object. With this purpose, databases are produced which contain radar characteristics of reference objects, such as impulse responses, transfer functions, resonance frequencies, etc.

The databases are generated from the results of numerical simulations and *in situ* measurements of the characteristics of the radiation scattered by objects. The IR functions estimated by processing the obtained data are smooth as a probe pulse is different in properties from a delta pulse. In this connection, when examining reference objects, the spectrum of the probe pulse should be as broad as or even broader than the spectrum of the pulse used to solve a similar radar problem in actual practice.

An object can be recognized either by its single characteristic or by a set of characteristics (signature). The latter method of object recognition is termed signature recognition. In particular, the set of resonance frequencies of an object can particularly be used as its signature.

Radar recognition deals mainly with objects having linear electromagnetic properties. There are parametric and nonparametric mathematical models of objects and associated methods of signal processing and recognition.

In nonparametric recognition of objects, the reflected signals and their spectra or the IRs estimated from measurements are used. The database in this case is presented by sets of reflected signal waveforms, spectra, and IRs obtained for various aspect angles of reference objects. The decision is made after scanning the data for all objects. As a discrimination parameter, the maximum correlation factor, pointing to the most closely resembling object, or the minimum residual between measurements and based data can be used.

This approach demands a large memory capacity to store reference data and high data processing rates. In this connection, investigations are being performed to attain better information compression compared with digitization of signals [68]. Besides, as radars vary in characteristics, it is necessary to obtain information about the reference objects for an actual radar system or to perform calculations using the database generated for objects probed with UWB pulse radiation having a wider frequency band [69].

The development of radar recognition methods is aimed mainly at elaborating parametric models of objects and reflected signals. This is necessary for creating automated recognition systems and reducing time demands, which is of critical importance for high-speed objects. Parametric models treat an object as a collection of scattering centers or describe it in terms of natural resonance scattering.

An object can be recognized if the radar database includes a reference object corresponding to the probed one; otherwise, it is necessary that an image of the object be constructed using appropriate measurements and then analyzed by a human operator.

In view of the importance and complexity of the problem of recognition of radar objects, it has been the objective of much research. The relevant studies are partly summarized in books [41, 70] and reviews [71–75, 53, 66]. Below we consider two methods for object recognition and imaging.

1.2.2.1   The E-pulse Method

Among the methods of object recognition, the E-pulse method is most theoretically developed and verified experimentally on airplane mockups [75–79]. The method is based on the resonance model of radiation scattering by objects. The model was theoretically substantiated in the context of the singularity expansion method proposed by Baum [80, 81] and developed in a number of subsequent studies.

The development of the singularity expansion method was stimulated by experimental studies on the action of a nuclear EMP on objects. It was observed that in typical nonstationary responses, damped sinusoidal oscillations were dominant. This served as a starting point to relate the characteristics of these responses to the analytical properties of their images (in the sense of a bilateral Laplace transform) in the complex frequency plane (s-plane).

It is important that the positions of simple (first order) poles on the complex frequency plane depend only on the geometry of the object and do not depend on its aspect angle. Therefore, having determined the pole positions, we obtain the parameters that characterize a given object subjected to any type of excitation, which is of importance in solving a recognition problem.

Let us briefly discuss the resonance model of radiation scattering by objects. Suppose that a perfectly conducting object is irradiated with a plane wave, polarized along a direction ς, that propagates in a direction specified by a vector $\mathbf{k}$ and its time structure is that of an electric field $E(t)$. The Laplace transform of this field at a point specified in the object-related coordinate system reads

$$\mathbf{E}^i(\mathbf{r},s) = \varsigma E(s)\exp(-s\mathbf{k}\mathbf{r}/c) \tag{1.25}$$

where $E(s)$ is the Laplace transform of the function $E(t)$.

The Laplace transform $\mathbf{J}(\mathbf{r}, s)$ of the induced surface current density at a point $\mathbf{r}$ on the object surface is determined by solving an integral and can be presented [76] as a singular series expansion:

$$\mathbf{J}(\mathbf{r},s) = \sum_{m=1}^{M} a_m \mathbf{J}_m(\mathbf{r})\left(s - s_m\right)^{-1} + \mathbf{W}(\mathbf{r},s) \tag{1.26}$$

where $\mathbf{W}(\mathbf{r}, s)$ is some integer function, $\mathbf{J}_m(\mathbf{r})$ is a complex function describing the mth mode of the induced current density, $s_m = \sigma_m + i\omega_m$ is the mth complex natural (resonance) frequency, and a_m is the mth complex coupling coefficient depending on the vectors **k** and ς. When probe pulses of finite bandwidth and energy are used, only a finite number of resonance modes are excited.

The transform of the backscattered electric field polarized along the direction ς at the point **r** located in a far-field region can be written as follows:

$$E_\varsigma^s(\mathbf{r}, s, \mathbf{k}) = \frac{\exp(-sr/c)}{r} E(s) \mathbf{H}(s, \mathbf{k}, \varsigma) \tag{1.27}$$

where $r = |\mathbf{r}|$ and $\mathbf{H}(s, \mathbf{k}, \varsigma)$ is a transfer function depending on the aspect angle of the object (vector **k**) and on the polarization of the signal (vector ς).

Using representation (1.26), we can derive a two-component expression for the transfer function. One component describes the contribution of the integer function $\mathbf{W}(\mathbf{r}, s)$ and corresponds to the induced response and the other is determined by a series expansion in natural modes and corresponds to the proper component of the scattered pulse. Thus, we have

$$\mathbf{H}(s, \mathbf{k}, \varsigma) = \mathbf{H}'(s, \mathbf{k}, \varsigma) + \mathbf{H}''(s, \mathbf{k}, \varsigma) \tag{1.28}$$

where $\mathbf{H}'(s, \mathbf{k}, \varsigma)$ is the induced component of the transfer function and $\mathbf{H}''(s, \mathbf{k}, \varsigma)$ is its proper component having a structure defined by (1.26).

The inverse Laplace transformation of the transfer function gives an expression for the impulse response of the object:

$$\mathbf{h}(t, \mathbf{k}, \varsigma) = \mathbf{h}'(t, \mathbf{k}, \varsigma) + \mathbf{h}''(t, \mathbf{k}, \varsigma) \tag{1.29}$$

The IR proper (late time) component represents the resonance model of the object and can be expressed as the sum of exponentially damped oscillations:

$$\mathbf{h}''(t, \mathbf{k}, \varsigma) = \sum_{n=1}^{N} a_n(\mathbf{k}, \varsigma) \exp\left(\sigma_n t\right) \cos\left(\omega_n t + \phi_n(\mathbf{k}, \varsigma)\right) \quad t > 2L/c \tag{1.30}$$

Here, a_n and ϕ_n are the amplitude and phase of the nth excited mode. Note that resonance frequencies exist as complex conjugate pairs ($s_{-n} = s_n^*$); therefore, N is an even number.

The response of an object, $y(t)$, to an arbitrary probe pulse $x(t)$ is determined by solving the convolution (1.7); it contains, as well as the IR defined by (1.29), a late-time component ($t > 2L/c$) being the sum of the complex exponents with the same resonance frequencies. When the probe pulse is substantially different from a delta pulse, the response $y(t)$ contains a significant region where its induced component $y'(t)$ and proper component $y''(t)$ overlap each other. This complicates the determination of the time segment of the response that is actually associated with the resonance component.

The essence of the E-pulse method is the following. A discrimination signal $e(t)$ of finite duration T_e is selected by a certain procedure using the resonance frequencies obtained from experimental measurements to match the response of the object. If the operation of convolution of the object response with the matched E pulse is performed, the convolution result should tend to zero in the late-time period, that is, the following condition should be satisfied:

$$c(t) = e(t) \otimes y(t) = 0 \quad \text{for} \quad 2L/c + T_e \leq t \leq T_y \qquad (1.31)$$

where the sign $\otimes$ denotes the operation of convolution and T_y is the duration of the response. The database of objects contains E pulses independent of aspect angle.

The discrimination parameter used for object recognition can be defined as follows:

$$\Psi = \frac{\int_{2L/c+T_e}^{T_y} c^2(t)\,dt}{\int_0^{T_e} e^2(t)\,dt} \qquad (1.32)$$

and it is equal to zero only for a true E pulse. However, the measurement noise and inexactly determined proper resonance frequencies used in constructing the E pulse for an object pose problems in obtaining zero convolution for the late-time period and result in a nonzero value of Ψ, which we denote by Ψ_{min}. Therefore, the discrimination parameter can be presented as follows:

$$\Lambda = 10\lg\left(\Psi/\Psi_{min}\right) \qquad (1.33)$$

Experimental investigations and numerical calculations have demonstrated a feasibility of recognition of airplane mockups. Given the difference in discrimination parameter between two objects equal to 10 dB, the probability of recognition can reach 0.9 for the SNR defined by (1.8) approximately equal to 25 dB [75].

The disadvantage of the E-pulse method using the late-time component of the reflected signal is due to that the amplitude of the early-time component is often substantially greater than that of the late-time component, and the higher the noise level, the more problematic the application of the method. In this connection, a generalized E-pulse method is being developed [82, 83] in which both parts of the reflected signal are considered. For the early-time component of a signal, which is represented as the sum of the pulses reflected from local scattering centers with corresponding time delays, the E pulse is synthesized based on the frequency spectrum of this signal component.

Thus, the combined E pulse consists of the spectral and the time part that are associated, respectively, with the early-time and the late-time component of the reflected pulse. In this case, the generalized E pulse depends on the aspect angle of the object, and therefore the database contains E pulses corresponding to different angular positions of the object. Investigations performed with the use of airplane mockups have confirmed the workability of the proposed approach. It has been demonstrated that the use of the combined method allows one to determine the aspect angle of an object. Note that the measurements were performed over a wide frequency range (0.2–18 GHz), which is difficult to realize by using simple UWB radiation pulses.

1.2.2.2 The genetic function method

Reconstruction of the shape of a radar object from measurements is among the most complicated inverse problems. Investigations [84–87] have shown the shape of an object can be reconstructed only if its spatial (angular) observation database is large enough. In this case, so-called multiangle projections can be obtained, and solving the problem is reduced to using a tomographic method based on the inverse Radon transform [88].

For small-base radar systems, tomographic methods of object shape reconstruction are inapplicable. The use of the Lewis–Boyarsky transform [89, 90] for the reconstruction of a two-dimensional object [87], an observation angle of 10° was achieved as a limit for the given approach used with a quadruply overlapped frequency band of the probe pulse. In doing this, the SNR defined by (1.10) should be no less than 54 dB.

The use of short UWB probe pulses ($L/\tau_{pc} \gg 1$) provides high time resolution of the reflected signals. The solution of the object shape reconstruction problem using a time-to-space transformation makes it possible to reduce the angle base of the receivers and, hence, to increase the radar range. Application of this approach [91] to object shape reconstruction relies on invoking so-called genetic functions (GFs) to describe temporal images of object fragments.

Let us briefly discuss the results of examinations of the proposed approach. The object shape reconstruction by the GF method was realized using the radar system shown schematically in Figure 1.5. The UWB pulse radiator and the receiver are placed at the center of a rectangular coordinate system. Three receivers are arranged on the coordinate axes, each at a distance b from the system center.

In the numerical experiments, a conventionalized model of an airplane with a perfectly conducting surface was used. The reflected signals were simulated using a code based on the Kirchhoff method intended for solving nonstationary diffraction problems [92] in the single scattering approximation. The object was probed by bipolar pulses of different duration τ_p.

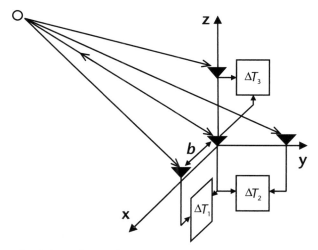

Figure 1.5 Schematic of the radar receiver-transmitter system. (With permission from Pleiades Publishing, Ltd.)

The database generation procedure is the following.

1. A complex object is represented as the sum of fragments, that is, simple geometric bodies.
2. For each fragment, the reflected signals are simulated for a given probe pulse waveform and for various sets of sizes and angles (θ and φ), and these are taken as genetic functions $g_n(t, \theta, \varphi)$.
3. A bank of genetic functions $g_n(t, \theta, \varphi)$ is generated with a chosen angular discretization step. The GFs are subdivided into classes to each there correspond characteristic fragments of the object and containing only their variations in shape. The size of the smallest fragment is limited by the spatial length of the radar pulse.
4. A signal $s(t)$ reflected from the complex object at known angles θ and φ is represented as the sum of genetic functions, each characterized by weight a_n and delay time τ_n:

$$s(t) = \sum_n a_n g_n\left(t - \tau_n\right) \tag{1.34}$$

The problem of shape reconstruction for a complex object is solved in the following order.

1. The GF set required for approximation of the signals scattered by the object at a given aspect angle is determined by solving the matrix

$$s_m = \sum_n a_n g_n\left(t_m - \tau_n\right) \tag{1.35}$$

    for the unknown quantities a_n and τ_n ($n = 1, 2, ..., N$). Here, s_m ($m = 1, 2, ..., M$; $M > N$) is the reflected signal received by the reference receiver of the receiver–transmitter system (Figure 1.5).
2. The coordinates of the fragments that correspond to the determined GFs are calculated by the formula

$$\tau_{n,j} = \frac{2c\Delta\tau_{n,j}R_n + b^2 - \left(c\Delta\tau_{n,j}\right)^2}{2b} \tag{1.36}$$

    where $\Delta\tau_{n,j}$ is the delay between the arrivals of the signal corresponding to the nth GF at the jth receiver and at the central (reference) receiver, and R_n is the distance to the nth fragment of the probed object determined by the delay between the pulse radiated by the transmitter and the nth local maximum of the reflected signal in the reference receiver.
3. The united fragments corresponding to the GFs that match coordinates calculated by (1.36) just constitute the reconstructed shape of the probed object.
    The accuracy of the object shape reconstruction was estimated by the equation

$$\eta = \frac{S - |\tilde{S} - S|}{S} \tag{1.37}$$

where S is the area of the object projection on a chosen plane and $\tilde{S}$ is the area of the reconstructed object projection on the same plane.

Investigations carried out for multipulse radar have shown that the accuracy of the shape reconstruction of an object substantially depends on the measurement noise, on the ratio of the object size L to the probe pulse spatial length $\tau_p c$, on the angular separation between the receivers in the measuring system, on the angular step of the GF database generation, and on the accuracy of determination of the object aspect angle. Note that the time intervals at which the signal is digitized should be shorter than the probe pulse duration ($\Delta t \leq \tau_p/20$).

Calculations performed to reconstruct the shape of a model airplane of length 4.5 m probed with a bipolar pulse of duration $\tau_p = 1$ ns at an angular separation between the receivers $\alpha = 2°$ and SNR $q_p = 20$ dB showed the reconstruction accuracy $\eta = 80\%$. Figure 1.6 presents the object shape reconstruction accuracy as a function of the ratio $L/\tau_p c$ for no error in the determination of the aspect angle. A reasonable accuracy is achieved at $L/\tau_p c \approx 20$. The projections of the object shape reconstructed by 10 implementations of the method are given in Figure 1.7.

For an object 50 m in size, the probe pulse duration $\tau_p = 1$ ns, SNR $q_p = 20$ dB, the angular step between the nearest aspect angles in the database $\Delta\alpha = 1°$, and the size of the receiving system $b = 50$ m, the distance to the object at which the reconstruction accuracy makes over 60% is 30 km. As the distance increases to 100 km, the reconstruction accuracy decreases to 17%, which is too low for recognition of the object by its reconstructed shape. In this case, the proposed method can be used to determine the set of GFs that would describe the signal reflected from the probed object. The information on the GF set can be used in solving the problem of recognition of the object.

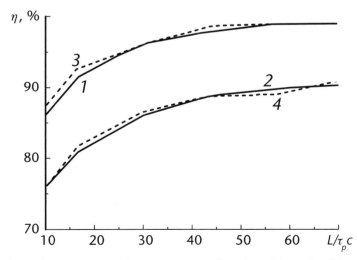

Figure 1.6 Object shape reconstruction accuracy as a function of the ratio of the object size to the probe pulse spatial length for $q_p = 26$ dB, $\tau_p = 1$ (1) and 2 ns (3) and for $q_p = 20$ dB, $\tau_p = 1$ (2) and 2 ns (4). (With permission from Pleiades Publishing, Ltd.)

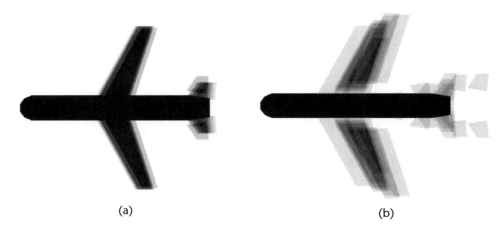

(a)   (b)

Figure 1.7 Projections of the reconstructed object shape obtained by averaging over 10 implementations of the method for SNR q_p = 26 (a) and 20 dB (b).

The simplest GF is a mirror response from a rectangular plate scatterer. In a numerical simulation [93], the signal reflected from an object was approximated by a set of six GFs of this type. The object was a 3D model of an airplane of length 10 m. To increase the angular base, inverse synthesis of the aperture was carried out for 10 s with a pulse repetition frequency of 2 Hz for the object moving at a constant altitude of 1 km with a velocity of 200 m/s. The calculations were performed for 21 aspect angles.

The coordinates of each scatterer were determined by the procedure described earlier. It was examined how the probe pulse duration τ_p, the distance between the receivers, b, and SNR q_p affect the accuracy of approximation of the object. The result for τ_p = 2 ns and b = 50 m is presented in Figure 1.8.

The sizes of the dots in Figure 1.8 are proportional to the sizes of the scatterers, and their coordinates are imposed on a common plane. This variety the GF method is reduced to a facet analysis [94]. Information about the local scattering centers of an airplane allows one to estimate the characteristic sizes of the object and its

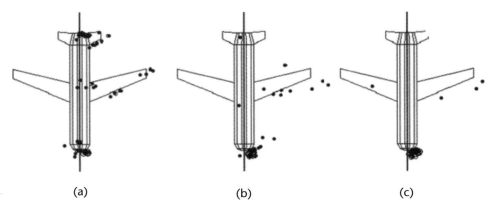

(a)   (b)   (c)

Figure 1.8 Approximation of an object by a set of dots for SNR q_p = ∞ (a), 26 (b), and 20 dB (c). (With permission from Springer.)

velocity of motion along a trajectory. These estimates can be used, either directly or in the context of the scattering center model, for recognition of the object [95].

1.3   Ultrawideband Communication Systems

The key parameter used to characterize a communication system is the rate of information transmission (information rate). According to the Shannon law (1.5), it is proportional to the frequency bandwidth Δf of the transmitted signal. There are three frequency ranges used in UWB communication systems: 3.1–10.6 GHz with a spectral density no more than −41.3 dBm/MHz [1], 57–66 GHz [96], and 275–300 GHz [97]. A frequency band is easy to widen by increasing the transmitted signal frequency.

In the period from 1984 to 2009, the frequency bandwidth for communication systems was doubled each 18 months [97]. As this took place, the information rate increased from 1 kb/s to 100 Mb/s. To meet the present-day requirements, data transmission rates should be as high as 5–10 Gb/s.

We shall briefly consider the UWB communication systems operating in the frequency range 3.1–10.6 GHz. Attention will be paid to the features required of the antennas in communication systems of this type. The UWB communication systems are subdivided into single-band (impulse radio) and multiband [50, 51, 98, 99]. At the lower frequency edge of the frequency range, they adjoin the UWB communication systems based on dynamic chaos [52, 100].

1.3.1   Single-band Ultrawideband Communications

Single-band UWB communications are developed on three lines related to the frequency ranges 3.1–10.6, 3–5, and 6–10 GHz. On the first line, the whole of the allocation band is utilized. This approach is most difficult to realize in view of stringent requirements for the antennas: The frequency band should be wider than 7.5 GHz (for the antenna could radiate and receive a pulse of duration 150–200 ps with small distortions) and the phase center should be stable within the frequency band of the pulse. Besides, it is necessary to ensure coexistence of UWB radio and narrowband communication systems in the frequency range between 5 and 6 GHz [101, 102].

To promote the realization of pulse radio, it was proposed to subdivide the frequency range into two ones: 3–5 and 6–10 GHz. Narrowing a frequency band moderates the requirements for antennas, increases the duration of the radiated pulse (consisting of several oscillations), and facilitates the realization of the analog part of a receiver–transmitter system. Besides, UWB communication systems of this type do not induce noises in the existing narrowband communication systems in the frequency range between 5 and 6 GHz.

In ultrawideband communications, information encoding in a pulse train is performed using pulse position modulation, pulse amplitude modulation, and bi-phase modulation. We shall consider only the third type of modulation, which is widely used and has many advantages. In bi-phase modulation, two types of pulses are used for information coding: direct and inverse (phase-shifted relative to the

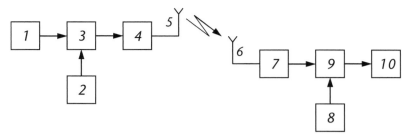

Figure 1.9 Structure of a receiver–transmitter system for single-band ultrawideband communication: transmitter: 1 – information source, 2 – pulse generator, 3 – modulator, 4 – pulse shaper, and 5 – antenna; receiver: 6 – antenna, 7 – amplifier, 8 – pulse generator, 9 – demodulator, and 10 – signal processing system.

direct pulse by 180°). These opposite-polarity pulses are used to code a logic zero and a logic unity. The information rate is one bit per pulse.

Conceptually, the operation of a single-band UWB communication system can be sketched as in Figure 1.9. Note that a pulse shaper is a device that restricts the bandwidth of a pulse to a specified limit. Among a great number of studies, only two that were aimed at increasing information rate for the frequency band 6–10 GHz deserve attention [103, 104]. In both studies, be-phase modulation was used. As a result, information rates of 750 Mb/s [103] and 2 Gb/s [104] were achieved.

1.3.2   Multiband Ultrawideband Communications

The idea of multiband UWB communication is to subdivide the entire frequency band (7.5 GHz) into 14 subbands of width 528 MHz, which comply with the definition of UWB radiation [1]. The subbands are grouped to form group subbands. This allows one to use a combination of subbands to optimize the operation of a communication system and reduce its power consumption.

Each subband has 128 frequency subcarriers (narrowband channels), which corresponds to a transmitted signal duration of 242.42 ns. In view of the cyclic prefix of duration 60.61 ns and the security interval of duration 9.47 ns, the duration of a symbol is 312.5 ns. Symbols can be radiated simultaneously (to obtain high information rate) or sequentially. A sequence of carrier frequencies is time-frequency coded.

A multiband communication system can support 10 information rates in the range 53.3–480 Mb/s [51]. The multiband UWB communication systems have the advantage of using narrow bands for information transmission. This makes the receiver-transmitter device less complicated (especially as regards analog-to-digital converters) and improves its spectral compliance (cutout of subbands occupied by narrowband communication systems) and fitting to requirements of various countries by using channel-tuning software.

In multiband communications, requirements for antennas are also moderated. Nevertheless, they should cover the full specified frequency range to provide simultaneous radiation of signals throughout the range. The compliance of a multiband communication system substantially reduces the information rate compared to a single-band system.

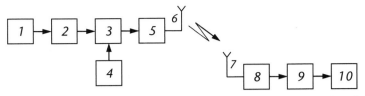

Figure 1.10   Structure of a receiver-transmitter system designed for ultrawideband direct chaotic communications: transmitter: 1 – ultrawideband chaos oscillator, 2 and 5 – amplifiers, 3 – modulator, 4 – driving pulse source, and 6 – antenna; receiver: 7 – antenna, 8 – amplifier, 9 – demodulator, and 10 – signal processing system.

1.3.3   Ultrawideband Direct Chaotic Communications

Dynamic chaos (random oscillations) [52] possesses a combination of properties that make it possible to utilize it as a data carrier in communication systems. In direct chaotic communication systems, the idea is realized to directly generate random oscillations carrying information and modulate them by an information signal. Pulse-amplitude modulation is used as a means for putting information in a chaotic signal. On the time axis, points are fixed, such that the presence of a pulse at these points implies that a logic unity is transmitted, whereas the absence of a pulse corresponds to transmission of a logic zero.

The structure of a receiver–transmitter system designed for ultrawideband direct chaotic communication is shown in Figure 1.10. In correspondence with this structure, a mockup of an information transmission system was built [100]. The chaos oscillator of the system (0.5–3.5 GHz) was capable of producing oscillations with a bandwidth of 3 GHz at a level of −20 dB.

A modulator is a device intended for switching a chaotic signal. If the voltage at the driving input of the modulator is at a level of 5 V (logic unity), the modulator transmits a chaotic signal. At an about zero signal level (logic zero), the modulator is locked and does not transmit a chaotic signal. Thus, when a two-level pulsed information signal is passed to the input of the modulator, a stream of chaotic radio pulses is formed at the output. The signals are radiated and received by UWB discone antennas.

Experiments on studying the transmission of digital information with the use of chaotic oscillations were carried out over a wide range of information rates, from 10 to 200 Mb/s. The maximum information rate was achieved with modulating pulses of duration 5 ns. A communication system was arranged between two computers of a local network. In a wireless mode at a data transmission rate of 10 Mb/s, transmission of video information was realized using the MPEG2 digital video standard. The quality of an image at the receiver side did not differ from that of the original one. The subsequent development of direct chaotic communications gave rise to a universal ultrawideband platform for household and office applications [105].

1.4   Susceptibility of Electronic Systems to Ultrawideband Electromagnetic Pulses

The actions of electromagnetic radiation pulses on hardware (circuit components, equipment, and systems) and on biological objects, including human beings, are subdivided into intentional and unintentional [106]. The unintentional action of electromagnetic fields on objects has been investigated for a long time by EMC experts. Investigations of the intentional action of strong electromagnetic fields on hardware, primarily on information systems, and on human beings (nonlethal weapons) are related to the progress in high-power microwave sources [3, 20, 107–109]. Great interest of EMC experts in intentional electromagnetic interference has quickened by the potential possibility of terrorist and illegal (criminal) activities using electromagnetic radiation sources.

We shall consider the investigations of the susceptibility of hardware to intentional electromagnetic interference with the only purpose to formulate technical requirements for high-power UWB radiation sources. Among the major tasks facing EMC experts are to study the mechanisms of the damaging action of radiation pulses, to determine criteria for distinguishing between different types of hardware failure, and to develop proper protection techniques and equipment. The final stage of the work is to subject the facilities of civic importance to susceptibility tests, as stringent as possible. Of civic importance are communication systems; ground, aerial, and sea transport; safety systems of power engineering objects (above all, nuclear reactors), computer systems of big banks, and life support medical equipment. Performing these tests calls for high-power UWB radiation sources.

Investigations of bioeffects caused by the action of UWB radiation are of importance as well [110, 111]. They are necessary for understanding nonthermal mechanisms of the interaction of short radiation pulses with biological media and objects and for developing safety standards for the population and, certainly, for operators of high-power UWB radiation sources.

We shall restrict our consideration to the results of investigations of the action of microwave narrowband and ultrawideband radiation pulses on hardware. The frequency range of interest is between 200 MHz and 5 GHz [106]. Sometimes this range is narrowed to 0.5–3 GHz [112]. The frequency range 0.2–5 GHz can be realized by exciting UWB radiators with bipolar voltage pulses of duration from 5 to 0.2 ns. For a narrower frequency band (0.5–3 GHz), it suffices to use UWB radiators excited with bipolar pulses of duration 0.3–2 ns. When considering the results of investigations of the susceptibility of electronic systems to microwave radiation, we shall also restrict ourselves, whenever possible, to short pulses, as the duration of UWB microwave radiation pulses is generally no more than 10 ns.

Experimental investigations were carried out on open areas for large-sized objects (airplanes, rockets, and cars), as well as in screened rooms, anechoic chambers, and TEM cells. Vertically and horizontally polarized radiation was used. Irradiation of objects was performed using single pulses of varied duration and pulse packets with the pulse repetition frequency varied from 1 to 1000 Hz. The peak field strength in the experiments reached ~100 kV/m. The experimental results are presented in monographs [20, 108] and reviews [113, 114].

The researchers distinguish two basic types of damage to a system: its transient malfunction, such that the system recovers its functions after termination of irradiation or reset (functional upset), and its physical damage requiring repair or replacement. For short pulses, the functional upset is associated with incomplete electrical breakdowns and the physical damage with thermal damage to the system components after completion of an electrical breakdown. The probability of both types of damage increases with peak electric field strength. Unfortunately, only one comprehensive study has been performed for bipolar UWB pulses with a central frequency of ~1 GHz [113]. In many experiments, monopolar UWB pulses of rise time ~100 ps with an FWHM of several nanoseconds were used. For such pulses, an appreciable portion of energy falls within the range 0.1–1 GHz, though the spectrum maximum corresponds to zero frequency. Note that a UWB radiation pulse contains no field component of near-zero frequency.

The results obtained allow some conclusions to be made. The critical field strength at which susceptibility of a system to the action of an electromagnetic pulse is detected increases with decreasing pulse duration and repetition frequency. However, the data obtained with bipolar pulses have shown that the frequency range for which the field–object interaction is efficient is more important. As the bipolar pulse duration was decreased from 1 ns (central frequency of 1 GHz) to 0.5 ns (central frequency of 2 GHz), the energy density required for the functional upset of a microprocessor board did not increase, but even decreased (to ~10^{-5} J/ m^2). The critical field strength of short pulses required for functional upset of various computer units ranges over wide limits: from 1 to 100 kV/m.

Note that the experimental investigations of the susceptibility of electronic systems to the action of UWB pulses were carried out for small-size objects and without screening. Metal screening of an explored object reduces the field strength inside the object by about −30 dB. The transfer function that determines the proportion between the voltage induced in the components of a system and the incident field strength in the given frequency range can be about −30 dB. It follows that for an induced voltage of 10–100 V, UWB radiation pulses with the field strength at an object ranging between 10 and 100 kV/m are necessary. For a source-object distance of 10 m (far-field region), the radiation effective potential should be 0.1–1 MV. The UWB sources radiating in the frequency range of interest are feasible with the use of a single antenna and with multielement antenna arrays excited with high-voltage bipolar pulses of different duration.

Comparative examinations have shown that the critical field strength for functional upset of objects is substantially lower when they are exposed to long pulses of narrowband microwave radiation rather than to short pulses of UWB radiation. This is due to that in an electronic system, resonance can readily be induced by long pulses, as the quality factor of the radiation source (number of field oscillations) can be greater than that of the irradiated system. When the frequency of narrowband radiation coincides with the resonance frequency of the irradiated electronic system, the critical field strength for functional upset of the system is minimal [106]. In contrast to narrowband radiation, which is able to induce resonance at one frequency, UWB radiation has the potential for giving rise to resonances over a wide frequency band. However, the efficiency of excitation of resonances is low, as the quality factor of a UWB radiator is, as a rule, substantially lower than that of the irradiated electronic system.

For estimation of critical fields for long electric circuits of large objects (cars, airplanes), theoretical models and computer codes are developed [112, 115]. The models are based on the topological approach, namely subdivision of large circuits into interlinking blocks. Among theoretical studies, worthy of mention is one by Vdovin et al. [116] who investigated analytically malfunctions (bit errors) in an information system exposed to a bipolar radiation pulse. Experimentally, similar malfunctions were investigated for a microprocessor board exposed to a monopolar UWB pulse [113].

One more task facing EMC experts is to examine the electromagnetic radiation emitted by various devices and objects. Unintentional emission of electromagnetic radiation can be used for unauthorized access to information [117] and for recognition of objects [118]. In addition, the obtained information about the frequency band (as a rule, ultrawideband) of a device allow one to properly choose the parameters of a radiation source to be used to check the device for vulnerability. Furthermore, the investigations aimed at reducing the emission of electromagnetic radiation by various devices should be helpful in increasing the critical field strength for their functional upset. It is clear that time-resolved investigations of the emission of electromagnetic radiation by devices and systems of varied purpose call for ultrawideband receiving antennas having wide dynamic ranges.

1.5   Ultrawideband Technology Applications

In this section, we briefly review the main ultrawideband technology applications without claiming completeness. The choice of the material was driven by the desire to outline intensely developing UWB technology applications and show their relationship with the research results presented in the book.

Real-time location systems. Detection and localization of people using UWB radar has many practical applications, including antiterrorist operations and search and rescue of victims affected by man-made accidents and natural disasters [119, 120]. In such situations, the actually achievable localization accuracy is of importance. Various aspects of the problem arising in time-difference-of-arrival measurements (timing jitter, thermal noise, multipath propagation, etc.) are discussed in detail in [119].

For UWB radar systems with small antenna arrays, data on the localization accuracy were obtained based on computer simulations and experimental diagnostics [121]. These data make it possible to perform advance development of an antenna system for a given controlled area or to decide whether a radar system is feasible for a desired application.

In the recent decade, UWB positioning and location technologies have received wide acceptance [122, 123]. Thus, Multispectral Solutions, Inc. has patented a UWB real time location system (RTLS) intended for real-time high-accuracy location or watching of various objects (e.g., personnel, equipment, etc.) in an arbitrary large room (or a system of connected or isolated rooms) under multipath and noise conditions [124]. The Eliko company [125] has announced the development of a high-accuracy RTLS UWB technology. The technology is expected to ensure 2D/3D micropositioning of objects through various obstacles, such as concrete walls, including indoor and outdoor.

Zebra Technologies Corporation [126] reasons that "ultra-wideband (UWB) wireless platforms provide the highest precision for any RTLS solution, and with the latest advancements and innovative technology by Redpoint Positioning, the solution is revolutionizing the RTLS market." It is stated that this technology provides a means for processing more than thousands of tags per second with 30-cm accurate tag location.

Among numerous methods for object location using UWB signals, the near-field RTLS technology [127–129] is progressing rapidly. The systems involved operate at low (generally about 1 MHz) frequencies and harness the features of the phase behavior of evanescent fields in the near-field region of a tag transmitter. The achievable tag location accuracy is 1–3 ft at a distance of 60–200 ft [127].

For some applications, low-frequency signals may be favored due to their good penetration into absorbing media. The presence of the field radial component, along with the transverse component, in the near-field region and the lack of synchronization between the components of the electric and magnetic fields make it possible to track a larger number of useful signal parameters [127, 128]. It is reported [127–129] that this technology is successfully harnessed in tracking and communications systems that are operated in and around standard cargo containers and in close conditions of signal propagation at nuclear facilities and in warehouses, in fire safety systems, and in systems for preventive warning of emergencies, including attacks on computer information networks.

Internet of Things. In recent years, the Internet of Things (IoT) and Machine-to-Machine (M2M) technologies became widespread which sometimes are combined with the IoT/M2M abbreviation. The Gartner Group analysts interpret the concept of Internet of Things as a network of physical objects incorporating a technology that allows the objects to measure the parameters of their own state or of the environment, and utilize and transfer this information [130]. Note that this, by the way, most frequently cited, definition does not contain the word Internet; that is, it is not told that the Internet of Things is necessarily part of the Internet [131].

To realize IoT technologies, it is rational to use navigation methods based on wireless data networks. Real-time locating systems are quite simple and allow display of the locations of all objects on an interactive map. The objects can be both things, such as goods or tools in the warehouse, and transport facilities (electric vehicles, automobiles), and even people. An RTLS can be deployed in a room, building, or on virtually any area where there is the opportunity to place the system infrastructure and organize its work. An RTLS is a network of base stations, whose coordinates are known. The objects that need to track are provided with active tags whose signals are picked up by the base stations. By processing data from all stations, the locations of all tags in the RTLS network coverage area are determined [132].

Where real-time tracking of objects, intelligent automation solutions and error-free, fast and reliable operation in severe conditions are desired, the RFID (radio frequency identification) technology is commonly used. RFID is a technology of wireless communication via a radio signal between the electronic tag placed on the object and a special radioelectronic device that picks up the tag signal and reads the data it carries [133]. The tag of an object may contain information about the object type, cost, mass, and temperature, logistics data, and any other information about the object, allowing for easy information readout. An RFID tag typically comprises a receiver, a transmitter, an antenna, and a memory unit. Receiving energy from a radio signal emitted by a permanently fixed reader or by a hand-held scanner, the tag responds by its own signal, containing useful information.

A topical problem is the development of UWB systems for digital wireless personal local and sensor networks that would ensure communication security [134]. Growing demand for high-speed communication channels is expected in the future. This is due to that the M2M technology calls for high-quality voice and video streaming. Internet of things will affect a growing variety of applications, including household appliances, climate control systems, automatic monitoring of children, telemedicine, automatic vehicles, etc. Automatic vehicles, which will be able to operate without a permanent driver action on the controls, will not, for sure, get along without the IoT. First prototypes of cars that communicate with each other and automatically call emergency services in the event of a traffic accident have already appeared.

Intelligent Transport Systems. In recent years, intelligent transport systems (ITSs) are intensely developing in different countries. With certain differences in interpretation of the ITS concept taken in different countries, a generalizing definition can be the following: ITS is a system-integrating modern information and communication technologies and automation aids with a transport infrastructure, transportation facilities and users, aimed to improve the safety and efficiency of the transport process and the comfort for drivers and transport users [135].

According to forecasts of the World Health Organization, Geneva, if urgent effective efforts to create safe transport and advanced control systems are not be taken, more than two million people a year will die on the roads by 2030, not taking into account other modes of transport. To reduce damage from the loss of life and injury to people justifies the costs for the development of new technologies and their introduction into practice.

A major trend in the ITS development in Europe, the United States, and Japan, which is actively promoted in the last 15 years, is toward the realization of the intelligent vehicle concept. The Enhanced Transport Security international program is already running. Even first tests of onboard intelligent systems have shown that they can reduce the number of road accidents by 40% and the number of road fatalities by 50%.

Currently, more than 10 types of onboard ITSs, such as systems for car interval control in heavy traffic, systems for holding a car on the runway, driver fatigue (nap) alarm systems, side collision avoidance systems, motorcycle detection systems, and others are already on sale or undergoing field tests.

To operate efficiently, the whole complex should share real time information and use UWB data transmission technologies. Leading companies have developed integrated circuits that use the 802.15.4–2011 UWB standard. For instance, the DecaWave DW1000 chip [136] operates in six sub-bands in the frequency range 3.5–6 GHz. A DW1000-based RTLS system is capable of determining the location of objects to within 10 cm and real time tracking of 11,000 RFID tags within a radius of 20 m. The PL3120 chip developed by Pulse~LINK have similar characteristics. The devices that use DW1000 or PL3120 chips require subtle UWB antennas. In Chapter 7 of this book, a UWB printed antenna 25 × 20 mm in size operating in the frequency range 3.1–10.6 GHz is described.

MIMO systems. Requirements on transmission capacity of mobile networks are permanently strengthened. Increasing channel bandwidth does not solve completely the problem of providing high throughput rates because the frequency range is still limited. One way to increase the capacity of wireless systems is the use of multiple transmitting and receiving antennas—Multiple Input Multiple Output (MIMO) technology—and special processing of the signal in this case [137]. The following is a classification and a brief description of MIMO systems.

Diversity reception (Rx Diversity) is a technology in which the number of receiving antennas is greater than that of transmitting ones. In reference to MIMO, the relevant system is called a single input multiple output (SIMO) system. In the simplest configuration, a SIMO system comprises one transmitting and two receiving antennas (1×2 SIMO system).

Diversity reception does not increase throughput rate; however, it enhances transmission reliability. In this case, two signals are received, and there are different ways of their processing. For instance, a signal with the best signal/noise ratio can be selected using the so-called switched diversity method, or the signals can be combined using the maximum ratio combining (MRC) method to increase the signal/noise ratio.

Transmission diversity (Tx Diversity) is a technology in which the number of transmitting antennas is greater than that of receiving ones. The relevant system is called a multiple input single output (MISO) system. In the simplest configuration, a MISO system comprises two transmitting and one receiving antennas (1×2 MISO system). SIMO and MISO systems are simple to implement, as they do not require special preparation of the signal.

Spatial multiplexing is a technology using several transmitting and several receiving antennas. Unlike MISO and SIMO, this technology is aimed not at improving transmission reliability but at increasing transmission rate. Therefore, MIMO is used for data transmission to mobile stations that are in good radio conditions, whereas SIMO and MISO are used to transmit data to mobile stations that are in poor radio conditions. To increase the data transmission rate in a MIMO case, the input data stream is divided into several streams, each transmitted independently by an individual antenna.

A new technology intensely developing in recent years is the so-called MIMO radar [138]. MIMO radars can be divided into two classes: radars with colocated antennas and coded signals and radars with widely separated antennas (Statistical MIMO radars).

The most significant feature of these radars is the large number of degrees of freedom that offers many important advantages such as high angular resolution and reduced side lobes of the antenna system, the possibility of simultaneous exact measurements of parameters of a large number of targets, and high adaptability, including the possibility of adaptive formation of the antenna system pattern during transmission. In addition, this radar technology allows efficient target search within a wide sector without spatial scanning [139]. It should be noted that MIMO radar with antenna diversity (Statistical MIMO radars) is not a new concept. This is a particular case of multiposition radar.

ITELITE company has developed various design solutions of MIMO antennas spanning various bands within the frequency range 0.9–5.9 GHz [140]. MIMO antennas for the 0.86–12.75 GHz frequency range have been developed by COBHAM. [141]

Dual-polarized UWB radars. Using UWB pulses with dual polarization enhances the object detection and recognition capabilities of radars. Balzovsky et al. [142] have shown that sequential probing of metal plates with linearly and elliptically polarized UWB pulses allows for estimation of the angle of inclination of the plate and its width. A study performed by the same authors [143] has shown that the polarization structure of a pulse having passed through a wall depends on the wall heterogeneity. Numerical simulations [144] predict that the use of radiation pulses with orthogonal polarizations makes more reliable detection of small cancerous tumors, which are usually overlooked in X-ray mammography.

UWB dual-polarized MIMO radar [145] intended for the detection of improvised explosive devices (IEDs) is being developed in cooperation by research institutions of Germany and Colombia. The project's executors have performed first experiments on probing IEDs with orthogonally polarized radiation pulses [146]. Our research team [142, 143] is also developing a radar system capable of transmitting and receiving orthogonally polarized UWB pulses. It differs from the project [145] by the use of other types of antennas and other pulse waveforms and by the possibility to scan the patterns of the transmitting and receiving antenna arrays. It is proposed [147] to use successive orthogonally polarized UWB pulses to detect metallic objects on the background of randomly rough earth's surface. Note that dual-polarized transmitting and receiving antenna arrays and pattern scanning are discussed in this book (Chapter 8).

Biomedical applications and biological effects. Interest in this area of application of UWB pulses [148] is dictated, above all, by concern for people's health. We first consider the biomedical applications of UWB radiation pulses. Note that in biomedicine, low-power pulses with high repetition frequencies are used to obtain required information within a short time. Among biomedical applications, measuring breathing and heartbeat rhythms of human and taking images of cancerous breast tumors, especially in women, are of primary importance.

Clinical trials of the UWB technology application to monitor chest movement, breathing, cardiac and motor activity of patients were carried out in Russia, China (Taiwan) [149] and Spain [150]. The cardiac rhythm and its variation and the breathing frequency were measured. The respiratory function of premature infants was monitored. The studies have shown positive results: UWB measurements coincided with data obtained by other methods. Note that the Novelda AS private company (Norway) has developed XeThru Sensor Technology for human vital sign monitoring [151].

UWB radars for respiratory motion detection of people through walls or rubble have been developed [152–155]. It should be noted that to probe objects behind obstacles, it is necessary to switch from the UWB pulses with central frequency $f_0 = 7$ GHz (3.1–10.6 GHz) to pulses with $f_0 = 1.3$ GHz or less (0.3–0.5 GHz), depending on the material and thickness of the obstacle.

UWB radar is a promising technology for the early detection and monitoring of breast cancer based on the difference in permittivity and conductivity between normal and cancerous tissues, and it is being developed as an attractive alternative to x-ray mammography. UWB radar imaging [156–158] includes breast irradiation with UWB pulses, receiving of backscattered signals, and use of these signals to detect and image dielectric scatterers. The authors of [158] propose to use an array of antennas of three types with different frequency bands to cover the entire frequency range 3.1–10.6 GHz. Note that in this book (Chapter 7), combined antennas of plane and volumetric design whose pulse radiation spectra include the whole of this frequency range are presented.

Studies of the biological effects of irradiation with nanosecond UWB pulses of high-field strength (up to 1 MV/m) are carried out on animals, mainly on mice. High field strengths are needed to assess the limits dangerous to humans. In addition, studies of interest are those with external fields comparable to the fields in a biological cell. High-field strengths can be most readily obtained in TEM cells. To do this, monopolar voltage pulses with the peak frequency corresponding to zero frequency are generally used. The results of the relevant studies presented in the review [111] are grouped into sections according to their effects on the cardiovascular system, central nervous system, behavior, genotoxicity, teratogenesis, carcinogenesis and other functions. The studies have shown that the depth of penetration of electromagnetic radiation and the loss of its energy in a biological matter depend on the pulse spectrum waveform (duration and voltage rise and fall times). In this regard, experiments in free space where a radiated UWB pulse contains no energy at a near-zero frequency are of interest.

Let us dwell only on two experiments [159, 160]. In the experiment [159], a four-element array of TEM antennas excited by a monopolar pulse of FWHM 220 ps was used. The UWB pulse at the antenna aperture had two lobes of different polarity. In this experiment, mice grafted with cancer were placed in a dielectric container and exposed to UWB pulses to investigate the inhibition of tumor growth after irradiation. The irradiation was performed at a pulse repetition frequency of 13, 16, 20, and 200 Hz. A therapeutic effect was observed only in mice irradiated at pulse repetition frequencies of 16 and 13 Hz and peak field strength of 220 kV/m. In the experiment [160], the mice were located far away from a multielement array excited by a bipolar voltage pulse of duration 2 ns. They were irradiated with UWB pulses of peak field strength 120 kV/m with a pulse repetition frequency of 100 Hz for 60 min. The liver functions and the kidney functions were investigated. Note that the designs of this type of UWB source are discussed in this book (Chapter 9).

Electronic warfare and car stop. Research on neutralization of electronic systems by high-power UWB pulses (non-lethal weapons) is being performed for a long time in relation to the solution of different problems. Another area of electronic warfare technology is the development of compact UWB devices for prompt detection, location, and identification of radio frequency (RF) signals.

Here, we only briefly discuss the studies on stopping cars. These studies were carried out at DIEHL [161], Eureka Aerospace [162, 163] and other companies [164, 165]. The results show that a malfunction of a car occurs when it is irradiated with pulses of critical field strength 20–40 kV/m at a pulse repetition rate of up to 100 Hz. To stop a car within a distance of 50 m, UWB radiation sources with effective potential rE_p = 1–2 MV are needed. The development and engineering of UWB sources of this type are described in this book (Chapter 9).

In the experiments earlier, UWB radiation pulses with linear and elliptical (circular) polarization were used. The sources of circularly polarized radiation offer the possibility of experimentation with varying continuously the angle between the electric field vector and the object within one pulse. High-power sources of elliptically polarized UWB radiation based on cylindrical helical antennas excited with short high-voltage pulses have been developed in the USA [166] and France [167], as well as at the Institute of High Current Electronics (Tomsk, Russia) [168, 169].

It is well known that the critical field strength for the malfunction of a system decreases with increasing pulse duration. However, the width of the radiation spectrum is usually also reduced. The pulse duration and the radiation spectrum width can be increased simultaneously by integrating radiation pulses in free space. Simple calculations [170] show that the use of four bipolar voltage pulses of different duration with optimum time delays between them for exciting an array of combined antennas can increase the ratio bandwidth $b = f_H/f_L$ of the radiation spectrum at a level of -10 dB to 22.7 (fivefold). In this case, the duration of the radiation pulse synthesized in free space increases approximately by half. To widen the pulse radiation spectrum is also of importance for UWB radars intended for object detection and recognition.

Other applications. The above analysis does not cover fully all application areas of UWB technology. Thus, development of ground-penetrating radars for the detection of mines [171], for permittivity estimation [172], for measuring soil water content in media [173], and for industrial applications [174–176] are being continued. Through the wall radars [177–179] are being developed for imaging objects behind a wall and for human detection. Considerable effort is put into the development of ground-based and air-based scanning UWB radars for remote sensing of objects and media [180,181], space-based radars systems for subsurface sounding of Mars and Moon soils [182], and borehole radars for examination of near-well soils [183, 184].

Conclusion

The period of the development of UWB radio systems can be conventionally subdivided into three basic stages. The first stage (1887–1913) was associated with the discovery that electromagnetic waves can be produced using spark generators and antennas of various types. During this period, radio and radar were invented and first sources of high-power UWB radiation capable to cause damage to the objects were developed and built. The second stage (1960–1990) featured, first of all, by the development and exploration of nuclear EMP simulators. At the same time, intense research and development work on subsurface UWB radars was carried out.

The contemporary stage, going back to the early 1990s of the twentieth century, is characterized by more stringent requirements on UWB systems of various applications. For antenna systems, this is, first of all, the requirement of small distortions of the radiated and received pulse waveforms. This is dictated by that the knowledge of the radiated and received pulse waveforms is important for obtaining information about an object in UWB radar and for signal coding in pulse radio communications.

In this period, the progress in pulse technologies has given impetus to the development of UWB radiation sources of various purposes with various parameter sets. Much attention has been given to UWB radio systems with controlled radiation characteristics. Theoretical methods and computer codes are being elaborated to form a basis for the refinement of all constituents of the UWB radio systems intended for applications.

To develop UWB radio systems for various applications requires knowledge of the physics of radiation of electromagnetic waves, their propagation in media, and scattering of short pulses by objects (Chapters 2 through 4). The accuracy of the results of relevant studies depends on the accuracy of estimation of the impulse responses of the radio system and its constituent elements (Chapter 5). The main elements of radio systems are receiving (Chapter 6) and transmitting (Chapter 7) antennas. Transmitting and receiving antenna arrays enable control of the radio system characteristics (Chapter 8). The use of UWB pulses with high electric field calls for the development of high-power radiation sources (Chapter 9).

Problems

1.1 Formulate the well-known methods of classification of radiations by frequency bands. What are the causes of differences in these methods?

1.2 What the concept of an ultrawideband radio system includes? Why the characteristics of the pulse propagation channel should be taken into account in elaborating technical requirements on the parameters of UWB radiation?

1.3 List the stages of development of UWB radio systems. Describe the basic achievements for each stage.

1.4 Describe the basic differences between UWB radar and narrowband radar.

References

[1]   Federal Communication Commission USA (FCC) 02-48, ET Docket 98-153, First Report and Order, April 2002.

[2]   Giri, D. V., and F. M. Tesche. "Classification of Intentional Electromagnetic Environments," *IEEE Trans. Electromagn. Compat.*, Vol. 46, No. 3, 2004, pp. 322–328.

[3]   Benford, J., J. A. Swegle, and E. Schamiloglu, *High Power Microwaves*. Second edition, New York: Taylor and Francis, 2007.

[4]   GOST R 51317.1.5–2009: *Electromagnetic Compatibility of Hardware. High-Power Electromagnetic Actions on Civil Systems. Main principles*, Moscow: Standard Publishers, 2009 (in Russian).

[5]   Rumsey, V. H., *Frequency Independent Antennas*, New York: Academic Press, 1966.

[6]   Kudryavtsev, P. S., *History of Physics*, Vol. 2, Moscow: Uchpedgiz, 1956 (in Russian).

[7]   Grigoryan, A. T., and A. N. Vyaltsev, Heinrich Hertz (1857–1894), Moscow: Nauka, 1968 (in Russian).

[8]   Rodionov, V. M., *Origin of Radio Engineering*, Moscow: Nauka, 1985 (in Russian).

[9]   Migulin, V. V., "*Origin of Radio and First Steps of Radio Engineering.*" In *100 Years of Radio. Collective papers*, pp. 7–24, V. V. Migulin and A. V. Gorokhovsky (eds.), Moscow: Radio i Svyaz, 1995 (in Russian).

[10] Schantz, H., *The Art and Science of Ultrawideband Antennas*, London: Artech House, 2005.

[11] Kobzarev, Yu. B., *Establishment of Domestic Radiolocation: Proceedings, Memoirs, Reminiscences*, Moscow: Nauka, 2007 (in Russian).

[12] Tesla, N., Colorado Springs. Diaries. 1899–1900 (Russian transl. of Nikola Tesla's Colorado Springs Notes, Samara: Agni Publishing House, 2008).

[13] Tesla, N., Papers, Second edition, Samara: Agni Publishing House, 2008 (Russian transl.).

[14] Schumann, W. O., "Über die Stralungslosen Eigenschwingungen einer Leitenden Kugel die von Luftschicht und einer Ionosphärenhülle Umgeben Ist," *Zeitschrift Naturforschung*, Vol. 7a, 1952, pp. 149–154.

[15] Bliokh, P. V., A. P. Nikolaenko, and Yu. F. Filippov. *Global Electromagnetic Resonances in the Earth–Ionosphere Cavity, Kiev: Naukova Dumka*, 1977 (in Russian).

[16] Shannon, C. A. "A Mathematical Theory of Communication," *Bell System Techn. J.*, Vol. 27, No. 3, 1948, pp. 379–423; Vol. 27, No. 4, 1948, pp. 623–656.

[17] Shannon, C. A., *Papers on Information Theory and Cybernetics*, Moscow: Izd. Inostr. Liter., 1963 (Russian transl., eds. R. L. Dobrushin and O. B. Lupanov).

[18] Medvedev, Yu. A., B. M. Stepanov, and G. V. Fedorovich, *Physics of Radiation Exitation of Electromagnetic Fields*, Moscow: Atomizdat, 1980 (in Russian).

[19] *Physics of the Nuclear Explosion, Vol.1: Development of an Explosion*, V. M. Loborev (Ed.), Moscow: Nauka, Fizmatlit, 1997 (in Russian).

[20] Balyuk, N. V., L. N. Kechiev, and P. V. Stepanov, *High-Power Electromagnetic Pulses: Action on Electronic Equipment and Methods of Protection*, Moscow: IDT Group Ltd., 2008 (in Russian).

[21] Golubev, A. I., et al., "Influence of Environmental Asymmetry on the Shape of an Electromagnetic Pulse from an Atmospheric Nuclear Explosion," *Plasma Physics Report*, Vol. 25, No. 5, 1999, pp. 387–392.

[22] Baum, C. E., "EMP Simulators for Various Types of Nuclear EMP Environments: an Interim Categorization," *IEEE Trans. Electromagn. Compat.*, Vol. 20, No. 1, 1978, pp. 35–53.

[23] Baum, C. E., "From the Electromagnetic Pulse to High-Power Electromagnetics," *Proc. IEEE*, Vol. 80, No. 6, 1992, pp. 789–817.

[24] Baum, C. E., "Radiation of Impulse-Like Transient Fields." In *Sensor and Simulation Notes*, No. 321, C. E. Baum (ed.), New Mexico, Kirtland: Air Force Research Laboratory, Directed Energy Directorate, 1989.

[25] Baum, C. E., and E. G. Farr. *"Impulse Radiating Antenna."* In *Ultra-Wideband, Short-Pulse Electromagnetics*, pp. 139–147, H. Bertoni et al. (eds.), New York: Plenum Press, 1993.

[26] Agee, F. J., et al., "Ultra-Wideband Transmitter Research," *IEEE Trans. Plasma Sci.*, Vol. 26, No. 3, 1998, pp. 860–872.

[27] Koshelev, V. I., *"Antenna Systems for Radiation of High-Power Ultrawideband Pulses," Proc. 3rd All-Russian Scientific and Technical Conference on Radar and Radio Communications*, Moscow, Oct. 26–30, 2009, Vol. 1, pp. 33–37 (in Russian).

[28] Baum, C. E., et al., "JOLT: A Highly Directive, Very Intensive, Impulse-Like Radiator," *Proc. IEEE*, Vol. 92, No. 7, 2004, pp. 1096–1109.

[29] Oicles, J. A., Grant J. R., and Herman M. H., "Realizing the Potential of Photoconductive Switching for HPM Applications," *Proc. SPIE*, Vol. 2557, 1995, pp. 225–236.

[30] Koshelev, V. I., et al., "High-Power Source of Ultrawideband Radiation Wave Beams with High Directivity," *Proc. 15th Inter. Symposium on High Current Electronics*, Tomsk, 2008, pp. 383–386.

[31] Efremov, A. M., et al., "Generation and Radiation of Ultra-Wideband Electromagnetic Pulses with High Stability and Effective Potential," *Laser Part. Beams*, Vol. 32, No. 3, 2014, pp. 413–418.

[32] Bessarab, A. V., et al., "An Ultrawideband Electromagnetic Pulse Transmitter Initiated by a Picosecond Laser," *Doklady Physics*, Vol. 51, No. 12, 2006, pp. 651–654.

[33] Kondrat'ev, A. A., et al., "Experimental Study of a Microwave-Radiation Generator Based on a Superlight Source," *Doklady Physics*, Vol. 56, No. 6, 2011, pp. 314–317.

[34] Cook, C. E., "Pulse Compression—Key to More Efficient Radar Transmission," *Proc. IRE*, Vol. 48, March 1960, pp. 310–316.

[35]   Cook, C. E., and M. Bernfield, *Radar Signals*, New York: Academic Press, 1967.

[36]   Shirman, Ya. D., et al., "The First Domestic Research on Superbroadband Radar," *Soviet J. Commun. Technol. Electron.*, Vol. 36, No. 1, 1991, pp. 21–23.

[37]   Cook, J. C., "Proposed Monocycle-Pulse Very-High-Frequency Radar for Air-Borne Ice and Snow Measurement," *Trans. AIEE Commun. Electron.*, Vol. 79, Nov. 1960, pp. 588–594.

[38]   Cook, J. C., "Radar Exploration Through Rock in Advance of Mining," *Trans. AIME*, Vol. 254, June 1973, pp. 140–146.

[39]   Finkelshtein, M. I., V. L. Mendelson, and V. A. Kutev, *Radiolocation of Schistose Ground Covers*, M. I. Finkelstein (ed.), Moscow: Sov. Radio, 1977 (in Russian).

[40]   Harmuth, H. F., *Nonsinusoidal Waves for Radar and Radio Communications*, New York: Academic Press, 1981.

[41]   Astanin, L. Yu., and A. A. Kostylev, *Fundamentals of Ultrawideband Radar Measurements*, Moscow: Radio i Svyaz, 1989 (in Russian).

[42]   *Introduction to Ultra-Wideband Radar System*, J. D. Taylor (ed.), London: CRC Press, 1995.

[43]   *Ultra-wideband Radar Technology*, J. D. Taylor (ed.), London: CRC Press, 2000.

[44]   Chen, V. C., and H. Ling, *Time-Frequency Transforms for Radar Imaging and Signal Analysis*, London: Artech House, 2002.

[45]   *Ground Penetrating Radar*. Second edition, D. J. Daniels (ed.), London: IEE, 2004.

[46]   *Problems of Prospective Radiolocation*, A. V. Sokolov (ed.), Moscow: Radiotekhnika, 2003 (in Russian).

[47]   *Problems of Subsurface Radiolocation*, A. Yu. Grinev (ed.), Moscow: Radiotekhnika, 2005 (in Russian).

[48]   *Detection and Identification of Radiolocation Objects*, A. V. Sokolov (ed.), Moscow: Radiotekhnika, 2006 (in Russian).

[49]   *Bioradiolocation*, A. S. Bugaev, S. I. Ivashov, and I. Ya. Immoreev (eds.), Moscow: BMSTU Publishers, 2010 (in Russian).

[50]   Ghavami, M., L. B. Michael, and R. Kohno, *Ultra Wideband Signals and Systems in Communication Engineering*, London: John Wiley and Sons, 2004.

[51]   Siriwongpairat, W. P., and K. J. R. Liu, *Ultra-Wideband Communications Systems: Multiband OFDM Approach*, New Jersey: John Wiley and Sons, 2008.

[52]   Dmitriev, A. S., and A. I. Panas, *Dynamic Chaos: New Information Carriers for Communication Systems*, Moscow: Izd. Fiz.-Mat. Lit., 2002 (in Russian).

[53]   Koshelev, V. I., "High-Power Pulses of Ultrawideband Radiation for Radar," In *Active phased antenna arrays*, pp. 428–454, D. I. Voskresensky, and A. I. Kanaschenkov (eds.), Moscow: Radiotekhnika, 2004 (in Russian).

[54]   Shtager, E. A. and E. V. Chaevsky, *Wave Scattering by Complex-Shape Bodies*, Moscow: Sov. Radio, 1974 (in Russian).

[55]   Stadnik, A. M., and G. V. Ermakov. "The Distortion of Ultrawideband Electromagnetic Pulses in the Earth's Atmosphere," *J. Commun. Technol. Electron.*, Vol. 40, No. 10, 1995, pp. 25–32.

[56]   Bunkin, B. V., and V. A. Kashin. "Features, Problems and Prospects of Subnanosecond Videopulsed Radars," *Radiotekh.*, No. 4–5, 1995, pp. 128–133.

[57]   Immoreev, I. Ya., "Ultrawideband Location: Basic Features and Differences from Conventional Radiolocation," *Elektromagn. volny i electron. sistemy*, Vol. 2, No. 1, 1997, pp. 81–88.

[58]   Immoreev, I. Ya., and D. V. Fedotov. "Optimum Processing of Radar Signals with Unknown Parameters," *Radiotekh.*, No. 10, 1998, pp. 84–88.

[59]   Immoreev, I. Ya., and V. S. Chernyak. "Detection of Ultrawideband Signals Reflected from Complex Targets," *Radiotekh.*, No. 4, 2008, pp. 3–10.

[60] Koshelev, V. I., V. T. Sarychev, and S. E. Shipilov. "Detection of Ultrawideband Pulsed Signals on the Background of Noise and Interferences," *Proc. 2nd Scientific Workshop on Ultrawideband Signals in Radar, Communication, and Acoustics*, Murom, July 4–7, 2006, pp. 332–336 (in Russian).

[61] Koshelev, V. I., V. T. Sarychev, and S. E. Shipilov. "Radar Target Detection at Noise and Interference Background." In *Ultra-Wideband, Short-Pulse Electromagnetics* 7, pp. 715–722, F. Sabath, et al. (eds.), New York: Springer, 2007.

[62] Koshelev, V. I., S. E. Shipilov, and V. T. Sarychev. "*Suboptimal Method of UWB Signal Detection at Noise, Interference and Clutter Environment*," Proc. 5th European Radar Conference, Amsterdam, Oct. 27–31, 2008, pp. 232–235.

[63] Harmuth, H. F., "Radar Equation for Nonsinusoidal Waves," *IEEE Trans. Electromagn. Compat.*, Vol. 31, No. 2, 1989, pp. 138–147.

[64] Lorber, H. W., "A Time Domain Radar Range Equation," In *Ultra-Wideband, Short-Pulse Electromagnetics* 2, pp. 355–364, L. Carin and L. B. Felsen (eds.), New York: Plenum Press, 1995.

[65] Briker, A. M., N. V. Zernov, and T. E. Martynova. "Scattering Properties of Antennas Illuminated by Nonharmonic Signals," *J. Commun. Technol. Electron.*, Vol. 45, No. 5, 2000, pp. 510–514.

[66] Koshelev, V. I., "Detection and Recognition of Radar Objects at Probing by High-Power Ultrawideband Pulses," *Proc. 2007 IEEE Inter. Conf. on Ultra-Wideband*, Singapore, Sep. 24–26, 2007, IEEE Catalog Number 07EX1479C.

[67] Balzovskii, E. V., I. Buyanov Yu, and V. I. Koshelev. "Dual Polarization Receiving Antenna Array for Recording of Ultra-Wideband Pulses," *J. Commun. Technol. Electron.*, Vol. 55, No. 2, 2010, pp. 172–180.

[68] Koshelev, V. I., et al., "Parametric Identification of Ultrabandwidth Signals on the Background of Noise," *Russ. Phys. J.*, Vol. 51, No. 6, 2008, pp. 601–609.

[69] Koshelev, V. I., et al., "Estimation of Informational Characteristics of Radar Objects in Ultrawideband Probing," *Zh. Radioelektron.*, No. 6, 2001, Available from: http:// jre. cplire.ru/jre/jun01/1/text.html.

[70] Nebabin, V. G., and V. V. Sergeev, *Methods and Procedure of Radar Identification*, Moscow: Radio i Svyaz, 1984 (in Russian).

[71] Kostylev, A. A., "Identification of Radar Targets with the Use of Ultrawideband Signals: Methods and Applications," *Zarubezh. Radioelektron.*, No. 4, 1984, pp. 75–104.

[72] Kostylev, A. A., "Identification and Use of Resonance Models of Scatterers and Antennas," *Zarubezh. Radioelektron.*, No. 1, 1991, pp. 23–34.

[73] Kononov, A. F., "Application of Tomographic Methods to Radar Imaging of Objects by Ultrawideband Signals," *Zarubezh. Radioelektron.*, No. 1, 1991, pp. 35–49.

[74] Kostylev, A. A., and N. Kalinin Yu. "Methods of Experimental Determination of Identification Signs Using Ultrawideband Signals," *Zarubezh. Radioelektron.*, No. 10, 1992, pp. 21–40.

[75] Kuznetsov, Yu. V., "Identification of Targets in Ultrawideband Radiolocation." In *Active Phased Antenna Arrays*, pp. 234–319, D. I. Voskresensky and A. I. Kanaschenkov (eds.), Moscow: Radiotekhnika, 2004 (in Russian).

[76] Rothwell, E., et al., "Radar Target Discrimination Using the Extinction-Pulse Technique," *IEEE Trans. Antennas Propogat.*, Vol. 33, No. 9, 1985, pp. 929–937.

[77] Rothwell, E. J., K. M. Chen, and D. P. Nyquist. "Extraction of the Natural Frequencies of a Radar Target from a Measured Response Using E-pulse Techniques," *IEEE Trans. Antennas Propogat.*, Vol. 35, No. 6, 1987, pp. 715–720.

[78] Baum, C. E., et al., "The Singularity Expansion Method and its Application to Target Identification," *Proc. IEEE*, Vol. 79, No. 10, 1991, pp. 1481–1492.

[79] Ilavarasan, P., et al., "Performance of an Automated Radar Target Discrimination Scheme Using E Pulses and S Pulses," *IEEE Trans. Antennas Propogat.*, Vol. 41, No. 5, 1993, pp. 582–588.

[80] Baum, C. E., "On the Singularity Expansion Method for the Solution of Electromagnetic Interaction Problem." In *Interaction Notes*, No. 88, C. E. Baum (ed.), New Mexico, Kirtland: Air Force Weapons Laboratory, 1971.

[81] Baum, C. E., "The Singularity Expansion Method." In *Transient Electromagnetic Fields*, pp. 129–179, L. B. Felsen (ed.), New York: Springer-Verlag, 1976.

[82] Rothwell, E. J., et al., "A General E-Pulse Scheme Arising from the Dual Early-Time/Late-Time Behavior of Radar Scatterers," *IEEE Trans. Antennas Propogat.*, Vol. 42, No. 9, 1994, pp. 1336–1341.

[83] Li, Q., et al., "Radar Target Identification Using a Combined Early-Time/Late-Time E-Pulse Technique," *IEEE Trans. Antennas Propogat.*, Vol. 46, No. 9, 1998, pp. 1272–1278.

[84] Das, Y., and W. M. Boerner., "On Radar Target Shape Estimation Using Algorithms for Reconstruction from Projections," *IEEE Trans. Antennas Propogat.*, Vol. 26, No. 2, 1978, pp. 274–279.

[85] Moffat, D. L., et al., "Transient Response Characteristics in Identification and Imaging," *IEEE Trans. Antennas Propogat.*, Vol. 29, No. 2, 1981, pp. 192–205.

[86] Dai, Y., et al., "Time-Domain Imaging of Radar Targets Using Algorithms for Reconstruction from Projections," *IEEE Trans. Antennas Propogat.*, Vol. 45, No. 8, 1997, pp. 1227–1235.

[87] Koshelev, V. I., S. E. Shipilov, and V. P. Yakubov. "Reconstructing the Shape of an Object in the Narrow Range-of-View Superwideband Radio Location," *J. Commun. Technol. Electron.*, Vol. 44, No. 3, 1999, pp. 281–284.

[88] Radon, J., "On the Determination of Function from their Integrals Along Certain Manifolds," *Ber. Saechs. Akad. Wiss. Leipzig, Math. Physics Kl.*, Vol. 69, No. 2, 1917, pp. 262–277.

[89] Lewis, R. M., "Physical Optics Inverse Diffraction," *IEEE Trans. Antennas Propogat.*, Vol. 17, No. 3, 1969, pp. 308–314.

[90] Bojarski, N. N., "Low Frequency Inverse Scattering," *IEEE Trans. Antennas Propogat.*, Vol. 30, No. 4, 1982, pp. 775–778.

[91] Koshelev, V. I., S. E. Shipilov, and V. P. Yakubov V. P. "Shape Reconstruction in Small-Angle Superwideband Radar Using Genetic Function," *J. Commun. Technol. Electron.*, Vol. 45, No. 12, 2000, pp. 1333–1338.

[92] Gutman, A. L., "The Kirchhoff Method of Calculating Pulsed Fields," *J. Commun. Technol. Electron.*, Vol. 42, No. 3, 1997, pp. 247–252.

[93] Koshelev, V. I., Shipilov S. E., and Yakubov V. P., "The Problems of Small Base Ultrawideband Radar." In *Ultra-Wideband, Short-Pulse Electromagnetics* 4, pp. 395–399, E. Heyman, B. Mandelbaum, and J. Shiloh (eds.), New York: Plenum Press, 1999.

[94] Bhalla, R., et al., "3D Scattering Center Representation of Complex Target Using the Shooting and Bouncing Ray Technique," *IEEE Trans. Antennas and Propagation Magazine*, Vol. 40, No. 5, 1998, pp. 30–39.

[95] Li, Q., et al., "Scattering Center Analysis of Radar Targets Using Fitting Scheme and Genetic Algorithm," *IEEE Trans. Antennas Propogat.*, Vol. 44, No. 2, 1996, pp. 198–207.

[96] "Special Issue on Antennas and Propagation Aspects of 60–90 GHz Wireless Communications," *IEEE Trans. Antennas Propagat.*, Vol. 57, No. 10, Part I of two Parts, 2009.

[97] Federici, J., and L. Moeller. "Review of Terahertz and Subterahertz Wireless Communications," *J. Appl. Phys.*, Vol. 107, No. 11, 2010, p. 111101.

[98] Aiello, G. R., and G. D. Rogerson. "Ultra-Wideband Wireless Systems," *IEEE Microwave Magazine*, Vol. 4, 2003, pp. 36–47.

[99]   Marenco, A. L., and R. Rice, *On Ultra Wideband (UWB) Technology and its Applications to Radar and Communications, Georgia: The Georgia Tech Ultra Wideband Center of Excellence, 2009,* http:// www.uwbtech.gatech.edu

[100]   Dmitriev, A. S., et al., "Experiments of Ultrawideband Directing Chaotic Data Transmission in the Microwave Band," *J. Commun. Technol. Electron.,* Vol. 47, No. 10, 2002, pp. 1112–1120.

[101]   Manzi, G., et al., "Coexistence Between Ultra-Wideband Radio and Narrow-Band Wireless LAN Communication System—Part I: Modeling and Measurement of UWB Radio Signals in Frequency and Time," *IEEE Trans. Electromagn. Compat.,* Vol. 51, No. 2, 2009, pp. 372–381.

[102]   Manzi, G., et al., "Coexistence Between Ultra-Wideband Radio and Narrow-Band Wireless LAN Communication System—Part II: EMI Evaluation," *IEEE Trans. Electromagn. Compat.,* Vol. 51, No. 2, 2009, pp. 382–390.

[103]   Kulkarni, V. V., et al., "A 750 Mb/s, 12 pJ/b, 6-to-10 GHz CMOS IR-UWB Transmitter with Embedded On-Chip Antenna," *IEEE J. Solid-State Circuits,* Vol. 44, No. 2, 2009, pp. 394–403.

[104]   Zhou, L., et al., "A 2-Gb/s 130 nm SMOS RF-Correlation-Based IR-UWB Transceiver Front-End," *IEEE Trans. Microw. Theory Tech.,* Vol. 59, No. 4, 2011, pp. 1117–1130.

[105]   Dmitriev, A. S., et al., "Ultrawideband Wireless Communications Based on Dynamic Chaos," *J. Commun. Technol. Electron.,* Vol. 51, No. 10, 2006, pp. 1126–1140.

[106]   Radasky, W. A., Baum C. E., and Wik M. W., "Introduction to the Special Issue on High-Power Electromagnetics (HPEM) and Intentional Electromagnetic Interference (IEMI)," *IEEE Trans. Electromagn. Compat.,* Vol. 46, No. 3, 2004, pp. 314–321.

[107]   Bugaev, S. P., et al., *Relativistic Multiwave UHF Generators,* Novosibirsk: Nauka, 1991 (in Russian).

[108]   Giri, D. V., *High-Power Electromagnetic Radiators: Nonlethal Weapons and Other Applications,* Cambridge: Harvard University Press, 2004.

[109]   Dobykin, V. D., et al., *Radioelectronic Struggle: Powered Destruction of Radioelectronic Systems,* Moscow: Vuzov. Kniga, 2007 (in Russian).

[110]   Holden, S., et al., "Ultra-Wideband (UWB) Radiofrequency (RF) Bioeffects Research in DERA," In *Ultra-Wideband, Short-Pulse Electromagnetics 5,* pp. 739–747, P. D. Smith and S. R. Cloude (eds.), New York: Plenum Press, 2002.

[111]   Schunck, T., et al., *"Penetration and Propagation into Biological Matter and Biological Effects of High-Power Ultra-Wideband Pulses: A Review,"* Electromagnetic Biology and Medicine, 2014 Informa Healthcare USA, Inc., http://informahealthcare.com/ebm.

[112]   Parmantier, J. P., "Numerical Coupling Models for Complex Systems and Results," *IEEE Trans. Electromagn. Compat.,* Vol. 46, No. 3, 2004, pp. 359–367.

[113]   Nitsch, D., et al., "Susceptibility of Some Electronic Equipment to HPEM Threats," *IEEE Trans. Electromagn. Compat.,* Vol. 46, No. 3, 2004, pp. 380–389.

[114]   Backstrom, M. G., and K. G. Lovstrand. "Susceptibility of Electronic Systems to High-Power Microwaves: Summary of Test Experience," *IEEE Trans. Electromagn. Compat.,* Vol. 46, No. 3, 2004, pp. 396–403.

[115]   Paletta, L., et al., "Susceptibility Analysis of Wiring in a Complex System Combining a 3-D Solver and a Transmission-Line Network Simulation," *IEEE Trans. Electromagn. Compat.,* Vol. 44, No. 2, 2002, pp. 309–317.

[116]   Vdovin, V. A., V. V. Kulagin, and V. A. Cherepenin. "Noises and Malfunctions Under Nonthermal Action of a Short Electromagnetic Pulse on Radioelectronic Devices," *Elektromagn. Volny i Elektron. Sistemy,* Vol. 8, No. 1, 2003, pp. 64–73.

[117]   Kuznetsov, Yu. V., et al., "Development of Methods for Analysis of Electromagnetic Radiations in a Wide Frequency Band," *Usp. Sovrem. Radioelektron.,* No. 1–2, 2009, pp. 132–139.

[118] Dong, X., et al., "Detection and Identification of Vehicles Based on Their Unintended Electromagnetic Emission," *IEEE Trans. Electromagn. Compat.*, Vol. 48, No. 4, 2006, pp. 752–759.

[119] Sachs, J. *Handbook of Ultra-Wideband Short-Range Sensing: Theory, Sensors, Applications*, New Jersey: Wiley-VCH, 2012.

[120] "*Ultra-Wideband Radio Technologies for Communications, Localization and Sensor Applications*," R. Thomä, et al. (eds.), Rijeka (Croatia): InTech, 2013.

[121] Rovňáková, J., D. Kocur, and P. Kažimír. "Investigation of Localization Accuracy for UWB Radar Operating in Complex Environment," *Acta Polytechnica Hungarica*, Vol. 10, No. 5, 2013, pp. 203–219.

[122]. Sahinoglu, Z., S. Gezici, and I. Guvenc, *Ultra-Wideband Positioning Systems: Theoretical Limits, Ranging Algorithms, and Protocols*, Cambridge: Cambridge University Press, 2008.

[123]. Zhang C., et al., "Real-Time Noncoherent UWB Positioning Radar With Millimeter Range Accuracy: Theory and Experiment," *IEEE Trans. Microwave Theory Technology*, Vol. 58, No. 1, 2010, pp. 9–20.

[124] Ameti, A., et al., "Extensible Object Location System and Method Using Multiple References," US Patent 8,149,169, April 3, 2012.

[125] Sensing the Future in the IoT, Elico. Available from: http://www.eliko.ee

[126] Zebra Makes Businesses as Smart and Connected as the World We Live in. Available from: http://www.zebra.com

[127] Schantz, H. G., et al., "Method of Near-Field Electromagnetic Ranging and Location," U.S. Patent 9,285,453, March 15, 2016.

[128] Schantz, H. G., and R. E. DePierre. "Directive, Electrically-Small UWB Antenna System and Method," U.S. Patent 9,209,525, December 8, 2015.

[129] Schantz, H. G., et al., "Multiple Phase State Near-Field Electromagnetic System and Method for Communication and Location," U.S. Patent 8,253,626, August 28, 2012.

[130] Sarwar, U. Internet of Things: The Next Technology Revolution. Available from: http://www.slideshare.net/usmanusb/the-internet-of-things-the-next-technology-revolution

[131] Metaxatos, P., and S. A. Nelson. The Internet of Things Needs Design, Not Just Technology. Available from: https://hbr.org/2016/04/the-internet-of-things-needs-design-not-just-technology

[132] Vijayakumar, P., and V. Vijayalakshm. RTLS Based Intelligent Transport System for BRTS Using RFID & IEEE 802.15.4 Modeled Wireless Mesh Networking. Available from: https://www.iimcal.ac.in/sites/all/files/sirg/6-5-analyzing-RTLS.PDF

[133] Bandakkanavar, R., Radio Frequency Identification Technology. Available from: http://krazytech.com/technical-papers/radio-frequency-identification-technology.

[134] Arslan, H., Z. N. Chen, and M. G. Di Benedetto. *Ultra Wideband Wireless Communication*, New Jersey: John Wiley & Son, 2006.

[135] Williams, B., *Intelligent Transport Systems Standards*, Boston/London: Artech House, 2008.

[136] ScenSor DWM1000 Module. Available from: http://www.decawave.com/products/dwm1000-module

[137] Chockalingam, A., and B. S. Rajan. *Large MIMO Systems*, Cambridge: Cambridge University Press, 2014.

[138] Li, J., and P. Stoica. *MIMO Radar Signal Processing*, New Jersey: John Wiley & Sons, 2009.

[139] Chernyak, V. S. "A New Direction in Radiolocation MIMO Radar," *Applied Radio Electronics*, Vol.8, No. 4, 2009, pp. 477–489.

[140] MIMO Antennas, ITELITE. Available from: http://www.itelite.net/en/Katalog/MIMO- 80211-n/

[141] MIMO Antennas, COBHAM. Available from: http://www.cobham.com/communications-and-connectivity/antenna-systems/microwave-antennas/mimo-antennas/product-brochure/

[142] Balzovsky, E. V., et al., "Dual Polarized Receiving Steering Antenna Array for Measurement of Ultrawideband Pulse Polarization Structure," *Rev. Sci. Instrum.*, Vol. 87, No. 3, 2016, p. 034703.

[143] Nekrasov, E. S., et al. "Investigation of the Polarization Characteristics of Ultrawideband Pulses Used for Sounding Metallic Objects Behind a Wall," *Proc. 10th All-Russian Science-and-Technology Conf. Radar and Radio Communications*, Moscow, Nov. 21–23, 2016, pp. 214–218 (in Russian).

[144] Hagness, S. C., A. Taflove, and J. E. Bridjes. "Three-Dimentional FDTD Analysis of a Pulsed Microwave Confocal System for Breast Cancer Detection: Design of an Antenna-Array Element," *IEEE Trans. Antennas Propagat.*, Vol. 47, No. 5, 1999, pp. 783–791.

[145] Sachs, J., et al., "Humanitarian Microwave Detection of Improvised Explosive Devices in Colombia (Project MEDICI)," *Proc. European Electromagnetics Symposium*, London, July 11–14, 2016, 5.a.2.pdf.

[146] Martinez, D., et al., "UWB Backscattering Characterization of Improvised Explosive Devices," *Proc. European Electromagnetics Symposium*, London, July 11–14, 2016, 6.a.1.pdf.

[147] Koshelev, V. I., A. A. Petkun, and V. M. Tarnovsky. "Detection of Metal Objects near a Random Rough Surface of Medium at Sounding by Orthogonally Polarized Ultrawideband Pulses," *Progress In Electromagnetic Res. B*, Vol. 70, 2016, pp. 27–40.

[148] Staderini, E. M., "UWB Radars in Medicine," *IEEE Aerospace and Electronic Systems Magazine*, No. 1, 2002, pp. 13–18.

[149] Immoreev, I., and Tao T. H., "UWB Radar for Patient Monitoring," *IEEE Aerospace and Electronic Systems Magazine*, No. 11, 2008, pp. 11–18.

[150] Lazano, A., D. Girbau, and R. Villarino. "Anylisis of Vital Signs Monitoring using an IR–UWB Radar," *Progress In Electromagnetic Res.*, Vol. 100, 2010, pp. 265–284.

[151] Single-chip radar sensors with sub-mm resolution–XeThru. Available from: http://www.xethru.com.

[152] Nezirovic, A., A. G. Yarovoy, and L. P. Lighart. "Signal Processing for Improved Detection of Trapped Victims Using UWB Radar," *IEEE Trans. Geosci. Remote Sensing*, Vol. 48, No. 4, 2010, pp. 2005–2014.

[153] Xu, Y., et al., "Vital Sign Detection Method Based on Multiple Higher Order Cumulant for Ultrawideband Radar," *IEEE Trans. Geosci. Remote Sensing*, Vol. 50, No. 4, 2012, pp. 1254–1265.

[154] Lv, H., et al., "Characterization and Identification of IR-UWB Respiratory-Motion Response of Trapped Victims," *IEEE Trans. Geosci. Remote Sensing*, Vol. 52, No. 11, 2014, pp. 7195–7204.

[155] Chen, T. C., et al., "Ultrawideband Syntetic Aperture Radar for Respiratory Motion Detection," *IEEE Trans. Geosci. Remote Sensing*, Vol. 53, No. 7, 2015, pp. 3749–3763.

[156] Klemm, I. J., et al., "Radar-Based Breast Cancer Detection Using a Hemispherical Antenna Array–Experimental Results," *IEEE Trans. Antennas Propagat.*, Vol. 57, No. 6, 2009, pp. 1692–1704.

[157] Conceicao, R. C., et al., "Comparison of Planar and Circular Antenna Configurations for Breast Cancer Detection Using Microwave Imaging," *Progress In Electromagnetic Res.*, Vol. 99, 2009, pp. 1–20.

[158] Wang, Y., et al., "Synthetic Bandwidth Radar for Ultra-Wideband Microwave Imaging Systems," *IEEE Trans. Antennas Propagat.*, Vol. 62, No. 2, 2014, pp. 698–705.

[159] Zinovev, S. V., et al., "Determination of Therapeutic Value of Ultra-Wideband Pulsed Electromagnetic Microwave Radiation on Models of Experimental Oncology," *J. Med. Phys.*, No. 3, 2015, pp. 62–67 (in Russian).

[160] LU, X., K. GUO, and Y. XIE. "The Acute UWB Pulse Exposure Hyperglycaemia and Hepatic Injury of KM Mouse," *Proc. European Electromagnetics Symposium*, London, July 11–14, 2016, 5.b.2.pdf.

[161] HPEM CarStop: Non-Violent System for Selective Stopping of Vehicles in Dynamic Scenarious, Diehl BGT Defence GmbH, Germany. Available from: http://www.diehl.com/fileadmin/diehl-defence/user_upload/flyer/HPEMcarStop_Flyer.pdf

[162] High-Power Compact Microwave Source for Vehicle Immobilization, Eureka Aerospace. Available from: https://www.ncjrc.gov/pdffiles1/nij/grants/236756.pdf

[163] Eureka Aerospace I Solutions to High-Tech Problems. Available from: http: www.eurekaaerospace.com

[164] Kuk, J. H., et al., "Analysis of HPEM Effects on an Automobile by Using Ultra-Wideband Pulse Generators," *Proc. European Electromagnetics Symposium*, London, July 11–14, 2016, P7.pdf.

[165] Urbancokova, H., J. Valouch, and S. Kovar. "Stopping of Transport Vehicles Using the Power Electromagnetic Pulses," *Przeglad Elektrotechniczny*, Vol. 91, No. 8, 2015, pp. 101–104.

[166] Morton, D., et al., "HPM WBTS, A Transportable High-Power Wide-Band Microwave Source," *Proc. IEEE Inter. Power Modulator and High Voltage Conf.*, Atlanta, May 23–27, 2010, pp. 186–189.

[167] Delmote, P., S. Pinguet, and F. Bieth. "Performances of a Compact, High-Power WB Source with Circular Polarization," In *Ultra-Wideband, Short-Pulse Electromagnetics* 10, pp. 239–250, S. Sabath and E. L. Mokole (eds.), New York: Springer, 2014.

[168] Andreev, Yu. A., et al., "A Source of High-Power Pulses of Elliptically Polarized Ultra-wideband Radiation," *Rev. Sci. Instrum.*, Vol. 85, No. 10, 2014, p. 104703.

[169] Andreev, Yu. A., et al., "Radiation of High-Power Ultrawideband Pulses With Elliptical Polarization by Four-Element Array of Cylindrical Helical Antennas," *Laser Part. Beams*, Vol. 33, No. 4, 2015, pp. 633–640.

[170] Koshelev, V. I., V. V. Plisko, and E. A. Sevostyanov. "Widening Radiation Spectra by Integration of Electromagnetic Pulses in Free Space," *Proc. 4nd All-Russian Microwave Conf.*, Moscow, Nov. 23–25, 2016, pp. 115–119 (in Russian).

[171] Giannakis, I., A. Giannopoulos, and A. Yarovoy. "Model-Based Evaluation of Signal-to-Clutter Ratio for Landmine Detection Using Ground-Penetrating Radar," *IEEE Trans. Geosci. Remote Sensing*, Vol. 54, No. 6, 2016, pp. 3564–3573.

[172] Hislop, G., "Permittivity Estimation Using Coupling of Commercial Ground-Penetrating Radars," *IEEE Trans. Geosci. Remote Sensing*, Vol. 53, No. 8, 2015, pp. 4157–4164.

[173] Serbin, G., and D. Or, "Ground-Penetrating Radar Measurement of Soil Water Content Dynamics Using a Suspended Horn Antenna," *IEEE Trans. Geosci. Remote Sensing*, Vol. 42, No. 8, 2004, pp. 1695–1705.

[174] Tan, A. E. C., K. Jhamb, and K. Rambabu, "Design of Transverse Electromagnetic Horn for Concrete Penetrating Ultrawideband Radar," *IEEE Trans. Antennas Propagat.*, Vol. 60, No. 4, 2012, pp. 1736–1743.

[175] Oloumi, D., et al., "Imaging of Oil-Wall Perforations Using UWB Synthetic Aperture Radar," *IEEE Trans. Geosci. Remote Sensing*, Vol. 53, No. 8, 2015, pp. 4510–4519.

[176] Shangguan, P., and I. L. Al-Qadi. "Calibration of FDTD Simulation of GPR Signal for Asphalt Pavement Compaction Monitoring," *IEEE Trans. Geosci. Remote Sensing*, Vol. 53, No. 3, 2015, pp. 1538–1548.

[177] Dogaru, T., C. Le, and L. Nguyen. "Synthetic Aperture Radar Images of a Simple Room Based on Computer Model," Army Research Laboratory Technical Report, Adelphi, MD, ARL-TR-5193, May 2010.

[178] Charvat, G. L., et al., "A Through-Dielectric Ultrawideband Switched-Antenna-Array Radar Imaging System," *IEEE Trans. Antennas Propagat.*, Vol. 60, No. 11, 2012, pp. 5495–5500.

[179] Ahmag, F., and M. G. Amin, "Through-the-Wall Human Motion Indication Using Sparsity-Driven Change Detection," *IEEE Trans. Geosci. Remote Sensing*, Vol. 51, No. 2, 2013, pp. 881–890.

[180] Negrier, R., et al., "High-PRF UWB Optoelectronic Radar System: A Clean-Type Algorithm to Overcome Depth Limitation," *IEEE Trans. Antennas Propagat.*, Vol. 64, No. 3, 2016, pp. 1080–1088.

[181] Yan, J. B., et al., "A Dual-Polarized 2-18-GHz Vivaldi Array for Airborne Radar Measurements of Snow," *IEEE Trans. Antennas Propagat.*, Vol. 64, No. 2, 2016, pp. 781–785.

[182] Smirnov, V. M., et al., "Space-Based Radar Systems Intended for Subsurface Sounding of Mars and Moon Soils," *J. Commun. Technol. Electron.*, Vol. 61, No. 2, 2016, pp. 112–119.

[183] Yang, H., et al., "Time-Gating-Based Time Reversal Imaging for Impulse Borehole Radar in Layered Media," *IEEE Trans. Geosci. Remote Sensing*, Vol. 54, No. 5, 2016, pp. 2695–2705.

[184] Koshelev, V. I., et al., "Object Detection by Three-Channel Antenna System of Ultrawideband Borehole Radar," *Proc. European Electromagnetics Symposium*, London, July 11–14, 2016, 4.a.3.pdf.

CHAPTER 2

Ultrawideband Pulse Radiation

Introduction

For a long period, in studying the characteristics of the electromagnetic radiation generated by a variety of sources, the researchers were interested mainly in stationary fields harmonically depending on time [1]. For these cases, rather efficient analytical and numerical methods were developed. If it was necessary to investigate nonstationary (time domain) processes, a frequency-domain to time–domain conversion was performed by using integral Laplace or Fourier transforms. However, it was found that this approach (formally quite reasonable) has a number of weak points when used to analyze essentially nonstationary processes characterized by very wide frequency ranges. Therefore, the current efforts of many researchers are focused on the development of new methods that would match the physical characteristics of the analyzed processes. However, the place and capabilities of new methods can hardly be appreciated without insight into the features and the specificity of implementation of well-established methods.

This chapter first describes the analytical methods that allow one to obtain formal solutions to the radiation problems for elementary radiators generating essentially nonstationary signals. Representations are given for the fields of elementary radiators, such as the electric and the magnetic Hertzian dipole and the slot radiator. The fields of finite-size (ring and disk) sources are found using an unconventional coordinate system related to the source. For each radiator type, a direct time-domain analysis is carried out.

Despite the apparent simplicity and academic nature of the above problems, they still find wide application. For instance, the problem of the nonstationary radiation of disk has several aspects. First, the interest in this problem is caused by the development of rational methods for calculating pulsed fields generated by various radiating systems [2–4]. Second, these radiators turned out to be attractive in developing electromagnetic missile and in investigating the feasibility of focusing UWB radiation [5–10]. Finally, these radiators became often used in elaborating the time-domain concepts of the reactive energy and Q factor of an antenna and in establishing fundamental limits on the antenna characteristics [11–13].

A considerable part of the chapter is devoted to the studies aimed at developing criteria for evaluating the field region boundaries for small-size radiators, such as electric monopoles, asymmetric electric dipoles, and bicone and combined antennas. The attention is focused on assessing the far-field boundaries of UWB radiation. To know the far-field boundaries for a UWB radiator, proper metrological procedures should be chosen for measuring the radiator characteristics. In the studies discussed, two approaches were used: one based on the determination of the far-field boundaries by the criterion of constancy of the radiation effective potential and the other based on the condition of the maximum (near field) and the minimum (far field) difference in the time dependences of the corresponding electromagnetic field components. In addition, the field zone boundaries for aperture radiators are characterized.

The advisability of studying experimentally a particular type of pattern as applied to a UWB radiator of particular purpose is discussed in detail.

A detailed description of the procedure of evaluating the efficiency of a radiator in relation to the field energy, peak power, and peak strength is given.

2.1   Elementary Sources of Ultrawideband Pulse Radiation

2.1.1   The Electric Hertzian Dipole

The simplest elementary radiator is an electric Hertzian dipole. It is an idealized model of an actual antenna composed of a short conductor with metal balls fixed at its ends [14, 15]. It is supposed that time-varying charges $+q(t)$ and $-q(t)$ are localized on the balls, providing a uniform distribution of current $I(t)$ along the conductor at any time. The current and the charges are related as

$$I = -\frac{dq}{dt} \qquad (2.1)$$

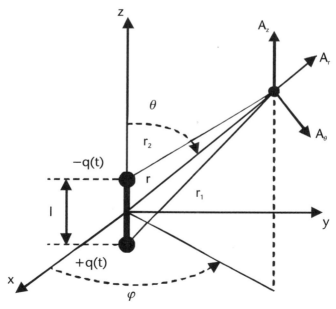

Figure 2.1   Sketch illustrating the calculation of the Hertzian dipole field.

Let us specify the dipole orientation and length by a vector l (Figure 2.1) and express the field in a space with material parameters ε and μ in terms of a vector potential $\mathbf{A}^e$ and a scalar potential φ^e:

$$\mathbf{H} = \frac{1}{\mu} \nabla \times \mathbf{A}^e, \quad \mathbf{E} = -\frac{\partial \mathbf{A}^e}{\partial t} - \nabla \varphi^e \tag{2.2}$$

As the medium is assumed to be nonconductive, the potentials satisfy equations:

$$\nabla^2 \mathbf{A}^e - \frac{1}{c^2} \frac{\partial^2 \mathbf{A}^e}{\partial t^2} = -\mu \mathbf{j}_i^e, \quad \nabla^2 \varphi^e - \frac{1}{c^2} \frac{\partial^2 \varphi^e}{\partial t^2} = -\frac{\rho^e}{\varepsilon} \tag{2.3}$$

where $c = 1/\sqrt{\varepsilon\mu}$ is the velocity of light in the space with the given material parameters. The particular solutions of these equations are determined by [14, 15]

$$\mathbf{A}^e(\mathbf{r}, t) = \frac{\mu}{4\pi} \int_V \frac{\mathbf{j}_i^e\left(\mathbf{r}', t - \dfrac{|\mathbf{r} - \mathbf{r}'|}{c}\right)}{|\mathbf{r} - \mathbf{r}'|} d^3\mathbf{r}' \tag{2.4}$$

$$\varphi^e(\mathbf{r}, t) = \frac{1}{4\pi\varepsilon} \int_V \frac{\rho^e\left(\mathbf{r}', t - \dfrac{|\mathbf{r} - \mathbf{r}'|}{c}\right)}{|\mathbf{r} - \mathbf{r}'|} d^3\mathbf{r}' \tag{2.5}$$

where the integration is performed over the volume occupied by extraneous currents (j_i^e) and charges (ρ^e).

Performing the integration in (2.4) over the conductor cross-section and assuming that the dipole dimensions are very small compared to the distance to the observation point ($l/r \ll 1$), we find

$$\mathbf{A}^e(\mathbf{r}, t) = \frac{\mu}{4\pi} \int_{-l/2}^{l/2} \frac{I\left(t - \dfrac{r}{c}\right)}{r} dl = \frac{\mu l}{4\pi r} I\left(t - \frac{r}{c}\right) \tag{2.6}$$

where $I(t - r/c)$ is the current in the dipole.

The scalar potential is found as the algebraic sum of the potentials induced by the charges $+q(t)$ and $-q(t)$ (Figure 2.1):

$$\varphi^e(\mathbf{r}, t) = \frac{1}{4\pi\varepsilon} \left[\frac{q\left(t - \dfrac{r_1}{c}\right)}{r_1} - \frac{q\left(t - \dfrac{r_2}{c}\right)}{r_2} \right] \tag{2.7}$$

Expanding each of the two functions in square brackets in (2.7) in powers of the small parameter $(l\cos\theta)/2$ and truncating the expansion to the terms linear with respect to this parameter, we obtain

$$\varphi^e(\mathbf{r},t) = \frac{1}{4\pi\varepsilon}\frac{\partial}{\partial r}\left[\frac{q\left(t-\dfrac{r}{c}\right)}{r}\right]l\cos\theta = \frac{1}{4\pi\varepsilon}\left[-\frac{q'\left(t-\dfrac{r}{c}\right)}{cr}-\frac{q\left(t-\dfrac{r}{c}\right)}{r^2}\right]l\cos\theta \quad (2.8)$$

where $q'(t - r/c)$ is the derivative of the function $q(t - r/c)$ with respect to the total argument or, which is the same, to the time t.

As can be seen from (2.6), the vector potential has only one component, A_z^e. In view of this, we obtain the following expressions for the potential components in spherical coordinates (see Figure 2.1):

$$A_r^e = A_z^e\cos\theta = \mu\frac{I\left(t-\dfrac{r}{c}\right)}{4\pi r}l\cos\theta \quad (2.9)$$

$$A_\theta^e = -A_z^e\sin\theta = -\mu\frac{I\left(t-\dfrac{r}{c}\right)}{4\pi r}l\sin\theta, \qquad A_\varphi^e = 0 \quad (2.10)$$

Hence, the only nonzero component of the magnetic field, H_φ, is determined by the expression

$$H_\varphi = \frac{1}{r}\left[\frac{\partial}{\partial r}\left(rA_\theta^e\right) - \frac{\partial}{\partial\theta}A_r^e\right] = \frac{l}{4\pi}\left[\frac{1}{cr}I'\left(t-\frac{r}{c}\right)+\frac{1}{r^2}I\left(t-\frac{r}{c}\right)\right]\sin\theta \quad (2.11)$$

The simplest way to determine the electric field strength [14] is by using Maxwell's first equation

$$\nabla \times \mathbf{H} = \varepsilon\frac{\partial\mathbf{E}}{\partial t} \quad (2.12)$$

$$\varepsilon\frac{\partial E_r}{\partial t} = \frac{1}{r^2\sin\theta}\cdot\frac{\partial}{\partial\theta}\left(r\sin\theta H_\varphi\right) = \frac{l}{2\pi}\left[\frac{1}{cr^2}I'\left(t-\frac{r}{c}\right)+\frac{1}{r^3}I\left(t-\frac{r}{c}\right)\right]\cos\theta \quad (2.13)$$

$$\varepsilon\frac{\partial E_\theta}{\partial t} = -\frac{1}{r\sin\theta}\cdot\frac{\partial}{\partial r}\left(r\sin\theta H_\varphi\right)$$
$$= \frac{l}{4\pi}\left[\frac{1}{c^2 r}I''\left(t-\frac{r}{c}\right)+\frac{1}{cr^2}I'\left(t-\frac{r}{c}\right)+\frac{1}{r^3}I\left(t-\frac{r}{c}\right)\right]\sin\theta \quad (2.14)$$

Integrating (2.13) and (2.14) with respect to time yields

$$E_r = \frac{l}{2\pi\varepsilon}\left[\frac{1}{cr^2}I\left(t-\frac{r}{c}\right)+\frac{1}{r^3}\int I\left(t-\frac{r}{c}\right)dt\right]\cos\theta \quad (2.15)$$

$$E_\theta = \frac{l}{4\pi\varepsilon}\left[\frac{1}{c^2 r}I'\left(t-\frac{r}{c}\right)+\frac{1}{cr^2}I\left(t-\frac{r}{c}\right)+\frac{1}{r^3}\int I\left(t-\frac{r}{c}\right)dt\right]\sin\theta \quad (2.16)$$

Thus, (2.11), (2.15), and (2.16) representing the electric and magnetic field components contain terms proportional to $1/r$, $1/r^2$, and $1/r^3$. This provides an opportunity to distinguish two specific zones: the near zone and the far one. The near zone is associated with the domination of the terms containing $1/r^2$ and $1/r^3$, whereas for the far zone, the dominating terms are those proportional to $1/r$. The boundaries of the near and the far zone are not fixed. With the chosen criteria of domination of some terms over the other, the positions of the boundaries depend on the time behavior of the functions $I'(t - r/c)$, $I(t - r/c)$, and $\int I(t - r/c)\, dt$.

It was noticed [16] that the retarded effect shows up at somewhat greater distances than it might seem at first glance. Actually, let us examine (2.11) in more detail. If the ratio r/c is small compared to the effective pulse duration and the functions $I(t)$ and $I'(t)$ do not take too large values at any time point in the time interval under consideration, it is admissible to use the following expansions:

$$I'\left(t - \frac{r}{c}\right) \approx I'(t) - I''(t)\left(\frac{r}{c}\right) + \frac{1}{2}I'''(t)\left(\frac{r}{c}\right)^2 \qquad (2.17)$$

$$I\left(t - \frac{r}{c}\right) \approx I(t) - I'(t)\left(\frac{r}{c}\right) + \frac{1}{2}I''(t)\left(\frac{r}{c}\right)^2 \qquad (2.18)$$

Therefore, we may write

$$H_\varphi \approx \frac{l}{4\pi}\left[\frac{1}{r^2}I(t) - \frac{1}{2c^2}I''(t) + \frac{1}{2}I'''(t)\frac{r}{c^3}\right]\sin\theta \qquad (2.19)$$

This equation shows that in a region near the dipole, the time behavior of the field follows the time variation of the dipole current with no delay. For the characteristic size of this region, different estimates are available in the literature [17–19]. For example, in [17, p. 303], it is stated that "... the effective source of the emitted radiation is not at the oscillator itself, but at a quarter wave-length in advance of it." This idea is actively promoted [19] later.

Let us briefly analyze the energy characteristics of the field radiated by an electric Hertzian dipole. The flux of the Poynting vector **S** through a sphere of radius r is determined by the expression [15]

$$P = \int_0^\pi \int_0^{2\pi} \frac{(\mathbf{S} \cdot \mathbf{r})}{r} r^2 \sin\theta\, d\theta\, d\varphi = \frac{l^2}{6\pi c^2}\sqrt{\frac{\mu}{\varepsilon}}$$

$$\times \left\{ \left[\frac{dI(t')}{dt}\right]^2 + \frac{2c}{r}I(t')\frac{dI(t')}{dt} + \frac{c^2}{r^2}\left[I^2(t') + \frac{dI(t')}{dt}\int I(t')\, dt'\right] + \frac{c^3}{r^3}I(t')\int I(t')\, dt' \right\} \qquad (2.20)$$

where $t' = t - r/c$, that represents the power radiated and accumulated in the space by the time t.

The radiated power is determined by

$$P = \frac{l^2}{6\pi c^2}\sqrt{\frac{\mu}{\varepsilon}}\left[\frac{dI(t')}{dt}\right]^2 \qquad (2.21)$$

Analysis of this equation shows that the power can be increased

1. by abruptly changing the current $I(t)$ (this, in particular, accounts for the detectable radiation of circuits containing fast-switched elements) and
2. for a given law of variation of the current $I(t)$, either by increasing the current amplitude or by increasing the dipole length l.

2.1.2 The Slot Radiator

Maxwell's equations for free space with no extraneous currents read

$$\nabla \times \mathbf{H} = \varepsilon \frac{\partial \mathbf{E}}{\partial t}, \quad \nabla \times \mathbf{E} = -\mu \frac{\partial \mathbf{H}}{\partial t} \tag{2.22}$$

In view of the symmetry of the equations (including the minus sign on the right side of the second equation), it is clear that to any their solution {E,H} there corresponds a dual solution {E′,H′} constructed by the rule

$$\mathbf{E}' = Z_0 \mathbf{H}, \ \mathbf{H}' = -\frac{\mathbf{E}}{Z_0}, \ Z_0 = \sqrt{\frac{\mu}{\varepsilon}} \tag{2.23}$$

For instance, the field of an electric Hertzian dipole is associated with the dual field [15]

$$E'_\varphi = \frac{Z_0 l}{4\pi}\left[\frac{1}{cr}I'\left(t - \frac{r}{c}\right) + \frac{1}{r^2}I\left(t - \frac{r}{c}\right)\right]\sin\theta\,, \tag{2.24}$$

$$H'_r = \frac{l}{2\pi}\left[\frac{1}{r^2}I\left(t - \frac{r}{c}\right) + \frac{c}{r^3}\int I\left(t - \frac{r}{c}\right)dt\right]\cos\theta \tag{2.25}$$

$$H'_\theta = -\frac{l}{4\pi}\left[\frac{1}{cr}I'\left(t - \frac{r}{c}\right) + \frac{1}{r^2}I\left(t - \frac{r}{c}\right) + \frac{c}{r^3}\int I\left(t - \frac{r}{c}\right)dt\right]\sin\theta \tag{2.26}$$

The dual field is generated by a slot, shaped like the dipole, cut in a conducting plane [15]. The slot is excited at the center by a voltage source. It is easy to verify directly that the Poynting vector and the power radiated through a sphere of infinite radius have the same values as the respective quantities for a Hertzian dipole.

2.1.3 The Magnetic Hertzian Dipole

The field of an electric Hertzian dipole can be described, using the formulas for vector and scalar products, by more compact expressions [15]:

$$\mathbf{H}(\mathbf{r},t) = \frac{l}{4\pi c}\left[\frac{1}{r}\frac{\ddot{\mathbf{p}} \times \mathbf{r}}{lr} + \frac{c}{r^2}\frac{\mathbf{p} \times \mathbf{r}}{lr}\right] \tag{2.27}$$

$$E(\mathbf{r},t) = \frac{Z_0 l}{4\pi c}\left\{\frac{1}{r}\frac{\mathbf{r}\times\mathbf{r}\times\ddot{\mathbf{p}}}{lr^2} + \frac{c}{r^2}\left[\frac{\mathbf{r}\times\mathbf{r}\times\dot{\mathbf{p}}}{lr^2} + \frac{2(\dot{\mathbf{p}}\cdot\mathbf{r})\mathbf{r}}{lr^2}\right] + \frac{c^2}{r^3}\left[\frac{\mathbf{r}\times\mathbf{r}\times\mathbf{p}}{lr^2} + \frac{2(\mathbf{p}\cdot\mathbf{r})\mathbf{r}}{lr^2}\right]\right\} \quad (2.28)$$

where $\mathbf{p} = q(t - r/c)\mathbf{l}$ is the dipole moment; the dotted symbols denote derivatives with respect to the argument.

We again consider the dual fields given by

$$E' = Z_0 H, \ H' = -\frac{E}{Z_0} \quad (2.29)$$

and the magnetic dipole moment given by

$$\mathbf{m} = Z_0\mathbf{p} \quad (2.30)$$

We next take into account that the magnetic dipole moment is determined by the equation

$$\mathbf{m} = \mu I(t)\mathbf{s} \quad (2.31)$$

where $I(t)$ is the current flowing in a small loop of vector area $\mathbf{s}$. The direction of $\mathbf{s}$ forms a right-handed system with the current $I(t)$.

The field of the magnetic Hertzian dipole is then determined by the equations

$$E'(\mathbf{r},t) = \frac{Z_0 s}{4\pi c^2}\left[\frac{1}{r}\frac{d^2 I(t')}{dt^2} + \frac{c}{r^2}\frac{dI(t')}{dt}\right]\frac{\mathbf{s}\times\mathbf{r}}{sr} \quad (2.32)$$

$$H'(\mathbf{r},t) = -\frac{s}{4\pi c^2}\left\{\frac{1}{r}\frac{d^2 I(t')}{dt^2}\frac{\mathbf{r}\times\mathbf{r}\times\mathbf{s}}{sr^2} + \left[\frac{c}{r^2}\frac{dI(t')}{dt} + \frac{c^2}{r^3}I(t')\right]\left[\frac{\mathbf{r}\times\mathbf{r}\times\mathbf{s}}{sr^2} + \frac{2(\mathbf{s}\cdot\mathbf{r})\mathbf{r}}{sr^2}\right]\right\} \quad (2.33)$$

The characteristic feature of this field is that, in the far zone, it is proportional to the second time derivative of the loop current. A magnetic dipole is a less efficient radiator than an electric dipole. Indeed, for the Poynting vector in the far zone, we have the expression

$$S' = E'\times H' = Z_0\left(\frac{s}{4\pi c}\right)^2\frac{1}{c^2 r^2}\left(\frac{d^2 I(t')}{dt^2}\right)^2\frac{\mathbf{r}}{r}\sin^2\theta \quad (2.34)$$

and for the instantaneous power radiated through a sphere of infinitely large radius, the expression

$$P = \frac{Z_0 s^2}{6\pi c^4}\left(\frac{d^2 I(t')}{dt^2}\right)^2 \quad (2.35)$$

2.2   Fields of Finite-size UWB Pulse Radiators

2.2.1   Radiation from Ring Sources

The problem of the nonstationary radiation generated by a circular loop (for the case of the current density symmetrically distributed over the loop) was solved analytically [20]. The geometry of the problem is shown in Figure 2.2. There is a circular loop of radius a with a uniformly distributed current arbitrarily varying with time. The loop lies in the plane xOy and its center coincides with the origin of the cylindrical coordinate system ρ,φ,z. The vector of the extraneous current density in the loop has a single nonzero component:

$$j_\varphi = I(\tau)\delta(z)\delta(\rho - a) \tag{2.36}$$

Here, $I(\tau) = \chi(\tau)u(\tau)$ is the current in the loop; $\tau = ct$, where c is the velocity of light; $\chi(\tau)$ is the Heaviside function, and $u(\tau)$ is a differentiable function. The time point at which the current starts flowing in the loop is taken for the time zero, $\tau = 0$.

The problem is to determine the nonstationary field at an arbitrary observation point P, which, by virtue of the problem symmetry, can be considered lying in the plane xOz. The distances from this point to the nearest and to the outermost point of the loop are determined as $R_2 = \sqrt{(\rho - a)^2 + z^2}$ and $R_1 = \sqrt{(\rho + a)^2 + z^2}$, respectively. The point $Q(\rho,0,0)$ is the projection of the point P onto the plane xOy in which the loop is located.

In solving the problem, the radial coordinate ρ entering into the inhomogeneous wave equation for the field component E_φ is eliminated by applying the Fourier–Bessel integral transform. The Klein–Gordon equation for transform is obtained whose solution for zero initial conditions is found by the Riemann method. The seeking field, on applying the inverse Fourier–Bessel integral transform, is presented in the form of a double integral. This integral, when the current is given by a Heaviside function, can be expressed in terms of an associated Legendre function having particular degree and order values, which can be reduced to elementary functions. As a result, the following representation for E_φ is obtained [20]:

$$E_\varphi = \frac{Z_0}{2\pi\rho}\frac{\cos\varphi}{\sin\varphi} \tag{2.37}$$

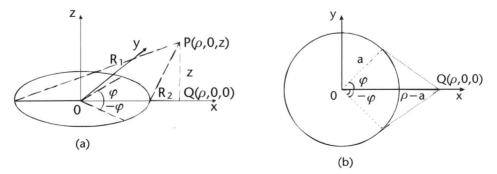

(a)

(b)

Figure 2.2   The problem geometry.

$$\cos\varphi = \frac{\rho^2 + z^2 + a^2 - \tau^2}{2\rho a} \tag{2.38}$$

$$\sin\varphi = \frac{1}{2\rho a}\left[(\rho + a)^2 + z^2 - \tau^2\right]^{1/2}\left[\tau^2 - z^2 - (\rho - a)^2\right]^{1/2} \tag{2.39}$$

$$z^2 + (\rho - a)^2 < \tau^2 < z^2 + (\rho + a)^2 \tag{2.40}$$

This result has a simple geometrical interpretation. The time interval during which the field at the observation point P is different from zero begins when the radiation pulse from the nearest point of the loop arrives at the point P and ends upon the arrival of the pulse from the most distant point of the loop. At the point P, at any fixed time, a field pulse is observed that has been generated by two loop elements symmetrical about the x-axis. The angular position of these elements is determined by the angle φ (Figure 2.2). The loop elements with angular coordinates $\varphi = 0$ and $\varphi = \pi$ radiate pulses of the Dirac delta function type. The more distant from the x-axis a pair of loop elements symmetrical about the axis, the less intense is their total radiation. Moreover, the pulses generated by the elements with coordinates $\varphi = \pm\pi/2$ quench each other. All this is in complete agreement with physical expectations.

The solution corresponding to an arbitrary time dependence of the loop current can easily be found, using (2.37), in the form of a convolution integral [20]:

$$E_\varphi = \frac{Z_0}{2\pi\rho}\int_{T_1}^{T_2}\frac{\partial I(\tau')}{\partial\tau'}\frac{\rho^2 + z^2 + a^2 - (\tau - \tau')^2}{\left[(\rho + a)^2 + z^2 - (\tau - \tau')^2\right]^{1/2}\left[(\tau - \tau')^2 - z^2 - (\rho - a)^2\right]^{1/2}}d\tau' \tag{2.41}$$

The integration limits are determined by the equations

$$T_2 = \tau - \sqrt{(\rho - a)^2 + z^2} \tag{2.42}$$

$$T_1 = \begin{cases} 0 & \text{for } \tau - \sqrt{(\rho + a)^2 + z^2} < 0 \\ \tau - \sqrt{(\rho + a)^2 + z^2} & \text{for } \tau - \sqrt{(\rho + a)^2 + z^2} > 0 \end{cases} \tag{2.43}$$

We present a quite simple approach to solving the problem. Note, first, that the vector potential of the field generated by a loop can be found by (2.4). As the current density in any cross section of the source is the same at a fixed time, the current will also be the same. Two current elements $I(t)ad\varphi'$ with angular coordinates $+\varphi'$ and $-\varphi'$ symmetrical about the x-axis (Figure 2.3) generate an electromagnetic field. This field at a point P of the plane xOz can be completely described by a vector potential having a single component:

$$dA_\varphi^e = \frac{\mu a}{2\pi}\frac{I\left(t - \dfrac{s}{c}\right)}{s}\cos\varphi' d\varphi' \tag{2.44}$$

To determine the vector potential of the total field using a conventional approach, this equation should be integrated with respect to φ' from 0 to π:

$$A_\varphi^e = \frac{\mu a}{2\pi} \int_0^\pi \frac{I\left(t - \frac{s}{c}\right)}{s} \cos\varphi' \, d\varphi' = \frac{\mu a}{2\pi} \int_0^\pi \frac{I\left(t - \frac{\sqrt{\xi^2 + z^2}}{c}\right)}{\sqrt{\xi^2 + z^2}} \cos\varphi' \, d\varphi' \quad (2.45)$$

However, the foregoing indicates that it is desirable to take the point Q for the origin and ξ for the integration variable (Figure 2.3).

Noting that

$$\cos\varphi' = \frac{\rho^2 + a^2 - \xi^2}{2a\rho}, \quad d\varphi' = \frac{2\xi d\xi}{\left[(\rho + a)^2 - \xi^2\right]^{1/2}\left[\xi^2 - (\rho - a)^2\right]^{1/2}} \quad (2.46)$$

we obtain

$$A_\varphi^e = \frac{\mu}{2\pi\rho} \int_{\rho - a}^{\rho + a} \frac{I\left(t - \frac{\sqrt{\xi^2 + z^2}}{c}\right)}{\sqrt{\xi^2 + z^2}} \frac{\left(\rho^2 + a^2 - \xi^2\right)}{\left[(\rho + a)^2 - \xi^2\right]^{1/2}\left[\xi^2 - (\rho - a)^2\right]^{1/2}} \xi d\xi \quad (2.47)$$

Next, we introduce into consideration a new integration variable t' by $\xi^2 + z^2 = (ct')^2$. Clearly, it has the meaning of the time it takes for an electromagnetic wave to pass (with the velocity of light) from a variable point of the loop to the observation point. Then we have

$$A_\varphi^e = \frac{Z_0}{2\pi\rho} \int_{R_2/c}^{R_1/c} I(t - t') \frac{\left(\rho^2 + a^2 + z^2 - (ct')^2\right)}{\left[(\rho + a)^2 + z^2 - (ct')^2\right]^{1/2}\left[(ct') - z^2 - (\rho - a)^2\right]^{1/2}} dt' \quad (2.48)$$

Finally, changing the integration variable by $t - t' = t''$, we obtain

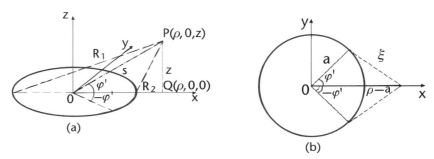

(a)   (b)

Figure 2.3 Sketch illustrating the calculation of the field of a loop source with nonstationary electric current.

$$A_\varphi^e(\rho,z,t) =$$

$$\frac{Z_0}{2\pi\rho}\int\limits_{t-R_1/c}^{t-R_2/c} I(t'')\frac{\left(\rho^2 + a^2 + z^2 - c^2(t-t'')^2\right)}{\left[(\rho + a)^2 + z^2 - c^2(t-t'')^2\right]^{1/2}\left[c^2(t-t'')^2 - z^2 - (\rho - a)^2\right]^{1/2}}\,dt'' \tag{2.49}$$

This representation of the total field, as can be seen from Figure 2.3, corresponds to the case $\rho > a$ where the projection of the observation point onto the plane xOy is outside the region covered by the loop in this plane. Similar considerations show that this representation is also valid for $\rho < a$. It should only be noted that for $t - R_1/c < 0$, the lower integration limit should be equal to zero in both cases. The expressions for the field components can be derived from the above representation by using well-known equations. For instance, taking into account the Lorentz gauge condition

$$\nabla \cdot \mathbf{A}^e + \frac{1}{c^2}\frac{\partial\varphi^e}{\partial t} = 0 \tag{2.50}$$

which relates the vector and the scalar potential, we obtain

$$\mathbf{E}(\rho,z,t) = c^2\int\nabla\nabla \cdot \mathbf{A}^e dt - \frac{\partial\mathbf{A}^e}{\partial t} \tag{2.51}$$

In our case, for $\mathbf{A}^e(0,0,A_\varphi^e)$, we have $\nabla \cdot \mathbf{A}^e = 0$ (as A_φ^e does not depend on φ), and, hence,

$$E_\varphi(\rho,z,t) = \frac{\partial A_\varphi^e}{\partial t} \tag{2.52}$$

Thus, according to (2.52), we have

$$E_\varphi(\rho,z,t) = -\frac{Z_0}{2\pi\rho}\frac{\partial}{\partial t}\int\limits_{t-R_1/c}^{t-R_2/c} I(t'')F(t,t'')\,dt'' \tag{2.53}$$

where

$$F(t,t'') = \frac{\left(\rho^2 + a^2 + z^2 - c^2(t-t'')^2\right)}{\left[(\rho + a)^2 + z^2 - c^2(t-t'')^2\right]^{1/2}\left[c^2(t-t'')^2 - z^2 - (\rho - a)^2\right]^{1/2}} \tag{2.54}$$

To perform the differentiation in (2.53), we use the formula for differentiation of a definite integral with variable integration limits,

$$\frac{\partial}{\partial y}\int\limits_{\varphi(y)}^{\psi(y)} f(x,y)\,dx = \int\limits_{\varphi(y)}^{\psi(y)}\frac{\partial f(x,y)}{\partial y}dx + \psi'(y)f\left(\psi(y),y\right) - \varphi'(y)f\left(\phi(y),y\right) \tag{2.55}$$

to obtain

$$E_{\varphi}(\rho,z,t) = -\frac{Z_0}{2\pi\rho}\left\{\int_{t-R_1/c}^{t-R_2/c} I(t'')\frac{\partial}{\partial t}F(t,t'')\,dt''\right.$$

$$\left. + I\left(t-\frac{R_2}{c}\right)F\left(t,t-\frac{R_2}{c}\right) - I\left(t-\frac{R_1}{c}\right)F\left(t,t-\frac{R_1}{c}\right)\right\} \quad (2.56)$$

Taking into account that

$$\frac{\partial}{\partial t}F(t,t'') = -\frac{\partial}{\partial t''}F(t,t'') \quad (2.57)$$

we integrate the integral in (2.56) by parts. Using the result of this integration,

$$\int_{t-R_1/c}^{t-R_2/c} I(t'')\frac{\partial}{\partial t}F(t,t'')\,dt'' = -\int_{t-R_1/c}^{t-R_2/c} I(t'')\frac{\partial}{\partial t''}F(t,t'')\,dt''$$

$$= -\left\{I\left(t-\frac{R_2}{c}\right)F\left(t,t-\frac{R_2}{c}\right) - I\left(t-\frac{R_1}{c}\right)F\left(t,t-\frac{R_1}{c}\right) - \int_{t-R_1/c}^{t-R_2/c}\frac{\partial I(t'')}{\partial t''}F(t,t'')\,dt''\right\} \quad (2.58)$$

in (2.56), we find

$$E_{\varphi}(\rho,z,t) = -\frac{Z_0}{2\pi\rho}\int_{t-R_1/c}^{t-R_2/c}\frac{\partial I(t'')}{\partial t''}F(t,t'')\,dt'' \quad (2.59)$$

Finally, putting $t = \tau/c$ in (2.59) and passing to a new integration variable by $t'' = \tau'/c$, we obtain a representation for the solution that coincide with (2.41).

2.2.2 Radiation from Disk and Circular Aperture Sources

Let there be a source shaped as a disk of radius a whose center is at the origin of a cylindrical coordinate system (ρ,φ,z). The disk is excited uniformly by a current arbitrarily varying with time and flowing along the x-axis. Assuming that the surface current density $\mathbf{J}^e$ is known, we determine the vector potential of the disk field at an arbitrary observation point $P(\rho,0,z)$ lying in the plane xOz (Figure 2.4). As the surface current density vector has only one nonzero component, $J_x^e = J$, the vector potential will also have a single component, A_x^e.

We again use an unconventional coordinate system, namely we put the origin of the coordinate system not at the center of the disk, but at the point Q that is the projection of the observation point onto the plane in which the disk is located (Figure 2.4).

For $\rho > a$, the disk element of angular dimension 2α enclosed by the arcs of circles with radii ξ and $\xi + d\xi$ (Figure 2.4), makes a contribution to the vector potential that can be described as

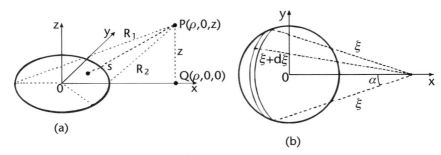

Figure 2.4   Sketch illustrating the calculation of the nonstationary field of a disk source.

$$
dA_x^e = \frac{\mu}{2\pi} \frac{J\left(t - \frac{\sqrt{\xi^2 + z^2}}{c}\right)}{\sqrt{\xi^2 + z^2}} \alpha \xi d\xi
\tag{2.60}
$$

As we have

$$
\alpha = \arccos\left(\frac{\xi^2 + \rho^2 - a^2}{2\rho\xi}\right)
\tag{2.61}
$$

then, integrating (2.60) with respect to ξ from $\xi = \rho - a$ to $\xi = \rho + a$, we obtain the vector potential of the field generated by the entire disk source:

$$
A_x^e = \frac{\mu}{2\pi} \int_{\rho-a}^{\rho+a} \frac{J\left(t - \frac{\sqrt{\xi^2 + z^2}}{c}\right)}{\sqrt{\xi^2 + z^2}} \arccos\left(\frac{\xi^2 + \rho^2 - a^2}{2\rho\xi}\right) \xi d\xi
\tag{2.62}
$$

The passage to a new integration variable t' by $\xi^2 + z^2 = (ct')^2$ yields the following representation of the vector potential:

$$
A_x^e = \frac{Z_0}{2\pi} \int_{R_2/c}^{R_1/c} J(t - t') \arccos\left(\frac{(ct')^2 - z^2 + \rho^2 - a^2}{2\rho\sqrt{(ct')^2 - z^2}}\right) dt'
\tag{2.63}
$$

Finally, one more change of the integration variable, $t - t' = t''$, results in the equation

$$
A_x^e = \frac{Z_0}{2\pi} \int_{t-R_1/c}^{t-R_2/c} J(t'') \arccos\left(\frac{c^2(t - t'')^2 - z^2 + \rho^2 - a^2}{2\rho\sqrt{c^2(t - t'')^2 - z^2}}\right) dt''
\tag{2.64}
$$

If $\rho < a$, that is, the projection of the observation point falls on the disk source, it is necessary to examine separately the contributions of the two disk parts to the total vector potential. The circle of radius $a - \rho$ centered at the point $(\rho,0,0)$ makes the following contribution:

$$A_{1x}^e = \frac{\mu}{2} \int_0^{a-\rho} \frac{J\left(t - \frac{\sqrt{\xi^2 + z^2}}{c}\right)}{\sqrt{\xi^2 + z^2}} \xi \, d\xi \tag{2.65}$$

After the passage to new integration variables t' and t'' by $(ct')^2 = \xi^2 + z^2$ and $t'' = t - t'$, this equation becomes

$$A_{1x}^e = \frac{Z_0}{2} \int_{t-R_2/c}^{t-z/c} J(t'') \, dt'' \tag{2.66}$$

The calculation of the contribution of the rest of the disk surface almost literally repeats the calculations performed for the case $\rho > a$. This is highlighted by the similarity between (2.64) and the resulting expression

$$A_{2x}^e = \frac{Z_0}{2\pi} \int_{t-R_1/c}^{t-R_2/c} J(t'') \arccos\left(\frac{c^2(t - t'')^2 - z^2 + \rho^2 - a^2}{2\rho\sqrt{c^2(t - t'')^2 - z^2}}\right) dt'' \tag{2.67}$$

Thus, for $\rho < a$, the total vector potential A_x^e is equal to $A_{1x}^e + A_{2x}^e$. The representations for the fields can be obtained using well-known equations just as it was made for the problem of the nonstationary radiation of a loop source. The solution of the problem of the radiation of an aperture source differs from the solution of the problem for a disk only by minor details.

The equations obtained offer a rather simple way to calculate the nonstationary fields of loop and disk sources. Figures 2.5–2.7 present examples of such calculations for a loop source. As can be seen from Figure 2.5, for a loop excited by a current pulse having the form of the Heaviside function, a significant contribution to the field is made only by the source parts nearest to or most distant from the observation point. The time behavior of the electric field component E_φ is consistent qualitatively with the features noted in Section 2.1. These features show up for

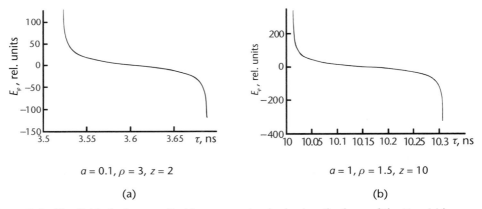

Figure 2.5   The field of a loop excited by a current pulse having the form of the Heaviside function

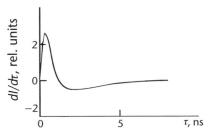

Figure 2.6 Behavior of the time derivative of the current in a loop.

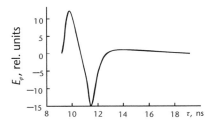

Figure 2.7 The field of a loop: $a = 1$, $\rho = 10$, $z = 2$.

both small and large sources; changing the position of the observation point does not affect the time behavior of the field.

Figure 2.6 shows the behavior of the time derivative of the source current given by the equation

$$\frac{dI(\tau)}{d\tau} = \tau^n \left[M^{n+1} e^{-M\tau} - e^{-\tau} \right] \chi(\tau), \; M = 5, \; n = 2 \tag{2.68}$$

and Figure 2.7 presents the time behavior of the field component E_φ corresponding to this pulsed excitation. Note that an increase in duration of the excitation pulse and the features of its behavior in the initial and final phases of the radiation process smooth the radiated pulse. It can also be seen that this is an equilibrium pulse.

2.3   The Structure of the Field of an Ultrawideband Radiator

2.3.1   The Boundaries of the Field Regions of a Short Radiator

For small-size antennas excited by harmonic oscillations, several field zones are distinguished depending on distance r: the near ($kr \ll 1$), the intermediate ($kr \approx 1$), and the far zone ($kr \gg 1$), where $k = 2\pi/\lambda$ and λ is the radiation wavelength [21]. The near zone features a $\pi/2$ phase difference between the electric and the magnetic field components; for the far zone, these components are in phase.

Apparently, the first attempt to determine the boundary of the far zone for an elementary radiator (electric Hertzian dipole) excited by a UWB or a time-domain current pulse was made by Harmuth [22]. The electric and the magnetic field region boundaries were determined by comparing the values of the components for fields varying as $1/r$ and $1/r^2$. The criterion for the electric field contains the integral and the time derivative of the current, and the criterion for the magnetic field contains the current and its time derivative:

$$r_E^2 >> \left| \frac{c^2 \int I(t)\, dt}{dI(t)/dt} \right|, \quad r_H >> \left| \frac{c I(t)}{dI(t)/dt} \right| \tag{2.69}$$

As rightly pointed out [23], the time derivative and integral of the current vary during the pulse. At some points of time, it may appear that the time derivative of the current is zero, and then, the boundary of the far zone will move to infinity. At other points of time, the current or the integral of the current may become zero, and then, the far zone will begin immediately at the radiator. The time dependence of the boundary position and its dependence on the field type (electric or magnetic) are the reasons why the criteria obtained by Harmuth [22] are not used in practice.

In this connection, a number of studies have been performed [24–27] to elaborate criteria for estimating the positions of field zone boundaries and their verification by means of numerical simulations and measurements for small-size radiators, such as electric monopoles [24], asymmetric electric dipoles [25], and bicone [26] and combined antennas [27]. The attention was focused on the estimation of the position of the far zone boundary of UWB radiation. In the present-day experimental studies, the far zone criterion $rE_p \approx const$ is widely used where E_p is the peak electric field at the distance r from the radiator. The quantity rE_p is called the effective radiation potential. Along with this criterion, it was proposed [24] to assess the boundaries of the field zones of UWB radiation using the criteria of the maximum (near field) and the minimum (far field) difference between the time functions of the electromagnetic field components E_z and H_φ. To quantify the difference between the field components, the RMSD is used which is calculated by $\sigma_f = \sqrt{\sum_{i=1}^{N} \left(E_{zi} - H_{\varphi i} \right)^2 / \sum_{i=1}^{N} H_{\varphi i}^2}$ where N is the length of the series that represents the sample function in time domain with a chosen sample spacing. In calculating σ_f, normalized dimensionless values of the field components are used.

First, we consider the results of studies of the electrical characteristics of a monopole [24]. To simulate the radiator, a code based on the finite-difference time-domain method was used. The geometry of the radiator is shown in Figure 2.8; the radiator parameters are the following: $a = 1$ mm, $b = 2.5$ mm, and $L = 10$ mm. The screen has unlimited dimensions. The simulation was performed for a Gaussian pulse $U(t) = U_0 \exp(-0.5(t/\tau_c)^2)$, a differentiated Gaussian (bipolar) pulse $U(t) = -U_0(t/\tau_c)\exp(-0.5(t/\tau_c)^2 + 0.5)$, double exponential pulses with the same rise time τ_r and different decay times τ_d (Figure 2.9), and double exponential pulses with the same pulse duration and different τ_r and τ_d (Figure 2.10).

For comparison, a harmonic signal $U(t) = U_0 \sin(\omega t)$ was also used. Here, U_0 is the pulse amplitude, τ_c is the characteristic length of the Gaussian pulse, and ω is the cyclic frequency. The pulse duration was determined by the base, and for the Gaussian pulse, it was $\tau_p = 8\tau_c$. With this duration, the pulse energy makes 99.99% of the energy of the mathematical Gaussian pulse. The grid step was chosen to satisfy the Courant condition $c\Delta t \leq \sqrt{\Delta r^2 \Delta z^2 / (\Delta r^2 + \Delta z^2)}$. In all calculations, the grid step was 0.25 mm, which provided a reasonable calculation accuracy.

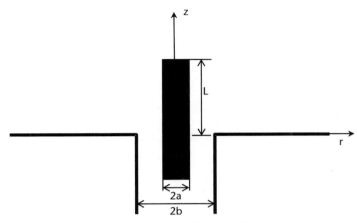

Figure 2.8   Geometry of an electric monopole. (With permission from Springer.)

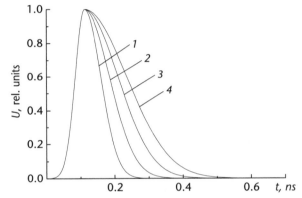

Figure 2.9   Voltage pulse waveforms $\tau_r = 0.11$ ns; $\tau_d = 0.3$ (1), 0.4 (2), 0.5 (3), and 0.6 ns (4). (With permission from Springer.)

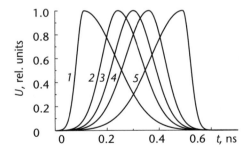

Figure 2.10   Voltage pulse waveforms: $\tau_p = 0.6$ ns; $\tau_r = 0.11$ (1), 0.24 (2), 0.3 (3), 0.36 (4), and 0.49 ns (5). (With permission from Springer.)

To estimate the positions of the boundaries of the field zones of a UWB radiator, the conditions of the maximum (near field) and minimum (far field) differences between the time functions of the electromagnetic field components E_z and H_φ were used. They are shown for illustration in Figure 2.11.

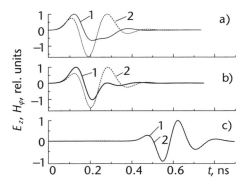

Figure 2.11 Time behavior of the field components E_z (1) and H_φ (2) for an exciting bipolar pulse of duration 0.333 ns at a distance $r = 2$ (*a*), 6 (*b*), and 120 mm (*c*). (With permission from Springer.)

The results obtained for the radiator excited by harmonic oscillations agree with the present physical notions. The phase difference $\Delta\varphi$ is about 89° for the near field and about 1° for the far field. These values are slightly different from the well-known theoretical estimates: $\Delta\varphi = 90°$ for the near field and $\Delta\varphi = 0$ for the far field. With the criteria $rE_p \approx const$ and $\Delta\varphi = 0$ used simultaneously, the position of the far zone boundary corresponding to the principal direction, $\theta = 90°$, was calculated as a function of L/λ (Figure 2.12). It was a minimum at $L/\lambda = 0.22$.

To determine the position of the far zone boundary for the radiator excited by monopolar and bipolar pulses, the criteria $rE_p \approx const$ and $\sigma_f \approx 0$ were used simultaneously. The corresponding dependence of this quantity on the ratio of the radiator length to the spatial length ($\tau_p c$) of the exciting bipolar (curve 1) and monopolar pulses (curve 2) is shown in Figure 2.13. Note that for the fields generated by the monopolar pulse and by the bipolar pulse whose length is twice that of the monopolar one, the far-zone boundaries almost coincide. Comparing these results with the results obtained for the radiator excited by harmonic oscillations (Figure 2.12), we see a moderated effect of resonant processes on UWB radiation pulses. As for

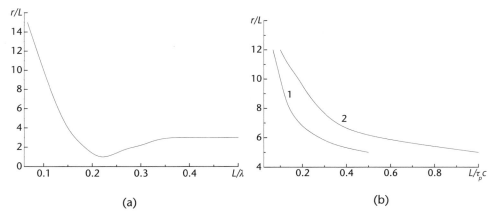

Figure 2.12 The position of the far zone of an electric monopole as a function of the harmonic oscillation frequency (a) and of the ratio of the radiator length to the spatial length of bipolar (1) and monopolar pulses (2) (b). (With permission from Springer.)

the near-zone boundary, in accordance with the criterion $\sigma_f \approx \sigma_{f\,max}$, it is located near the surface of the radiator.

The above criteria were used to estimate the position of the far-zone boundary for a radiator excited by pulses whose waveform is described by a double exponential function. In the first series of calculations, pulses with the same rise time τ_r, but with different decay times τ_d were used (Figure 2.9); accordingly, the pulse duration was $\tau_p = \tau_r + \tau_d$. Numerical examinations have shown that the estimated quantity was almost independent of the pulse duration, which was varied from 0.3 to 0.6 ns at a 0.11-ns rise time. It turned out that the far-zone boundary was at a distance of about 100 mm from the radiator, and it was close to the far-zone boundary for a symmetric Gaussian pulse of duration 0.22 ns. Similar calculations were performed for double exponential pulses of the same duration τ_p equal to 0.6 ns with varied τ_r and τ_d. Numerical examinations have shown that the maximum distance of the far-zone boundary from the radiator was obtained for a symmetric Gaussian pulse ($\tau_r = \tau_d$). Reducing τ_r or τ_d relative to the respective parameters of the symmetric pulse reduced the distance to the far-zone boundary from 135 to 100 mm. These distances were the same for pulses 1 and 5, and, respectively, for pulses 2 and 4 (Figure 2.10). The results obtained are easy to understand by examining the amplitude spectra of the exciting pulses. They show that a decrease in τ_r or τ_d increases the proportion of high-frequency components in the spectrum compared to the spectrum of a symmetric Gaussian pulse, and this results in a decrease in the distance to the far field zone. This also explains the independence of the position of the far-zone boundary on the pulse duration at a constant rise time: it is the rise time of a pulse that determines the proportion of high-frequency components in the pulse amplitude spectrum.

Consider next the results of investigations of an asymmetric electric dipole [25] as applied to the estimation of the position of the far-zone boundary. In contrast to the electric monopole with an infinite screen (Figure 2.8), the asymmetric dipole had a circular screen of finite size D. Investigations have shown that the characteristics of the radiators with an infinite screen are almost the same as those of the radiators with a screen of diameter $D/L = 20$. The difference shows up for a smaller screen diameter ($D/L = 10$) and becomes substantial for $D/L = 2$. For small screen diameters, the current flows onto the outer surface of the feeder, and a wave is formed and propagates along the surface. The ratio of the wave energy to the energy radiated into the upper half-space has a maximum at $D/L = 1$. For $D/L \geq 4$, the energy of the wave propagating near the feeder is low (~1%), and for $D/L = 10$, it is almost zero.

Using the criteria $rE_p \approx const$ and $\sigma_f \approx 0$, the position of the far-zone boundary corresponding to the principal direction of the peak power (E_p^2) pattern was determined for a radiator excited by bipolar pulses of different duration on varying the screen diameter of the screen (Figure 2.13). The position of the far-zone boundary was estimated for the pattern principal direction as indicated in Figure 2.14. The angle θ in the pattern was counted from the z-axis (Figure 2.8). As can be seen from Figure 2.13, the far-zone boundaries for the radiators with $D/L = 20$ and $D/L = \infty$ almost coincide. The distance to the far-zone boundary decreases with decreasing D/L and does not depend on $L/\tau_p c$ at $L/\tau_p c \geq 0.13$ for $D/L = 2$.

A bicone antenna has a wide bandwidth, allowing one to estimate the position of the far-zone boundary for a wide range of the frequencies and durations of

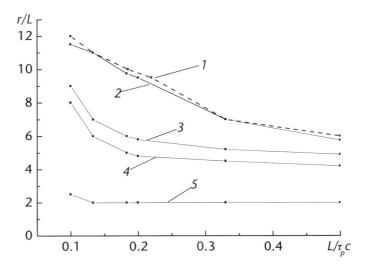

Figure 2.13 Position of the far-zone boundary for an asymmetric dipole as a function of the bipolar pulse duration for $D/L = \infty$ (1), 20 (2), 10 (3), 4 (4), and 2 (5). (With permission from Science & Technology Publishing House, Ltd.)

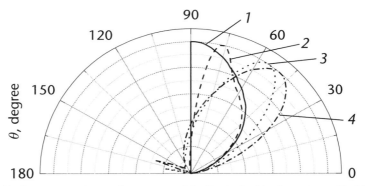

Figure 2.14 Peak-power patterns of an asymmetric dipole for bipolar pulses with $L/\tau_p c = 0.222$ and $D/L = 20$ (1), 10 (2), 4 (3), and 2 (4). (With permission from Science & Technology Publishing House, Ltd.)

bipolar pulses [26]. The geometry of the bicone antenna with a cone angle $2\theta_0 = 120°$ and an internal feeder is shown in Figure 2.15. The length of the cone generator, L, is 60 mm. The characteristic impedance of the feeder, as in the previous calculations, is 50 Ω.

Based on the previously chosen criteria, estimates were obtained for the position of the far-zone boundary corresponding to the principal direction, $\theta = 90°$, for harmonic oscillations and bipolar pulses. For harmonic oscillations, the criteria were the equality to zero of the phase difference between the electric and the magnetic field component, $\Delta\varphi \approx 0$, and the constancy of the effective radiation potential, $rE_p \approx const$. For a bipolar pulse, the criteria were the equality to zero of the RMS difference between the waveforms of the electric and magnetic field components, $\sigma_f \approx$

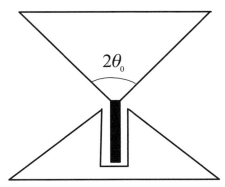

Figure 2.15   Geometry of a bicone antenna.

0, and $rE_p \approx const$. The obtained estimates are plotted in Figure 2.16 as functions of the oscillation frequency and of the bipolar pulse duration. For a bicone antenna with dimensions $L/\tau_p c < 0.5$, the distance to the far-zone boundary increases with $\tau_p c$ like for an electric monopole (Figure 2.12, curve 1). For large bicone antennas ($L/\tau_p c > 2$), it is almost a constant and equals $3L$. The results show that the estimates of the position of the far-zone boundary obtained for antennas excited by bipolar pulses and by harmonic oscillations almost coincide.

The above results of numerical simulations suggest that the position of the far-zone boundary for radiators excited by bipolar pulses can be estimated using the analytical equations obtained for radiators excited by harmonic oscillations. The relevant experimental investigations were performed for a combined antenna [27]. The antenna was optimized for excitation by a bipolar pulse of duration 3 ns. The transverse dimension of the antenna was half the central wavelength of the spectrum of the exciting pulse, $\lambda_0 = \tau_p c$. The quantity rE_p was measured as a function of the distance along the principal direction of the peak-power pattern. The distance at which this function is saturated ($rE_p \approx const$) corresponds to the far-zone boundary. An UWB dipole with resistive arms [28] was used for the receiving antenna. The

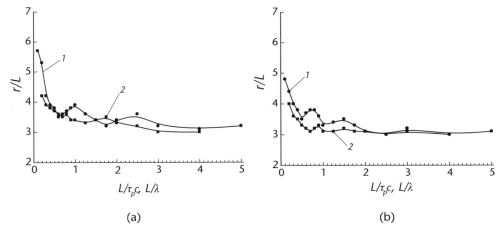

(a)   (b)

Figure 2.16   Estimates of the position of the far-zone boundary for a bicone antenna with $2\theta_0 = 120°$ as functions of the harmonic oscillation frequency L/λ (1) and of the bipolar pulse duration $L/\tau_p c$ (2), obtained using the criteria $\Delta\varphi \approx 0$ (1) and $\sigma_f \approx 0$ (2) (a) and the criterion $rE_p \approx const$ (b).

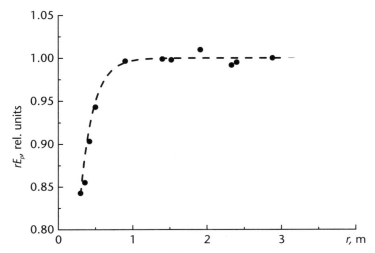

Figure2.17 The quantity rE_p as a function of distance for a combined antenna excited by a bipolar pulse of duration 3 ns. (With permission from David Publishing Company.)

measurements are presented in Figure 2.17. The experimentally evaluated distance from the radiation center [29] of the combined antenna, which, to within the measurement error, corresponds to its geometric center [30], to the far-zone boundary is about 1 m. The theoretical estimate ($r \gg \lambda_0/2\pi$) for a small-size radiator excited by a harmonic oscillation of wavelength λ_0 corresponds to the distance obtained in the experiment: $r \approx \lambda_0$.

2.3.2 The Boundaries of the Field Regions of Aperture Radiators

Among ultrawideband aperture radiators, the IRA antenna should be mentioned first. It is based on a parabolic reflector of diameter D that is excited by a pulse with a short rise time τ_r using two V-shaped feeders whose common center is located at the reflector focus [31, 32]. The formation of the radiation regions of IRA antennas was studied by the authors of [33–35].

The position of the near-field boundary is defined as the distance from the antenna for which the duration of the radiated pulse from quasistep signal is equal to the rise time of this signal and can be estimated by [35]

$$r = \frac{D^2}{8c\tau_r} \tag{2.70}$$

In the near field, the electric field amplitude remains almost unchanged [34, 35], that is, the so-called searchlight effect is realized. Outside the near zone, the field amplitude decreases. For the far zone, the condition $\Delta r/c < \tau_r$ should be satisfied. Here, Δr is the difference between the distances from the reflector center and from its edge to the observation point, and τ_r is determined by the 0.1–0.9 level of the voltage pulse amplitude. Experiments [33] have shown that for $\Delta r/c \approx \tau_r/5$, a far radiation zone is formed. Detailed theoretical studies [35] have revealed that the far zone, which is characterized by the decrease of the field strength as $1/r$, is formed even if $\Delta r/c \approx \tau_r/4$. Thus, to estimate the position of the boundary of this zone, we can use

$$r = \frac{D^2}{2c\tau_r} \tag{2.71}$$

In the intermediate zone, the field strength decreases more slowly than as $1/r$ [35]. Moreover, as follows from (2.70) and (2.71), the smaller τ_r, the more distant are the near and the intermediate zone from the radiator aperture.

Antenna arrays with simultaneously excited elements also are among the aperture radiators. The position of the far-zone boundary was estimated for a 4×4 square array of combined antennas with an aperture of 34.5×34.5 cm that was excited by a bipolar pulse of duration $\tau_p = 0.5$ ns [36]. The distance to the far-zone boundary was estimated by the well-known formula for an aperture antenna of arbitrary geometry with maximum transverse dimension D excited by a harmonic oscillation:

$$r = \frac{2D^2}{\lambda_0} \tag{2.72}$$

where $\lambda_0 = \tau_p c$, like in the above estimation of this quantity for a single combined antenna. For the considered array, (2.72) gives $r \approx 3$ m. Figure 2.18 shows the experimental dependence of the quantity rE_p on the distance between the array and the receiving antenna in the principal direction of the peak-power pattern. The receiving antenna was a combined antenna similar to an array element. The beginning of the horizontal portion of the curve corresponds to the boundary of the far zone. The theoretical estimates obtained by (2.72) comply with the measurements. Thus, the above results [27, 36] confirm the validity of using the formulas obtained for

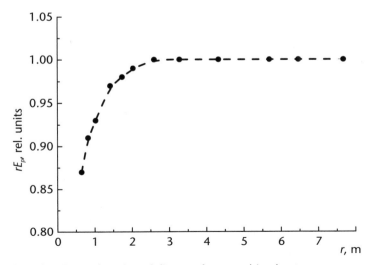

Figure 2.18 Quantity rE_p as a function of distance for a combined antenna array excited by a bipolar pulse of duration 0.5 ns (With permission from Pleiades Publishing, Ltd.)

harmonic oscillations to estimate the position of the far-zone boundary for both small-size antennas and arrays excited by a bipolar voltage pulse. This is due to that the energy of the exciting pulse is concentrated near the center frequency of the spectrum, which corresponds to the wavelength λ_0. Also note that for a bipolar pulse we have $\lambda_0 \approx 4\tau_r c$, and the substitution of this equation in (2.72) yields (2.71). Thus, (2.70)–(2.72) estimating the zones boundaries of UWB radiation fields can be used for wide-aperture antennas excited by pulses of different waveform.

2.4 Efficiency of the Generation of Electromagnetic Pulse Radiation

2.4.1 Radiation Patterns

One of the principal characteristics of a radiator is its radiation pattern. The waveform and duration of a radiated UWB pulse generally depend on the direction, unlike the same parameters of harmonic oscillations. Therefore, several pattern types have been proposed to date that can be used depending on the practical problem to be solved. They are distinguished by (1) peak field strength (E_p), (2) peak power (E_p^2), (3) peak field strength difference (E_{pp}), (4) squared peak field strength difference (E_{pp}^2), (5), average power (E_m^2), and (6) energy (W). Patterns of types 2 and 5 are discussed in [37] as applied to dipole antenna arrays. A pattern of type 4 is used in [38] to interpret the results of experimental investigations of the UWB radiation of a TEM antenna. A pattern of type 6 is useful for estimating the proportion of energy related to different pattern sectors [39]. Patterns of type 1 and type 3 are useful for detailed analysis of the angular dependence of the field, as they are based on the peak field strength. However, the most widely used in calculations and experiments is the peak-power pattern. As applied to harmonic signals, to this type of pattern there corresponds the power pattern, which is also widely used by researchers. Note that when plotted on the decibel scale, patterns of type 1 and type 2, and patterns of type 3 and type 4 do not differ from each other.

Radiation patterns are plotted, as a rule, in spherical coordinates. The center of the coordinate system is taken to be the phase center (or a partial phase center), if it exists and is invariable for all frequencies and angles (or for some ranges of frequencies and angles) of the radiated pulse, or the radiation center that is associated with the geometric center of the radiator. In general terms, an $F(\theta,\varphi)$ or $F^2(\theta,\varphi)$ pattern is a function of the angles θ and φ normalized for the maximum value of the parameter, for example, E_p or E_p^2. In some cases, the pattern is convenient to plot using the angle of elevation, δ, measured from the horizontal plane. In this case, the angles are measured from $\delta = \varphi = 0$. An important feature of the of patterns for UWB radiators is that the parameter by which a pattern is plotted lies in a spherical layer whose thickness is dictated by the maximum pulse duration within a given range of angles. For a sinusoidal pulse radiator, due to the independence of the pulse waveform and duration of the angles, the pattern is plotted based on a sphere of constant radius. Note that most of the patterns are plotted in two planes: vertical [$F(\theta,\varphi = 0)$] and horizontal [$F(\theta = 0,\varphi)$]. If the cross-section plane of a pattern corresponds to the polarization of the electric or the magnetic field, it is called an E-plane or an H-plane, respectively.

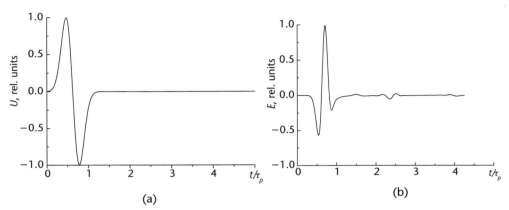

Figure 2.19 Waveforms of an exciting bipolar pulse of duration τ_p = 150 ps (a) and of the radiation pulse in the far zone (b).

As an example, we present the results of calculations of various types of pattern for a bicone antenna [26]. The geometry of the antenna (Figure 2.15) and its parameters are given above. The antenna is excited by a bipolar pulse of duration 150 ps. The waveforms of the exciting pulse and of the radiation pulse in the far zone are shown in Figure 2.19.

Figure 2.20 presents normalized patterns of type E_p (1), E_p^2 (2), E_{pp} (3), and W (4). It can be seen that for this UWB radiator, the four patterns are smooth and similar in shape, and their half-maximum width decreases sequentially.

2.4.2 The Energy, the Peak-power, and the Peak-field-strength Efficiency of a UWB Radiator

The efficiency of a UWB radiator is assessed using parameters that characterize its efficiency by energy, k_w, peak power, k_p, and peak field strength, k_E. Consider

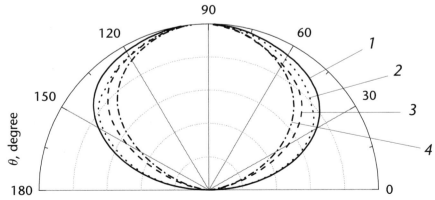

Figure 2.20 Radiation patterns of a bicone antenna plotted by peak electric field (1), peak power (2), maximum field strength difference (3), and energy (4) for a bipolar pulse of duration τ_p = 150 ps.

some methods for estimating the parameters k_w and k_p [39]. The energy efficiency of a radiator is defined as

$$k_w = \frac{W_{rad}}{W_g} \tag{2.73}$$

where W_{rad} is the energy radiated by the antenna and W_g is the energy in the voltage pulse at the antenna input. The radiated energy is experimentally determined as the difference between the energy in the voltage pulse at the antenna input and the energy in the pulse reflected from the antenna: $W_{rad} = W_g - W_{ref}$. The small energy losses due to the finite conductivity of the antenna are not taken into account. The value of k_w can also be estimated [40] from the spectrum of the pulse exciting the antenna or from the measured voltage factor of the standing wave in the feeder as a function of frequency (8.7). Both methods give nearly the same values of k_w.

Knowing the characteristics of the voltage pulse at the antenna input, the value of k_w and the space–time characteristics of the radiation of an antenna, we can find the peak-power efficiency of the antenna as

$$k_p = \frac{P_{rad}}{P_g} \tag{2.74}$$

where P_{rad} is the peak power of the radiation and P_g is the peak power in the pulse at the antenna input. The peak pulse power of the voltage generator is estimated as $P_g = U_{gmax}^2/\rho_f$, where U_{gmax} is the maximum absolute value of the generator pulsed voltage, and ρ_f is the wave impedance of the feeder. The radiation peak power is determined for the time corresponding to the maximum amplitude of the signal received by the antenna in the principal direction of the pattern. To account for the dependence of the radiated pulse waveform on the observation angle, the surface of the sphere is subdivided into sectors of area S_i (i = 1–12), such that the waveform of the radiated pulse in each sector could be considered invariable. Initially, the efficiency k_{pi} is estimated for each ith sector. The total peak-power efficiency is determined as the sum of all k_{pi}:

$$k_p = \sum_{i=1}^{12} k_{pi} = \frac{\sum_{i=1}^{12} \left(U_{a\max i}^2 S_i \right) k_w \int_0^{\tau_p} U_g^2(t)\, dt}{U_{g\max}^2 \sum_{i=1}^{12} \left(S_i \int_0^{\tau_{radi}} U_{ai}^2(t)\, dt \right)} \tag{2.75}$$

Here, $U_{ai}(t)$ is the time-depending pulsed voltage at the output of the receiving antenna, measured in the ith sector; $U_{a\max i}^2$ is the squared maximum value of $U_{ai}(t)$, and τ_{radi} is the duration of the respective radiation pulse. A detailed description of the method for estimating k_p is given in [39].

The peak-power patterns of an antenna plotted by measurements in the E-plane and H-plane also allow one to estimate the directivity D_0 of the antenna for the direction corresponding to $\delta = \varphi = 0$:

$$D_0 = \frac{4\pi F^2(0,0)}{\sum_{i=1}^{12} S_i F_i^2(\delta,\varphi)} \qquad (2.76)$$

where $F_i^2(\delta,\varphi)$ is the average value of the normalized pattern in the ith sector and $F^2(0,0) = 1$ is the value of the normalized pattern for the principal direction ($\delta = \varphi = 0$). Knowing k_p and D_0, one can calculate not only the radiation peak power, but also the peak strength of the far field at a given distance r:

$$E_p = \frac{1}{r}\sqrt{30 P_g k_p D_0} \qquad (2.77)$$

The peak electric field efficiency is estimated by the measurements in the far field for the pattern principal direction as

$$k_E = \frac{r E_p}{U_{g\max}} \qquad (2.78)$$

In contrast to the effective radiation potential $r E_p$ that characterizes a UWB radiation source, including the generator of the voltage pulse exciting the antenna, k_E characterizes only the radiator (antenna). This parameter is commonly used for comparative assessment of various antennas. In the frequency domain, to this parameter there corresponds the gain, which is equal to the product of the efficiency by the directivity parameter. The efficiency k_E also increases with k_w and D_0.

Conclusion

In this chapter, we have considered, first, the methods for solving the problems of the nonstationary radiation produced by elementary radiators. The possibilities of solving the problems directly in time domain have been demonstrated from the methodological point of view. It has been demonstrated that the duality principle can be helpful in deriving formulas for the fields of magnetic radiators. These problems have long been considered classics, and their solutions can be found in many publications [14–16, 18, 19]. However, not all these studies involved a thorough analysis of the solutions.

The solutions of the problems for finite-size radiators are remarkable in the sense that they are represented by integrals with finite integration limits. The structure of the solutions admits a clear physical interpretation, and the integrals are convenient for numerical computation regardless of the exciting pulse waveform.

Practically important criteria have been elaborated for estimating the positions of the field zone boundaries for UWB radiators. They have been verified by numerical simulations and measurements for a number of small-size radiators. The features of the patterns of antennas excited by a bipolar voltage pulse have been revealed. A complex method of evaluating the efficiency of a radiator by energy, peak power, and peak field strength has been proposed.

Problems

2.1 The complex amplitudes of the field components for an electric Hertzian dipole can be determined by [14]

$$E_r(\omega) = Z_0 \frac{p(\omega)\exp\left(\dfrac{-i\omega r}{c}\right)}{2\pi r}\left[-\frac{i\omega}{r} + \frac{c}{r^2}\right]\cos\theta \qquad (2.79)$$

$$E_\theta(\omega) = Z_0 \frac{p(\omega)\exp\left(\dfrac{-i\omega r}{c}\right)}{4\pi r}\left[-\frac{(i\omega)^2}{c} + \frac{i\omega}{r} + \frac{c}{r^2}\right]\sin\theta \qquad (2.80)$$

$$H_\varphi(\omega) = \frac{p(\omega)\exp\left(\dfrac{-i\omega r}{c}\right)}{4\pi r}\left[-\frac{(i\omega)^2}{c} + \frac{i\omega}{r}\right]\sin\theta \qquad (2.81)$$

where $p(\omega) = I(\omega)L$ is the dipole moment, and in contrast to [14] the time dependence $\exp(i\omega t)$ here is used.

Apply the inverse Fourier transform to these expressions to derive expressions for the components of the dipole field in time domain.

Solution: In view of the well-known properties of the transform, we can write

$$p(t) = \frac{1}{2\pi}\int_{-\infty}^{\infty} p(\omega)e^{i\omega t}\,d\omega, \quad p'(t) = i\omega\frac{1}{2\pi}\int_{-\infty}^{\infty} p(\omega)e^{i\omega t}\,d\omega \qquad (2.82)$$

$$p''(t) = (i\omega)^2\frac{1}{2\pi}\int_{-\infty}^{\infty} p(\omega)e^{i\omega t}\,d\omega \qquad (2.83)$$

In view of these equations, we find

$$E_r(t) = \frac{Z_0}{2\pi r}\left[\frac{p'(t')}{r} + \frac{cp(t')}{r^2}\right]\cos\theta \qquad (2.84)$$

$$E_\theta(t) = \frac{Z_0}{4\pi r}\left[\frac{p''(t')}{c} + \frac{p'(t')}{r} + \frac{cp(t')}{r^2}\right]\sin\theta \qquad (2.85)$$

$$H_\varphi(t) = \frac{1}{4\pi r}\left[-\frac{p''(t')}{c} + \frac{p'(t')}{r}\right]\sin\theta \qquad (2.86)$$

It can easily be seen that these expressions are equivalent to (2.11), (2.15), and (2.16).

2.2 Use (2.11), (2.15), and (2.16) to determine the far field ($r \gg c\tau_p$). Show that in this case, we have

$$\int_{-\infty}^{\infty} E_\theta(t)\, dt = 0 \tag{2.87}$$

Give a physical interpretation of the result obtained.

2.3 Use the principle of duality to find a solution to the problem of the radiation generated from a narrow annular slit in a conducting screen.

References

[1] Felsen, L. B., and N. Marcuvitz, *Radiation and Scattering of Waves*, New Jersey: Prentice Hall, Inc., 1973.

[2] Balakirev, V. A., and L. G. Sidel'nikov. "Formation of Electromagnetic Pulse by Aperture Antennas," *J. Commun. Technol. Electron.*, Vol. 44, No. 8, 1999, pp. 866–869.

[3] Skulkin, S. P., "On Some Features of the Pulsed Fields of Aperture Antennas," *Radiophysics and Quantum Electronics*, Vol. 42, No. 2, 1999, pp. 131–140.

[4] Dumin, O., and O. Tretyakov. "Radiation of Arbitrary Signals by Plane Disk," *Proc. 6th Int. Conf. Math. Meth. Electromagn. Theory*, Lviv, Ukraine, Sept. 19–24, 1996, pp. 248–251.

[5] Wu, T. T., "Electromagnetic Missiles," *J. Appl. Phys.*, Vol. 57, No. 7, 1985, pp. 2370–2373.

[6] Sodin, L. G., "The Characteristics of Pulse Radiation of Antennas (An Electromagnetic Missile)," *J. Commun. Technol. Electron.*, Vol. 37, No. 12, 1992, pp. 69–77.

[7] Sodin, L. G., "Antenna Pulse Radiation," *J. Commun. Technol. Electron.*, Vol. 43, No. 2, 1998, pp. 154–162.

[8] Kir'yashkin, V. V., and N. P. Chubinskii. "Studying of Possibility Collimating Electromagnetic-Wave Beams of Ultrawideband Signals," *J. Commun. Technol. Electron.*, Vol. 47, No. 1, 2002, pp. 18–26.

[9] Gutman, A. L. "The Kirchhoff Method of Calculating Pulsed Fields," *J. Commun. Technol. Electron.*, Vol. 42, No. 3, 1997, pp. 247–252.

[10] Mikhailov, E. M., and P. A. Golovinskii P. A. "Description of Diffraction and Focusing of Ultrashort Pulses on the Basis on a Nonstationary Kirchhoff –Sommerfeld Method," *J. Exp. Theor. Phys.*, Vol. 90, No. 2, 2000, pp. 240–249.

[11] Heyman, E., and T. Melamed. "Certain Considerations in Aperture Synthesis of Ultrawideband/Short-Pulse Radiation," *IEEE Trans. Antennas Propagat.*, Vol. 42, No. 4, 1994, pp. 518–525.

[12] Shlivinski, A., and E. Heyman. "Energy Consideration in Space-Time Synthesis of Collimated Pulsed Apertures," In *Ultra-Wideband, Short-Pulse Electromagnetics 4*, pp. 65–75, E. Heyman et al. (eds.), New York: Academic Press, 1999.

[13] Shlivinski, A., and E. Heyman. "Time-Domain Near-Field Analysis of Short-Pulse Antennas. Part I: Spherical Wave (Multiple) Expansion," *IEEE Trans. Antennas Propagat.*, Vol. 47, No. 2, 1999, pp. 271–279.

[14] Goldshtein, L. D., and N. V. Zernov. *Electromagnetic Fields and Waves*, Moscow: Sov. Radio, 1971 (in Russian).

[15] Harmuth, H. F., *Antennas and Waveguides for Nonsinusoidal Waves*, Orlando: Academic Press, 1984.

[16] Feynman, R. P., R. B. Leighton, and M. Sands, *The Feynman Lectures on Physics*, Vol. 2. Reading, Massachusetts: Addison-Wesley, 1964.

[17] Lodge, Oliver, *Modern Views of Electricity*, London: Macmillan and Co., 1907.

[18] Franceschetti, G., and C. H. Papas. "Pulsed Antennas," *IEEE Trans. Antennas Propagat.* Vol. 22, No. 5, 1974, pp. 651–661.

[19]   Schantz, H. G., "Electromagnetic Energy Around Hertzian Dipoles," *IEEE Antennas Propagat. Magazine*, Vol. 43, No. 2, 2001, pp. 50–62.

[20]   Borisov, V. V., "Radiation of Electromagnetic Signal by a Circular Current," *Proc. 10th All-Union Symp. Diffraction and Propagation of Waves*, Vinnitsa, Ukraine, 1990, pp. 171–174.

[21]   Fradin, A. Z., *Antenna-Feeder Devices*, Moscow: Svyaz, 1977 (in Russian).

[22]   Harmuth, H. F. *Transmission of Information by Orthogonal Functions*, New York: Academic Press, 1970.

[23]   Krymskii, V. V., V. A. Bukharin, and V. I. Zalyapin, *Theory of Nonsinusoidal Electromagnetic Waves*, Chelyabinsk: ChSTU Publishers, 1995 (in Russian).

[24]   Koshelev, V. I., S. Liu , and A. A. Petkun , "Criteria for the Field Boundaries of an Axially Symmetric Ultrawideband Radiator," *Russ. Phys. J.*, Vol. 51, No. 9, 2008, pp. 930–935.

[25]   Koshelev, V. I., S. Liu, and A. A. Petkun. "Effect of the Screen Diameter on the Characteristics of a Short Electrical Radiator," *Izv. Vyssh. Uchebn. Zaved., Fiz.*, Vol. 53, No. 9/2, 2010, pp. 49–53.

[26]   Koshelev, V. I., et al., "Frequency-Domain and Time-Domain Characteristics of Conical TEM Antennas," *Proc. 4th All-Russian Sci. and Tech. Conf. Radar and Radio*, Moscow, Russia, Nov. 29–Dec. 3, 2010, pp. 336–340. (in Russian).

[27]   Koshelev, V. I., et al., "Study on Stability and Efficiency of High-Power Ultrawideband Radiation Source," *J. Energy Power Eng.*, Vol. 6, No. 5, 2012, pp. 771–776.

[28]   Balzovsky, E. V., Yu. I. Buyanov, and V. I. Koshelev. "An Ultrawideband Dipole Antenna With Resistive Arms," *J. Commun. Technol. Electron.*, Vol. 49, No. 4, 2004, pp. 426–431.

[29]   Markov, G. T., and D. M. Sazonov, *Antennas*, Moscow: Energia, 1975 (in Russian).

[30]   Balzovskii, E. V., V. I. Koshelev, and S. E. Shipilov. "Ultrawideband Probing of Objects Hidden Behind Radio-Transparent Obstacles," *Izv. Vyssh. Uchebn. Zaved., Fiz.*, Vol. 53, No. 9/2, 2010, pp. 83–87.

[31]   Baum, C. E., "Radiation of Impulse-Like Transient Fields." In *Sensor and Simulation Notes*, No. 321, C. E. Baum (ed.), Kirtland, NM: Air Force Res. Lab., Directed Energy Directorate, 1989.

[32]   Baum, C. E., and E. G. Farr. "Impulse Radiating Antennas," In *Ultra-Wideband, Short-Pulse Electromagnetics*, Edited by H. Bertoni, et al., New York: Plenum Press, 1993, pp. 139–147.

[33]   Smith, I. D., et al., "Design, Fabrication and Testing of Paraboloidal Reflector Antenna and Pulser System for Impulse-Like Waveforms," *Proc. 10th Inter. Pulsed Power Conference*, Albuquerque, NM, 1995, Vol. 1, pp. 56–64.

[34]   Mikheev, O. V., et al., "New Method for Calculating Pulse Radiation From an Antenna With a Reflector," *IEEE Trans. Electomagn. Compat.*, Vol. 39, No. 1, 1997, pp. 48–54.

[35]   Giri, D. V., et al., "Intermediate and Far Fields of a Reflector Antenna Energized by a Hydrogen Spark-Gap Switched Pulser," *IEEE Trans. Plasma Sci.*, Vol. 28, No. 5, 2000, pp. 1631–1636.

[36]   Efremov, A. M., et al., "High-Power Sources of Ultra-Wideband Radiation With Subnanosecond Pulse Lengths," *Instrum. Exp. Tech.*, Vol. 54, No. 1, 2011, pp. 70–76.

[37]   Harmuth, H. F., *Nonsinusoidal Waves for Radar and Radio Communications*, New York: Academic Press, 1981.

[38]   Theodorou, E. A., et al., "Broadband Pulse-Optimized Antenna," *IEE Proc.*, Vol. 128, Pt. H, No. 3, 1981, pp. 124–130.

[39]   Andreev, Yu. A., I. Buyanov Yu, and V. I. Koshelev. "A Combined Antenna With Extended Bandwidth," *J. Commun. Technol. Electron.*, Vol. 50, No. 5, 2005, pp. 535–543.

[40]   Koshelev, V. I., and V. V. Plisko. "Energy Characteristics of Four-Element Arrays of Combined Antennas," *Izv. Vyssh. Uchebn. Zaved., Fiz.*, Vol. 56, No. 8/2, 2013, pp. 134–138.

CHAPTER 3

Propagation of Ultrawideband Pulses

Introduction

Any solution of Maxwell's equations, unique and continuous at all points of a homogeneous, linear, isotropic, and stationary medium, describing the transfer of a finite energy, represents a physically possible electromagnetic field. The simplest solutions of Maxwell's equations are those that depend on time and on a single spatial coordinate (so-called plane waves). Solutions of this type describe the waves that transfer infinitely great amounts of energy and therefore are physically impossible. However, the factors that determine the propagation of plane waves occur also in the propagation of more complex, realistic fields typical of practical applications. In addition, a proper superposition of plane waves with corresponding weights makes feasible a synthesis field satisfying a finite energy condition.

The chapter considers various factors affecting the propagation of plane waves in unbounded conducting media. The physical phenomena are discussed that occur during the propagation of UWB electromagnetic pulses in linear isotropic media without dispersion and in the presence of dispersion and that result from the non-linear interaction of a high-power radiation pulse with the propagation medium.

Methodologically, finding a solution to the problem of a UWB pulse propagating in a medium without dispersion is considered using a time-domain and a frequency-domain approach. The combining idea is to obtain a solution not separated in space and time variables as most adequate to describe the phenomena under investigation. On the other hand, the possibilities and complexity of the mathematical approach used to solve the problem are demonstrated.

The dispersion of a medium can be efficiently taken into account by using a rather accurate semiempirical model for the complex refractive index of the Earth's atmosphere. In solving the problem of a high-power electromagnetic pulse propagating through the lower atmosphere, a model is used in which Maxwell's equations and the Boltzmann equation for free electrons are solved self-consistently.

In reality, rather many natural media with different parameters form well-defined interfaces with each other. Reconstruction of the medium parameters from the measurements or calculations of the transmitted (or reflected) electromagnetic field is one of the most important problems of radiophysics, acoustics, and geophysics. However, the parameters of a medium cannot be reconstructed efficiently by solving the corresponding inverse problem if adequate methods for solving direct problems of the interaction of pulsed radiation with a layered medium are not available. Therefore, the chapter also presents several solutions of relevant direct problems.

To solve the problem of a plane pulse wave reflecting from or passing through the boundary of a conducting half-space, several methods are proposed. In one of them, the time-domain calculation of the reflection coefficient is of importance. Therefore, the method for calculating this coefficient is described in detail. Separately, the structure of the pulsed field of an infinite line source located parallel to the interface of two media is considered for the case where the displacement currents are negligible compared to the conduction currents in the medium.

The final section of the chapter is presented with somewhat greater attention. It is devoted to the study of the transient electromagnetic field of a horizontal electric dipole located at the upper boundary of a multilayered dielectric medium that is excited by a short current pulse. It is assumed that the medium is nondispersive. The structure of the solution obtained is such that it clearly illustrates the different types of waves (waveguide, leaky, and lateral) generated in the system and shows up their distinctive features. This provides for identification of these waves in the overall response of the layered structure and, thus, determination of the geometric parameters of the medium.

3.1 Propagation of Ultrawideband Electromagnetic Pulses in Conducting Media

3.1.1 Propagation of Ultrawideband Pulses in Unbounded Media

3.1.1.1 The Frequency-domain Solution

If the propagation medium is unbounded and nondispersive, the permittivity ε, permeability μ, and conductivity σ^e are constant scalars [1, 2], so that the constitutive equations read

$$\mathbf{D} = \varepsilon\mathbf{E}, \ \mathbf{B} = \mu\mathbf{H}, \ \mathbf{j} = \sigma^e\mathbf{E} \tag{3.1}$$

Maxwell's equations for such media with no sources can be written as

$$\nabla \times \mathbf{H} = \varepsilon\frac{\partial\mathbf{E}}{\partial t} + \sigma^e\mathbf{E} \tag{3.2}$$

$$\nabla \times \mathbf{E} = -\mu\frac{\partial\mathbf{H}}{\partial t} \tag{3.3}$$

$$\nabla \cdot \mathbf{E} = 0 \tag{3.4}$$

$$\nabla \cdot \mathbf{H} = 0 \tag{3.5}$$

Using routinely (3.2)–(3.5), we find that the electric field $\mathbf{E}$ and the magnetic field $\mathbf{H}$ satisfy the same second-order partial differential [1, 2]:

$$\varepsilon\mu\frac{\partial^2\mathbf{F}}{\partial t^2} + \nabla \times \nabla \times \mathbf{F} + \mu\sigma^e\frac{\partial\mathbf{F}}{\partial t} = 0, \ \ \mathbf{F} = \mathbf{E} \ \text{or} \ \mathbf{H} \tag{3.6}$$

By virtue of a well-known vector identity, we have

$$\nabla \times \nabla \times \mathbf{F} = \nabla \nabla \cdot \mathbf{F} - \nabla^2 \mathbf{F} = -\nabla^2 \mathbf{F} \tag{3.7}$$

Therefore, supposing that a plane wave propagates along the positive direction of the z-axis of a cartesian coordinate system and denoting the component F_x or F_y of the vector $\mathbf{F}$ by F, we obtain from (3.6) a simpler equation for the components $E_x(z,t)$ and $H_y(z,t)$:

$$\frac{\partial^2 F}{\partial z^2} - \varepsilon\mu\frac{\partial^2 F}{\partial t^2} - \mu\sigma^e\frac{\partial F}{\partial t} = 0 \tag{3.8}$$

The general solution of this equation for a time-harmonic plane wave is

$$F = \left[A\exp(ikz) + B\exp(-ikz)\right]\exp(-i\omega t) \tag{3.9}$$

where A, B, and k are the amplitudes and the wave number corresponding to a given frequency ω. In the case under consideration, the wave number is complex. Substitution of the representation (3.9) in (3.8) yields the following expression for the wave number:

$$A\exp(ikz - i\omega t)\left[-k^2 + \omega^2\varepsilon\mu + i\omega\mu\sigma^e\right] + B\exp(-ikz - i\omega t)\left[-k^2 + \omega^2\varepsilon\mu + i\omega\mu\sigma^e\right] = 0 \tag{3.10}$$

whence it follows that

$$k = \sqrt{\omega^2\varepsilon\mu + i\omega\mu\sigma^e} \tag{3.11}$$

For $\sigma^e = 0$, (3.11) becomes the well-known expression for the wave number of a nonconducting medium.

For an arbitrarily time-varying wave, the general solution of (3.8) can be represented by a Fourier expansion in a continuous spectrum of plane waves:

$$F(z,t) = \int_{-\infty}^{\infty} \left[A(\omega)\exp(ikz) + B(\omega)\exp(-ikz)\right]\exp(-i\omega t)\,d\omega \tag{3.12}$$

Let the function F and its derivative with respect to z be specified in the plane $z = 0$:

$$F(0,t) = \varphi(t), \quad \left.\frac{\partial F(z,t)}{\partial z}\right|_{z=0} = \psi(t) \tag{3.13}$$

Let us explore how the characteristics of a plane wave vary during its propagation in a conducting medium. Substitution of (3.9) in (3.13) yields

$$\varphi(t) = \int_{-\infty}^{\infty} [A(\omega) + B(\omega)] \exp(-i\omega t)\, d\omega$$

$$\psi(t) = \int_{-\infty}^{\infty} ik[A(\omega) - B(\omega)] \exp(-i\omega t)\, d\omega \tag{3.14}$$

Expressions (3.14) are nothing but the Fourier integral expansions of the known functions $\varphi(t)$ and $\psi(t)$. Applying a direct Fourier transform to (3.14), we find explicit expressions for the spectral densities of these expansions from which it follows that

$$A(\omega) = \frac{1}{4\pi} \int_{-\infty}^{\infty} \left[\varphi(\xi) - \frac{i}{k}\psi(\xi) \right] \exp(i\omega\xi)\, d\xi$$

$$B(\omega) = \frac{1}{4\pi} \int_{-\infty}^{\infty} \left[\varphi(\xi) + \frac{i}{k}\psi(\xi) \right] \exp(i\omega\xi)\, d\xi \tag{3.15}$$

Substitution of $A(\omega)$ and $B(\omega)$ in (3.12) yields

$$F(z,t) = F_1(z,t) + F_2(z,t) \tag{3.16}$$

where

$$F_1(z,t) = \frac{1}{2\pi} \int_{-\infty}^{\infty} \varphi(\xi)\, d\xi \int_{-\infty}^{\infty} \cos kz \exp[i\omega(\xi - t)]\, d\omega \tag{3.17}$$

$$F_2(z,t) = \frac{1}{2\pi} \int_{-\infty}^{\infty} \psi(\xi)\, d\xi \int_{-\infty}^{\infty} \frac{\sin kz}{k} \exp[i\omega(\xi - t)]\, d\omega \tag{3.18}$$

Equations (3.17) and (3.18) represent the formal solution of the problem. However, this solution is of little use in investigating the evolution of the parameters of a pulse plane wave during its propagation. This is especially clear for pulses of short duration (wide frequency band). The main reason for this is that the space and the time variable in the solution are separated (contained in different multipliers of the integrand), which is inconsistent with the physics of the problem.

Let us transform the frequency-domain representations (3.17) and (3.18) to time-domain representations. The idea of this transformation [1, 3, 4] is to use an integral representation for the function $\sin kz/k$ in (3.18):

$$\frac{\sin kz}{k} = \frac{c}{2} \int_{-z/c}^{z/c} e^{i\eta(\omega + ib)} J_0\left(\frac{b}{c}\sqrt{z^2 - c^2\eta^2} \right) d\eta \tag{3.19}$$

where $c = 1/\sqrt{\mu\varepsilon}$, $b = \sigma^e/2\varepsilon$, and $J_0(\cdot)$ is a zero-order Bessel function.

Substituting (3.19) in (3.18) and changing the order of integration, we obtain

$$F_2(z,t) = \frac{c}{2} \int\limits_{-z/c}^{z/c} d\eta \exp(-b\eta) \left[\frac{1}{2\pi} \int\limits_{-\infty}^{\infty} d\omega \int\limits_{-\infty}^{\infty} \psi(\xi) J_0 \left(\frac{b}{c} \sqrt{z^2 - c^2\eta^2} \right) \exp[i\omega(\xi + \eta - t)] d\xi \right] \quad (3.20)$$

However, according to the Fourier integral theorem, the expression in square brackets in the integrand is

$$\psi(t - \eta) J_0 \left(\frac{b}{c} \sqrt{z^2 - c^2\eta^2} \right) \quad (3.21)$$

Hence, we have

$$F_2(z,t) = \frac{c}{2} \int\limits_{-z/c}^{z/c} d\eta \exp(-b\eta) \psi(t - \eta) J_0 \left(\frac{b}{c} \sqrt{z^2 - c^2\eta^2} \right) \quad (3.22)$$

and, changing the integration variable by putting $t - \eta = \varsigma$, obtain

$$F_2(z,t) = \frac{c}{2} \int\limits_{t-z/c}^{t+z/c} \psi(\varsigma) J_0 \left(\frac{b}{c} \sqrt{z^2 - c^2(t - \varsigma)^2} \right) \exp[-b(t - \varsigma)] d\varsigma \quad (3.23)$$

A similar expression can be obtained for $F_1(z,t)$ if we take into account that (3.17) follows from (3.18) on changing $\psi(\xi)$ by $\varphi(\xi)$ and differentiating with respect to z. Making this change in (3.23), differentiating the resulting expression with respect to z as an integral with variable limits, and combining the resulting expression for $F_1(z,t)$ with the expression for $F_2(z,t)$, we finally find

$$F(z,t) = F_1(z,t) + F_2(z,t) = \frac{1}{2} \exp \left(\frac{b}{c} z \right) \varphi \left(t + \frac{z}{c} \right) + \frac{1}{2} \exp \left(-\frac{b}{c} z \right) \varphi \left(t - \frac{z}{c} \right)$$

$$+ \frac{c}{2} \exp(-bt) \int\limits_{t-z/c}^{t+z/c} \varphi(\varsigma) \frac{\partial}{\partial z} J_0 \left(\frac{b}{c} \sqrt{z^2 - c^2(t - \varsigma)^2} \right) \exp(b\varsigma) d\varsigma \quad (3.24)$$

$$+ \frac{c}{2} \exp(-bt) \int\limits_{t-z/c}^{t+z/c} \psi(\varsigma) J_0 \left(\frac{b}{c} \sqrt{z^2 - c^2(t - \varsigma)^2} \right) \exp(b\varsigma) d\varsigma$$

Consider, first, a particular case of a nonconducting medium ($\sigma^e = 0$). For this case, we have

$$b = 0, J_0(0) = 1, \left. \frac{\partial}{\partial z} J_0(z) \right|_{z=0} = 0 \quad (3.25)$$

and obtain for $F(z,t)$ the well-known d'Alembert formula

$$F(z,t) = \frac{\varphi \left(t + \dfrac{z}{c} \right) + \varphi \left(t - \dfrac{z}{c} \right)}{2} + \frac{c}{2} \int\limits_{t-z/c}^{t+z/c} \psi(\varsigma) d\varsigma \quad (3.26)$$

This formula describes two undumped plane waves propagating with a constant velocity c from the plane $z = 0$ in the directions $z \to -\infty$ and $z \to \infty$, respectively. The waves retain their original shape during the propagation.

In a medium with $\sigma^e \neq 0$, two waves propagate in the directions $z \to -\infty$ and $z \to \infty$. Their velocity c is independent of σ^e and ω, but the amplitudes decrease exponentially during the propagation. The dumping coefficient is given by $b/c = (\sigma^e/2)\sqrt{\mu/\varepsilon}$, as in the case of a plane harmonic wave. In this case, the waveform of the initial perturbation specified in the plane $z = 0$ is not retained due to the integral contributions decreasing exponentially with time. This gives rise to a "tail" following each wave, which decreases exponentially with time.

3.1.1.2   The Time-domain Solution

Let us find a solution to (3.8), that is, determine a field depending on the (single) coordinate z and time t [1]. Suppose that there is no field before the zero time, and at $t = 0$, a source located in the plane $z = 0$ is switched on and radiates a linearly polarized plane wave with components $E_x(z,t)$ and $H_y(z,t)$ in the direction $z > 0$. Also, suppose that the function $f(t)$ that describes the time variation of the electric field $E_x(z,t)$ in the plane $z = 0$ is known.

Applying the direct Laplace transform

$$G(z,s) = \int_0^\infty F(z,t)\exp(-st)\,dt = L[F(z,t)] \tag{3.27}$$

to (3.8), we obtain an ordinary differential equation for the transform $G(z,s)$:

$$\frac{d^2G}{dz^2} - h^2G = -\frac{h^2}{s}F(z,0) + \mu\varepsilon\frac{\partial F(z,t)}{\partial t}\bigg|_{t=0} \tag{3.28}$$

where $h^2 = \mu\varepsilon s^2 + \mu\sigma^e s$. However, as the field and, hence, its derivatives are everywhere equal to zero up to $t = 0$, (3.28) will actually be homogeneous:

$$\frac{d^2G}{dz^2} - h^2G = 0 \tag{3.29}$$

The general solution of (3.29) reads

$$G(z,s) = A\exp(-hz) + B\exp(hz) \tag{3.30}$$

where h refers to the branch of the double-valued function $\sqrt{h^2}$ that takes positive values when h^2 is real and positive.

The Laplace transformation of the condition for the electric field at $t = 0$ yields

$$G(0,s) = L[f(t)] \tag{3.31}$$

It is clear that the constant B should be put equal to zero, as the radiated field propagates in the direction $z > 0$. Actually, for the particular case of a nonconducting medium ($\sigma^e = 0$), we have $h = s/c$, $c = 1/\sqrt{\varepsilon\mu}$, and, according to the displacement theorem for a Laplace transform,

$$G(z,s) = G(0,s)\exp\left(-\frac{s}{c}z\right) = L\left[f\left(t - \frac{z}{c}\right)\right] \tag{3.32}$$

This equation implies that

$$E_x(z,t) = E_x\left(0, t - \frac{z}{c}\right) = f\left(t - \frac{z}{c}\right) \tag{3.33}$$

that is, the wave propagates in the positive-z direction.

The key to the subsequent procedure of seeking a solution is to find the function-original for the function $\exp(-hz)$ in the sense of the Laplace transform. It can be found based on the following integral representation [1]:

$$\frac{\exp(-hz)}{h} = c\int_{z/c}^{\infty} \exp(-bt)J_0\left(\frac{b}{c}\sqrt{z^2 - c^2t^2}\right)\exp(-st)\,dt \tag{3.34}$$

On differentiation, we obtain

$$\exp(-hz) = \exp\left(-\frac{b}{c}z\right)\exp\left(-s\frac{z}{c}\right) - c\int_{z/c}^{\infty} \exp(-bt)\frac{\partial}{\partial z}J_0\left(\frac{b}{c}\sqrt{z^2 - c^2t^2}\right)\exp(-st)\,dt \tag{3.35}$$

or

$$\exp(-hz) = \exp\left(-\frac{b}{c}z\right)\exp\left(-s\frac{z}{c}\right) - cL[\varphi(z,t)] \tag{3.36}$$

where

$$\varphi(z,t) = \begin{cases} 0 & \text{for } 0 < t < z/c \\[2mm] \exp(-bt)\dfrac{\partial}{\partial z}J_0\left(\dfrac{b}{c}\sqrt{z^2 - c^2t^2}\right) & \text{for } t \geq z/c \end{cases} \tag{3.37}$$

Proceeding from the above calculations, we can write

$$G(z,s) = G(0,s)\exp(-hz) = L[f(t)]\exp\left(-\frac{b}{c}z\right)\exp\left(-s\frac{z}{c}\right) - cL[f(t)]L[\varphi(z,t)] \tag{3.38}$$

In view of the displacement theorem and the convolution theorem for a Laplace transform, we find

$$E_x(z,t) = \exp\left(-\frac{b}{c}z\right)f\left(t - \frac{z}{c}\right) - c\int_{z/c}^{t} f(t - \xi)\exp(-b\xi)\frac{\partial}{\partial z}J_0\left(\frac{b}{c}\sqrt{z^2 - c^2\xi^2}\right)d\xi \quad (3.39)$$

or, changing the integration variable by putting $t - \xi = \beta$,

$$E_x(z,t) = \exp\left(-\frac{b}{c}z\right)f\left(t - \frac{z}{c}\right)$$
$$- c\exp(-bt)\int_{0}^{t-z/c} f(\beta)\exp(b\beta)\frac{\partial}{\partial z}J_0\left(\frac{b}{c}\sqrt{z^2 - c^2(t - \beta)^2}\right)d\beta \quad (3.40)$$

The solution obtained satisfies (3.8) and describes a wave propagating in the positive-z direction. In the plane $z = 0$, it coincides with the function $f(t)$ for all $t > 0$. From the computational viewpoint, it is advisable to express the Bessel function in the representation (3.40) in terms of a Bessel function of imaginary argument by using the well-known equations [5]

$$\frac{d}{dz}J_0(z) = -J_1(z), \quad J_1(iz) = iI_1(z) \quad (3.41)$$

Then we have

$$E_x(z,t) = \exp\left(-\frac{b}{c}z\right)f\left(t - \frac{z}{c}\right)$$
$$+ \frac{bz}{c}\exp(-bt)\int_{0}^{t-z/c} f(\beta)\exp(b\beta)\frac{I_1\left(b\sqrt{(t - \beta)^2 - z^2/c^2}\right)}{\sqrt{(t - \beta)^2 - z^2/c^2}}d\beta \quad (3.42)$$

In particular, if $f(t)$ is the Dirac delta function, $f(t) = \delta(t)$, we obtain the following expression for the space-time Green function (impulse response) of a conducting medium:

$$E_x(z,t) = \left[\delta\left(t - \frac{z}{c}\right) + \frac{bz}{c}\frac{I_1\left(b\sqrt{t^2 - z^2/c^2}\right)}{\sqrt{t^2 - z^2/c^2}}\right]\exp(-bt), \quad t \geq \frac{z}{c} \quad (3.43)$$

Expression (3.43) for the space–time Green function was obtained in a different way and examined in detail for narrowband pulses [6]. Using this expression, the solution (3.42) was obtained [7]. This solution was also found [8] using the Riemann method [9].

Taking into account that for $x \gg 1$ the following asymptotic formula is valid [5]:

$$I_1(x) \approx \frac{\exp x}{(2\pi x)^{1/2}} \quad (3.44)$$

we obtain for large t

$$\frac{I_1\left(b\sqrt{t^2 - z^2/c^2}\right)}{\sqrt{t^2 - z^2/c^2}} \approx \frac{\exp\left[\left(b\sqrt{t^2 - z^2/c^2}\right)\right]}{\sqrt{2\pi b}\sqrt{t^2 - z^2/c^2}} \approx \frac{\exp(bt)}{\sqrt{2\pi b}\sqrt{t}} \tag{3.45}$$

Thus, we see that the impulse response of a conducting medium at a point with a coordinate z is a pulse of amplitude $\exp(-bz/c)$ arrived at the point z at the time z/c, which is followed by a tail slowly decreasing with time. The greater the medium conductivity (b) or the distance from the source to the point z, the lower the pulse amplitude. The tail decreases with time as $1/\sqrt{t}$.

The procedure of evaluation of the integral entering into (3.42) strongly depends on the function $f(t)$ that describes the pulse waveform. Consider three waveform functions:

$$f_1(t) = E_0 \left(\frac{t}{T}\right)^n \left[M^{n+1}\exp\left(-M\frac{t}{T}\right) - \exp\left(-\frac{t}{T}\right)\right]\chi(t) \tag{3.46}$$

$$f_2(t) = E_0 \frac{t}{T}\exp\left(-\frac{t}{2T}\right)\left[\frac{1}{6}\left(\frac{t}{T}\right)^2 - \frac{3}{2}\frac{t}{T} + 2\right]\chi(t) \tag{3.47}$$

$$f_3(t) = E_0 \frac{t - t_0}{T}\exp\left(-\left(\frac{t - t_0}{T}\right)^2\right)\chi(t - t_0) \tag{3.48}$$

where the parameters n, T, M, and t_0 determine the pulse waveform and effective duration, E_0 is a constant, and

$$\chi(t) = \begin{cases} 0, & t < 0 \\ 1, & t \geq 0 \end{cases} \tag{3.49}$$

In a computer simulation performed using these expressions, equations were obtained that describe the variations in pulse waveform at a given distance from the plane $z = 0$ with increasing the medium conductivity. The plots of these waveforms are presented in Figures 3.1–3.3. They show that an increase in conductivity results in a more pronounced decay of the pulse, in an increase its effective duration, and in a distortion of its initial waveform. This is because the high-frequency components of the energy spectrum of a pulse, which are responsible for the pulse edges, are suppressed by a conducting medium more efficiently than the low-frequency components.

3.1.2   Earth's Atmosphere

In developing promising UWB location, radar, and communication systems, it is important to clearly understand the qualitative and quantitative impact of various

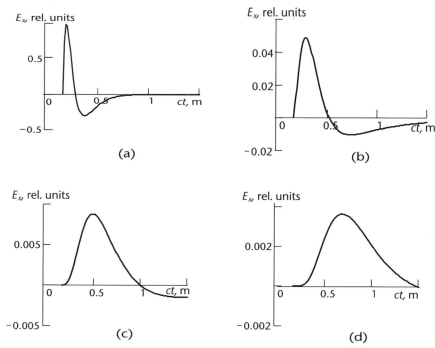

Figure 3.1 Distortion of the pulse waveform $f_1(t)$ depending on the medium conductivity: $E_0 = 2.2$, $n = 1$, $T = 0.3$, $z = 0.5$; $b = 0$ (a), 10 (b), 30 (c), and 50 (d).

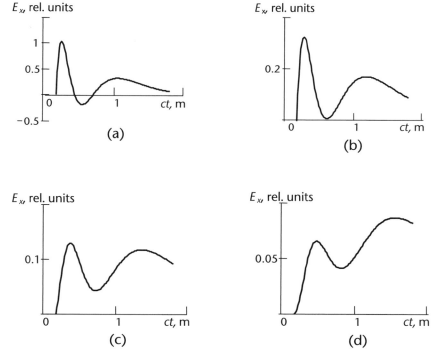

Figure 3.2 Distortion of the pulse waveform $f_2(t)$ depending on the medium conductivity: $E_0 = 1$, $n = 1$, $T = 0.3$, $z = 0.5$; $b = 0$ (a), 3 (b), 8 (c), and 15 (d).

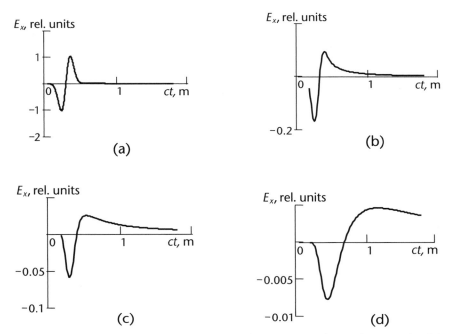

Figure 3.3 Distortion of the pulse waveform $f_3(t)$ depending on the medium conductivity: $E_0 =$ 2.4, $T = 0.3$, $t_0 = 0.4$, $z = 0.5$; $b = 0$ (a), 5 (b), 10 (c), and 30 (d).

factors on a radiation pulse propagating in a medium [10, 11]. As the mechanisms of the interaction of a short-pulse radiation with the propagation medium are highly intricate, the researchers have to restrict their analyses to simple model problems. But even in these cases, it is necessary to use complicated semiempirical models to calculate the complex refractive index of the medium, n_c. It is impossible to calculate this parameter exactly for a wide frequency band using the quantum-mechanical approach, as it requires taking into account the contributions of the wings of a great many spectral lines. However, a rather complete and accurate engineering model has been developed for estimating n_c of the Earth's atmosphere for frequency bands of width up to 1000 GHz [12, 13]. Models of this kind for the Earth's ionosphere, soil, and seawater are not available.

Let us consider the distortions of a plane electromagnetic pulse wave propagating in an unbounded dispersive and absorbing medium in the positive direction of the z-axis [14–16]. Represent the wave component $E_x = E(z,t)$ as a Fourier integral:

$$E(z,t) = \frac{1}{2\pi} \int_{-\infty}^{\infty} E(z,\omega)\exp(-i\omega t)\,d\omega \qquad (3.50)$$

For a homogeneous, isotropic, and dispersive medium, the spectral density of the radiated pulse, $E(z,\omega)$, can be expressed in terms of its spectral density in the plane $z = 0$ as $E(z,\omega) = E(0,\omega)\exp[ik(\omega)z]$, where $k(\omega)$ is the wave number defined as $k(\omega) = \omega n_c(\omega)/c$. The complex refractive index n_c is conventionally subdivided into the frequency-independent real part and the frequency-dependent real and imaginary parts (n_0, $n(\omega)$, and $\kappa(\omega)$, respectively):

$$n_c(\omega) = n_0 + n(\omega) + i\kappa(\omega) \tag{3.51}$$

In a model proposed by Liebe [13], the frequency-independent part n_0 of $n_c(\omega)$ is specified by the standard formula

$$n_0 = 1 + 10^{-6}\frac{77.6}{T}\left(p + \frac{4810}{T}e\right) + n_v \tag{3.52}$$

where the additional term n_v describes the contribution of the rotational spectrum of water vapor molecules. The atmospheric pressure p and the partial pressure of water vapor, e, are expressed in millibars and the temperature T is expressed in kelvins. The frequency-dependent part combines the contributions of the discrete absorption lines and continuum from the spectral windows for dry air, water vapor, and hydrosols. The model is complicated in that its spectroscopic data contain more than 450 parameters for describing the resonance lines of O_2 and H_2O for atmospheric conditions at altitudes of up to 30 km and their contribution to the continuous spectrum, whereas only p, T, e, and the hydrosol concentration serve as input data.

Thus, the calculation of the distortion of a pulse propagating in a dispersive and absorptive medium using a given model of the refractive index $n_c(\omega)$ is reduced to a numerical calculation of the inverse Fourier transform

$$E(z,t) = \frac{1}{2\pi}\int\limits_{-\infty}^{\infty} E(0,\omega)\exp\left\{\omega[in(\omega) - \kappa(\omega)]\frac{z}{c} - i\omega t'\right\}d\omega \tag{3.53}$$

where $t' = t - zn_0/c$ is the time count out from the arrival of the pulse spectral component propagating with phase velocity c/n_0 at the observation point z (the good reason for this counting the time t' will become apparent in the subsequent discussion).

Quantitative estimates for the pulse distortions can be obtained [14] using the temporal moments defined as

$$\overline{t^k(z)} = \int\limits_{-\infty}^{\infty} t^k w(z,t)\,dt, \quad k = 1,2,\ldots \tag{3.54}$$

where the function

$$w(z,t) = \frac{E^2(z,t)}{\int_{-\infty}^{\infty} E^2(z,t)\,dt} \tag{3.55}$$

plays the role of probability density. For instance, the first moment $\overline{t(z)}$ describes the displacement of the pulse "center of gravity" with distance z, and the quantity expressed in terms of this moment,

$$v_{eff}(z) = \frac{z}{\overline{t(z)} - \overline{t(0)}} \tag{3.56}$$

describes the "effective" velocity of propagation of the main portion (so-called "body") of the pulse. The quantity $\delta t(z)$, defined as

$$\delta t(z) = \sqrt{\overline{t^2(z)} - \overline{t(z)}^2} \tag{3.57}$$

describes the root mean square width (effective duration) of the pulse observed at the distance z from the plane $z = 0$.

In solving some problems, the first two moments are inadequate to completely characterize UWB pulse radiation [15]. Therefore, the asymmetry ratio $\gamma_3 = \mu_3/\mu_2^{3/2}$ and the degree of excess $\gamma_4 = \mu_4/\mu_2^2$, associated with the third and the fourth temporal moment, respectively, are introduced into consideration. These parameters are both expressed in terms of the so-called central moments $\mu_k = \overline{(t - \overline{t(z)})^k}$.

It is reasonable to characterize the decay of a pulse by its energy density normalized to the initial value, which is of interest for practical applications:

$$w(z) = \frac{\int_{-\infty}^{\infty} E^2(z,t)\, dt}{\int_{-\infty}^{\infty} E^2(0,t)\, dt} = \frac{\int_{-\infty}^{\infty} (E(z,\omega))^2\, d\omega}{\int_{-\infty}^{\infty} (E(0,\omega))^2\, d\omega} \tag{3.58}$$

Equations (3.54), (3.57), and (3.58) form the basis for numerical calculations [14] of the distortion and power losses of UWB pulses in the Earth's atmosphere. The initial waveform of the pulse $E(0,t)$ (or, equivalently, its spectral density) and the model description of the refractive index $n_c(\omega)$ in a specified frequency range serve as input data. In the numerical calculations [14], the UWB pulse waveform was simulated by a function of the form:

$$F(0,t) = F_0 \left(\frac{t}{T}\right)^n \left[M^{n+1} \exp\left(-\frac{Mt}{T}\right) - \exp\left(-\frac{t}{T}\right)\right] \chi(t) \tag{3.59}$$

Figures 3.4–3.7 present the final waveform of the UWB pulse calculated by (3.59), the pulse root mean square width calculated by (3.57), and the function of the form (3.58) that describes the pulse decay [14–16]. The integrals were evaluated using fast Fourier transforms. The complex refractive index was calculated using standard values of the air pressure ($p = 1013.25$ mbar) and temperature ($T = 273$ K), and of the water vapor density ($\rho = 7.5$ g/m^3).

Figure 3.4 illustrates the distortion of a pulse with the initial waveform described by (3.59) with the parameters $M = 5$, $n = 2$, and $T = 0.8$ ns, corresponding to the duration $\delta t(0) = 1$ ns, at a distance $z = 0$ (curve 1), 250 (curve 2), and 500 km (curve 3). It can be seen that the final pulse waveform is qualitatively the same as its original waveform. Characteristic of the pulse distortion is the flattening of its final waveform: the positive peak becomes lower depending on the parameters n, M, and T, and the negative peak decreases in absolute value, to which there corresponds a

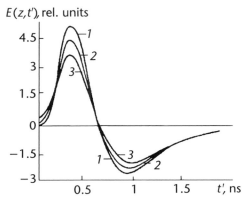

Figure 3.4 Typical distortion of a pulse described by (3.59) with duration $\delta t(0) = 1$ ns at a distance $z = 0$ (curve 1), 250 (curve 2), and 500 km (curve 3). (With permission from Nauka Publishing House.)

shift of the spectra toward the lower frequencies. This character of the waveform distortion of the nanosecond UWB pulse in the Earth's atmosphere is due to that in its spectral band (of width about 1 GHz), the damping coefficient of the spectral components increases monotonically with frequency. Therefore, as the pulse propagates in the atmosphere, its high-frequency harmonics decay faster than the low-frequency ones. The increase in the proportion of the spectrum low-frequency components in time domain corresponds to its broadening, which is the dominant effect in the distortion of nanosecond UWB pulse waveforms.

Calculations have shown that the body, that is, the main portion of a nanosecond UWB pulse, propagates with a velocity approximately equal to $v_0 = c/n_0$. On the other hand, v_0 can be treated as the quantity $(\partial k(\omega)/\partial \omega)^{-1}$ at some frequency ω_0 within the UWB pulse spectrum, that is, as a group velocity of a narrowband signal with carrier frequency ω_0 propagating in a medium whose dispersion coefficient is described by the same law $k = k(\omega)$.

Figure 3.5 shows the distortion of a picosecond ($\delta t(0) = 50$ ps) UWB pulse. The spectrum of the picosecond pulse is much broader than that of a nanosecond pulse. It includes the resonance absorption line of water vapor at a frequency of 22.2 GHz, which is responsible for much more substantial changes in pulse waveform even at small distances ($z < 5$ km). The leading and trailing edges of the UWB picosecond pulse are diffused by oscillations whose amplitude is comparable to the main peak, making the radar processing of signals of this duration practically impossible.

Thus, as water in various aggregation states is one of the most widespread components of natural environments, the waveform of a UWB signal of duration less than 0.1 ns propagating in such media will degrade rather quickly.

Figure 3.6 shows the variations in root mean square width of UWB pulses with parameters $M = 2$, $n = 5$, and different T with distance from the radiation source. The run of the $\delta t(z)$ curve remains qualitatively the same, but the pulse width increment strongly depends on the parameter T.

Figure 3.7 presents the calculated energy lost by UWB pulses of different duration (the initial parameters of the pulses are the same as for Figure 3.6).

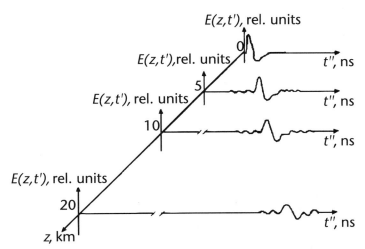

Figure 3.5 Typical distortion of the waveform of a picosecond pulse described by (3.59) during its propagation. (With permission from Nauka Publishing House.)

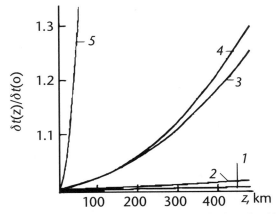

Figure 3.6 Variation in root mean square width of UWB pulses described by (3.59) with distance from the radiation source. Pulse parameters: $M = 2$, $n = 5$, and $T = 3$ (curve 1), 1.5 (curve 2), 0.3 (curve 3), 0.15 (curve 4), and 0.03 ns (curve 5). (With permission from Nauka Publishing House.)

Thus, the numerical calculations show that the main consequence of the distortion of nanosecond pulses in the Earth's atmosphere is their broadening (with the initial waveform qualitatively retained) and decay. For distances of up to 500 km, the decay of signals of this type is not too strong to preclude their radar processing.

The distortions of picosecond (up to 0.1 ns) UWB pulses, even at small (to 5 km) distances, are so strong that the extraction of radar information about the objects from the reflected signal waveforms is impossible. This is a fundamental limitation to the use of UWB electromagnetic pulses of duration less than 0.1 ns caused by the intense absorbing effect of the atmospheric water vapor at a frequency of 22.2 GHz.

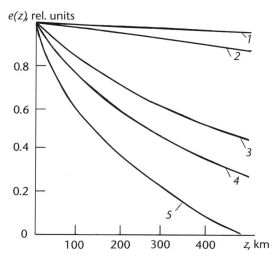

Figure 3.7 Energy lost by UWB pulses described by (3.59) with distance from the radiation source. Pulse parameters: $M = 2$, $n = 5$, and $T = 3$ (curve 1), 1.5 (curve 2), 0.3 (curve 3), 0.15 (curve 4), and 0.03 ns (curve 5). (With permission from Nauka Publishing House.)

3.1.3 Distortions of High-power Pulses in the Earth's Lower Atmosphere

It is well known that nonlinear effects set an upper power density limit (several megawatts per square meter) on microsecond electromagnetic pulses capable of propagating in the atmosphere. This corresponds to the pulse amplitudes of several hundreds of kilovolts per meter. A possible way to overcome this limitation is to use pulses of nanosecond duration, which is comparable to the time required for air ionization. In this connection, of considerable interest is to estimate the limiting intensity of nanosecond pulses that could propagate in the atmosphere with insignificant energy loss or without distortion.

A mathematical model of the pulse propagation process [17] is based on a kinetic description of the free electrons in the plasma and involves solving self-consistently Maxwell's equations for the electromagnetic field and the Boltzmann formula for the free electrons. To simplify the extensive kinetic description, three assumptions about the electron distribution function are used. One of them is that the ratio of the mean free path of the electrons to the characteristic spatial scale of the electromagnetic field is small. For this reason, the consideration is limited to altitudes of 50 km, as for an altitude of about 100 km, this ratio is close to unity. On the other hand, this assumption allows taking into account the local relationship between the electromagnetic pulse electric field and the electron distribution function and neglecting the effects of convection and diffusion of electrons from the discharge region.

The second assumption is that the ratio of the time of an electron-molecule collision to the characteristic time required to ionize the air is small. As electron collisions make their distribution function isotropic, it can be assumed that this function is weakly anisotropic, and, hence, the following approximation can be used:

$$f(\mathbf{v},t) = f_0(\varepsilon,t) + f_1(\varepsilon,t)\cos\theta, \qquad (3.60)$$

where θ is the angle between the electric field vector $\mathbf{E}$ and the electron velocity vector $\mathbf{v}$; $\varepsilon = mv^2/2$ is the kinetic energy of the electron.

Finally, it is assumed that the ratio of the free electron density to the air density is small. Then, we may neglect electron-electron collisions and take into account only elastic and inelastic collisions of electrons with ground-state air molecules.

With the above assumptions, the functions f_0 and f_1 satisfy the following kinetic equations [17]:

$$\frac{\varepsilon}{vN}\frac{\partial f_0}{\partial t} - \frac{eE}{3N}\frac{\partial \varepsilon f_1}{\partial \varepsilon} = \tilde{J}[f_0] + S_0 \tag{3.61}$$

$$\frac{\partial f_1}{\partial t} - eEv\frac{\partial f_0}{\partial \varepsilon} = -NvQ(\varepsilon)f_1(t) \tag{3.62}$$

where N is the density of the gas medium, e is the electron charge, and E is the electric field strength.

The collision integral $\tilde{J}[f_0]$ in (3.61) is a linear integro-differential operator, which describes the contributions of the elastic electron scattering by molecules, of the inelastic interactions of electrons with molecules whose rotational, vibrational, and electron energy levels are excited, and of the ionization and attachment of electrons to oxygen and water molecules. The structure of this operator is discussed in detail in Golubev et al. [17]. The function $Q(\varepsilon)$ in (3.62) describes the transport cross-section for all the above processes.

From (3.61), in view of the structure of the operator $\tilde{J}[f_0]$, it follows [17] that the electron density

$$n_e = 2\pi\left(\frac{2}{m}\right)^{3/2}\int\sqrt{\varepsilon}f_0\,d\varepsilon \tag{3.63}$$

varies, due to ionization and attachment, as

$$\frac{\partial n_e}{\partial t} = (v_i - v_\alpha)n_e \tag{3.64}$$

where v_i and v_α are, respectively, the ionization and the attachment frequencies, which are determined by the equation

$$v_b = \frac{8\pi N}{m^2 n_e}\int_0^\infty \varepsilon Q_b f_0\,d\varepsilon \tag{3.65}$$

Here, Q_b is the cross section of the corresponding process.

For the description of the propagation of electromagnetic waves be self-consistent, the kinetic equations should be complemented with Maxwell's equations. The propagating electromagnetic wave is plane and polarized along the x-axis, and the direction of its propagation coincides with the direction of the z-axis. The equations for the field components E_x and H_y are simplified by using a retarded reference system with $\tau = t - z/c$. In this system, the Faraday law is formulated as

$$c\frac{\partial E_x}{\partial z} = -\frac{\partial}{\partial \tau}(E_x + H_y) \tag{3.66}$$

The left side of this equation is small if the spatial length of the pulse is much less than the distance over which the pulse amplitude varies appreciably, $c\tau_p \ll L$. Then, we have $E_x = -H_y$, and the Ampere law is reduced to a single equation:

$$\frac{\partial E_x}{\partial z} + \frac{2\pi}{c}j_x = 0 \tag{3.67}$$

which corresponds to the so-called high-frequency approximation. This time-independent equation is convenient to perform numerical calculations. The current density given by

$$j_x(z,\tau) = -\left(\frac{8\pi e}{3m}\right)\int \varepsilon f_1 d\varepsilon \tag{3.68}$$

can be calculated using (3.61) and (3.62) in the retarded reference system.

An electromagnetic pulse propagating from the Earth vertically upward is considered. The air is assumed to consist of nitrogen, whose density distribution is described by an exponential law:

$$N(z) = N_0\exp\left(-\frac{z}{L}\right) \tag{3.69}$$

The parameters $N_0 = 2.7 \cdot 10^{19}$ cm^{-3} and $L = 7$ km correspond to the typical atmospheric conditions. The effect of oxygen, water vapors, and other impurities in air is described by second-order quantities.

It is assumed that before the arrival of the pulse, the air contains a small amount of free electrons of density $n_0 = 10^3$ cm^{-3} and temperature $T_0 = 0.1$ eV, and their energy distribution is Maxwellian. These electrons can be generated by the leading edge of the pulse due to the detachment of electrons from the negative ions of molecular oxygen. This occurs if the pulse amplitude is about one-tenth of the breakdown threshold. The presence of these electrons is considered as an initial condition because oxygen molecules are not taken into account in the expression for the collision integral. The data on the cross sections involved in the calculations are available in the publications cited in Golubev et al [17].

The initial pulse waveform $E_0(\tau)$ is chosen based on the natural limit of no charge transfer after the pulse propagation:

$$\int E_0(\tau)d\tau = 0 \tag{3.70}$$

Therefore, the propagation of a bipolar pulse with a short positive portion and a long negative portion is considered. The initial pulse waveform is described by the formula

$$E_0(\tau) = E(z = 0, \tau) = E_{\max} \frac{\varphi(\tau)}{\varphi_{\max}} \tag{3.71}$$

where

$$\varphi(\tau) = \frac{d}{d\tau} \frac{\exp \alpha(\tau - \tau_0)}{\beta + \alpha \exp(\alpha + \beta)(\tau - \tau_0)} \tag{3.72}$$

and $\varphi_{\max}$ denotes the maximum value of the function φ.

The calculations were performed for $\beta = 0.1\alpha$ and $\alpha\tau_0 = 10$. Only two pulse parameters were varied: $E_{\max}$ and the parameter α, which determines the pulse duration τ_p. The FWHM duration of the pulse positive portion equals approximately $3/\alpha$.

In the calculations [17], $E_{\max}$ and α were varied in the ranges 1–10 MV/m and 1–1000 ns^{-1}, respectively. For $\alpha > 100$ ns^{-1}, no decay of the pulse was observed up to an altitude of 50 km. Therefore, only pulses of longer duration were considered.

Figure 3.8 presents the pulse waveforms at different altitudes for $\alpha = 10$ ns^{-1} and initial pulse amplitudes of 1, 4, 7, and 10 MV/m. For the pulse with the 1 MV/m initial amplitude, its amplitude and waveform vary insignificantly with altitude. The pulse duration (~1 ns) is less than the time to the onset of avalanche-like ionization, although the amplitude is greater than the ionization threshold for altitudes above 10 km. Hence, the negative portion of the pulse produces a very small amount of plasma, so that this plasma has an insignificant effect on the pulse amplitude and waveform.

Strong changes can be seen for the pulse of initial amplitude 4 MV/m. As the pulse reaches an altitude of 20 km, its negative portion is completely absorbed, and the absorption of the positive portion begins and intensifies with increasing altitude. Note that the respective portion of the pulse waveform becomes very steep. A similar pattern is observed for the pulses with large initial amplitudes, although the absorption begins earlier and proceeds at higher rates.

The calculations have shown that the pulse duration is the most important factor because it determines the absorption of the pulse energy. Figure 3.9 illustrates the altitude dependence of the pulse amplitude for different pulse durations and initial amplitudes. The amplitudes of all pulses with $\alpha > 100$ ns^{-1} are independent of altitude. The absorption of the pulses of longer duration depends on their duration and initial amplitude. Thus, the pulse of amplitude 4 MV/m with $\alpha > 10$ ns^{-1} can propagate to altitudes above 30 km. On the other hand, a similar pulse of tenfold duration is absorbed almost completely at an altitude of 10 km. However, if the initial amplitude of a pulse is 10 MV/m and its duration is not changed ($\alpha > 10$ ns^{-1}), the pulse amplitude only halves at an altitude of 20 km.

The pulses with durations defined by $\alpha = 1$–10 ns^{-1} give up not all their energy into plasma formation. Examination of Figures 3.8 and 3.9 shows that the amplitude

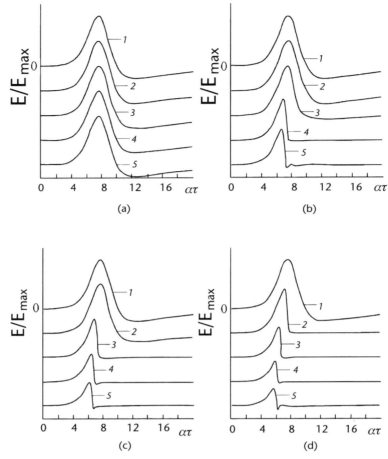

Figure 3.8 Waveforms of electric field pulses of different initial amplitude $E_{max} = 1$ (a), 4 (b), 7 (c), and 10 MV/m (d) at different altitudes: $z = 0$ (curve 1), 8 (curve 2), 16 (curve 3), 24 (curve 4), and 32 (curve 5) for $\alpha = 10$ ns^{-1}. (With permission from IEEE, Inc.)

and duration of a pulse propagating in the atmosphere depend weakly on the pulse initial amplitude but strongly depend on its initial duration. For instance, the pulse with $\alpha = 10$ ns^{-1} has amplitude of about 3 MV/m and duration of 0.2 ns, which is almost independent of the initial amplitude.

The altitude at which the absorption of the pulse energy is complete and the range of pulse durations corresponding to weak absorption can be estimate analytically. The maximum pulse duration is approximately inversely proportional to its maximum amplitude [17]:

$$\tau_{max}[\text{ns}] \approx 2.1/E_{max}[\text{MV/m}] \qquad (3.73)$$

For pulses of nanosecond duration, the absorption is significant only if the pulse amplitude is over the breakdown threshold.

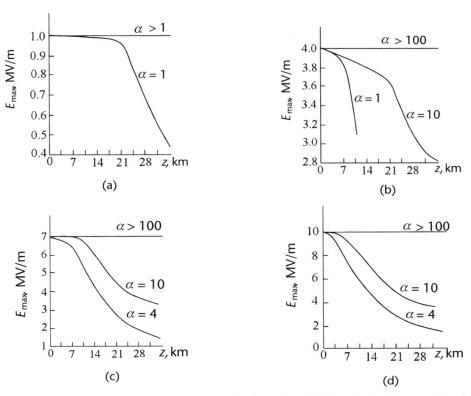

Figure 3.9 Altitude dependence of the amplitude of an electric field pulse for the pulse duration α = 1, 4, 10, and 100 ns^{-1} and the initial amplitude E_{max} = 1 (a), 4 (b), 7 (c), and 10 MV/m (d). (With permission from IEEE, Inc.)

3.2   Layered Media

3.2.1   Propagation of an Ultrawideband Pulse through an Interface between Two Media

3.2.1.1   Reflection of a Plane Electromagnetic Wave

This section describes an approach [18, 19] that allows one to obtain meaningful information about the behavior of the transient reflection coefficient of electromagnetic waves from, in general, a multilayered medium. This information is important, for instance, in determining the parameters of materials from the results of their probing with UWB radiation.

Suppose that a plane harmonic electromagnetic wave is incident from a free space onto plane interface with a conducting medium at an angle of θ_0. Assume that the electric field is polarized parallel to the interface (TE polarization) and the medium parameters $\varepsilon = \varepsilon_r \varepsilon_0$, μ_0, and σ^e are independent of frequency.

For the reflection coefficient, the following expression is known [18]:

$$\Gamma(\omega) = \frac{\cos\theta_0 - \sqrt{\varepsilon_r - \sin^2\theta_0 - i\dfrac{\sigma^e}{\omega\varepsilon_0}}}{\cos\theta_0 + \sqrt{\varepsilon_r - \sin^2\theta_0 - i\dfrac{\sigma^e}{\omega\varepsilon_0}}} \tag{3.74}$$

For the subsequent calculations, it is convenient to represent it as

$$\Gamma(s) = \frac{\sqrt{s} - \sqrt{Ds + B}}{\sqrt{s} + \sqrt{Ds + B}} \tag{3.75}$$

where

$$s = i\omega, \ D = \frac{\varepsilon_r - \sin^2 \theta_0}{\cos^2 \theta_0}, \ \text{and} \ B = \frac{\sigma^e}{\varepsilon_0 \cos^2 \theta_0} \tag{3.76}$$

and perform the regularization

$$\Gamma(s) = \Gamma_\infty(s) + \Gamma_0(s) \tag{3.77}$$

by selecting explicitly the limiting value of $\Gamma(s)$ as

$$\lim_{s \to \infty} \Gamma(s) = \Gamma_\infty(s) = \frac{1 - \sqrt{D}}{1 + \sqrt{D}} \tag{3.78}$$

and the additional term as

$$\Gamma_0(s) = 2\frac{\sqrt{D}}{1 + \sqrt{D}} \left[\frac{\sqrt{s} - \sqrt{s + B/D}}{\sqrt{s} + \sqrt{D}\sqrt{s + B/D}} \right] \tag{3.79}$$

Performing algebraic transformation of (3.79),

$$\Gamma_0(s) = -\frac{2\sqrt{D}}{D - 1} \left[\frac{s}{s + B/(D - 1)} \right] \left[1 - \sqrt{\frac{s + B/D}{s}} \right] - \frac{2B}{(1 + \sqrt{D})(D - 1)} \left[\frac{1}{s + B/(D - 1)} \right] \tag{3.80}$$

and using the well-known properties of the Laplace transform, we obtain a time-domain representation of the reflection coefficient:

$$\Gamma(t) = \frac{1 - \sqrt{D}}{1 + \sqrt{D}} \delta(t) + \frac{Be^{-Bt/2D}}{\sqrt{D}(D - 1)} \left[I_0\left(\frac{Bt}{2D}\right) + I_1\left(\frac{Bt}{2D}\right) \right] \chi(t) - \frac{2Be^{-Bt/(D-1)}}{(1 + \sqrt{D})(D - 1)} \chi(t)$$
$$- \frac{B^2 e^{-Bt/(D-1)}}{\sqrt{D}(D - 1)^2} \chi(t) \int_0^t e^{B(1+D)x/2D(D-1)} \left[I_0\left(\frac{Bx}{2D}\right) + I_1\left(\frac{Bx}{2D}\right) \right] dx \tag{3.81}$$

where $\delta(t)$ is the Dirac delta function, $\chi(t)$ is the Heaviside function, and $I_n(\cdot)$ is a modified Bessel function.

The first term in (3.81) describes the coefficient of reflection of a wave from a lossless half-space ($\sigma^e = 0$). The sum of the rest terms, $R(t)$, determines the effect of conductance on the wave reflection process.

Direct use of (3.81) in the calculation of $R(t)$ is problematic for large times t, as the integrand of the integral entering into (3.81) increases exponentially with x.

For this reason, a special transformation of (3.81) into a rapidly converging infinite series, convenient in performing numerical calculations, was proposed [20]. The transformation steps are as follows. First, the integral is written as

$$\int_0^t [\ldots]\, dx = \int_{-\infty}^t [\ldots]\, dx - \int_{-\infty}^0 [\ldots]\, dx \tag{3.82}$$

The use of the tabulated integral [21]

$$\int_0^\infty \exp(-\alpha x) I_n(\beta x)\, dx = \frac{\beta^n}{\sqrt{\alpha^2 - \beta^2}\left(\alpha + \sqrt{\alpha^2 - \beta^2}\right)^n} \tag{3.83}$$

shows that the integration over the interval $(-\infty, 0)$ compensates for the third term in (3.81), and, therefore, $R(t)$ can be represented as

$$
\begin{aligned}
R(t) = {}& \frac{B}{\sqrt{D}(D-1)}\left[f_0\!\left(\frac{Bt}{2D}\right) + f_1\!\left(\frac{Bt}{2D}\right)\right]\chi(t) \\[4pt]
& - \frac{B^2}{\sqrt{D}(D-1)^2}\,\chi(t)\int_{-\infty}^t \exp\!\left[\frac{-B(t-x)}{(D-1)}\right]\left[f_0\!\left(\frac{Bx}{2D}\right) + f_1\!\left(\frac{Bx}{2D}\right)\right]dx
\end{aligned}
\tag{3.84}
$$

where $f_n(x) = \exp(-x)I_n(x)$.

Using a new integration variable $u = B(t - x)/(D - 1)$ and a function $Q(x) = f_0(x) + f_1(x)$ and taking into account that

$$\int_0^\infty \exp(-u)\, du = 1 \tag{3.85}$$

we can rewrite (3.84) as

$$R(t) = -\frac{B}{\sqrt{D}(D-1)}\,\chi(t)\int_0^\infty \exp(-u)\left[Q\!\left(\frac{Bt}{2D} - \frac{D-1}{2D}u\right) - Q\!\left(\frac{Bt}{2D}\right)\right]du \tag{3.86}$$

Equation (3.86) has two noteworthy features. First, it involves the multiplier $\exp(-u)$ that decreases exponentially with increasing u, providing for rapid convergence of the integral. Second, it is clear that for large t, the square-bracketed expression is a good approximation for the derivative of the function $Q(x)$, and, therefore, the coefficient $R(t)$ should be proportional to $Q'(Bt/2D)$. However, this approximation is inadequate for small t.

The final transformation step includes the expansion of the function $Q(x)$ in a Taylor series:

$$Q(a + bu) = Q(a) + \sum_{n=1}^\infty b^n \frac{u^n}{n!} Q^{(n)}(a) \tag{3.87}$$

substitution of this expansion in (3.86), and term-by-term integration of the resulting expression in view of the equation

$$\int_0^\infty u^n \exp(-u)\,du = n! \tag{3.88}$$

The expansion of $R(t)$ in an infinite series reads

$$R(t) = -\frac{B}{\sqrt{D}(D-1)}\chi(t)\sum_{n=1}^\infty\left(\frac{1-D}{2D}\right)^n Q^{(n)}\left(\frac{Bt}{2D}\right) \tag{3.89}$$

Turning to the discussion of the algorithm of calculation by (3.89), we note some points important for performing numerical calculations [20]. First, the derivatives of the function $Q(x)$ can be calculated using the recurrence equation for the derivatives of the functions $f_n(x)$:

$$f_n'(x) = \frac{1}{2}f_{n-1}(x) - f_n(x) + \frac{1}{2}f_{n+1}(x) \tag{3.90}$$

Then, for instance, we have

$$Q'(x) = -\frac{1}{2}f_0(x) + \frac{1}{2}f_2(x) \tag{3.91}$$

$$Q''(x) = \frac{1}{2}f_0(x) - \frac{1}{4}f_1(x) - \frac{1}{2}f_2(x) + \frac{1}{4}f_3(x) \tag{3.92}$$

$$Q'''(x) = -\frac{5}{8}f_0(x) + \frac{1}{2}f_1(x) + \frac{1}{2}f_2(x) - \frac{1}{2}f_3(x) + \frac{1}{8}f_4(x) \tag{3.93}$$

The general calculation formula is represented as

$$Q^{(n)}(x) = \sum_{m=0}^{n+1} a_m^n f_m(x) \tag{3.94}$$

where $a_0^1 = -1/2$, $a_1^1 = 0$, $a_2^1 = 1/2$, and

$$a_0^n = -a_0^{n-1} + \frac{1}{2}a_1^{n-1}, \; a_1^n = a_0^{n-1} - a_1^{n-1} + \frac{1}{2}a_2^{n-1}, \; a_m^n = \frac{1}{2}a_{m-1}^{n-1} - a_m^{n-1} + \frac{1}{2}a_{m+1}^{n-1} \tag{3.95}$$

$$a_n^n = \frac{1}{2}a_{n-1}^{n-1} - a_n^{n-1}, \; a_{n+1}^n = \frac{1}{2}a_n^{n-1} \tag{3.96}$$

If the upper limit of summation (∞) in (3.89) is replaced by N, a number appropriate in view of calculation accuracy, it will suffice to calculate only two Bessel functions, I_N and I_{N+1}, for each t, as the values of the rest functions can be found using recurrence equations. It was especially emphasized [20] that the calculation of $Q'(x)$ is stable. However, to prevent overflows and underflows, care is required when (3.94) is used for large n.

Thus, the above analysis suggests that the calculation of $R(t)$ with different values of ε and σ^e can be performed using the approximate expression

$$R(t) \approx \tilde{R}(t) = -\frac{B}{\sqrt{D}(D-1)} \chi(t) \sum_{n=1}^{N} \left(\frac{1-D}{2D}\right)^n Q^{(n)}\left(\frac{Bt}{2D}\right) \tag{3.97}$$

For large t, the coefficient $R(t)$ is proportional to Q', and the calculations are rather accurate even if only the first term of the series in (3.97) is used. For small t, it may appear that much more series terms must be used. However, Rothwell and Suk [20] give an alternative approximate expression for $R(t)$, which holds for small t:

$$R(t) \approx R_s(t) = R(0)\exp\left(-\frac{Bt}{2D}\right)\chi(t) = -\frac{B}{\sqrt{D}(1+\sqrt{D})^2}\exp\left(-\frac{Bt}{2D}\right)\chi(t) \tag{3.98}$$

With this in mind, to improve the quality of the approximation (3.89), Rothwell and Suk [20] normalized the function $\tilde{R}(t)$ using the known behavior of $R(t)$ at small t:

$$R(t) \approx R_{app}(t) = \left\{\left[\frac{R_s(0)}{\tilde{R}(0)} - 1\right]\exp\left(-\frac{Bt}{2D}\right) + 1\right\}\tilde{R}(t) \tag{3.99}$$

Examining this expression, we see that $R_{app}(t)$ has the required value at $t = 0$ and tends to $R(t)$ as $N \to \infty$. In Rothwell and Suk [20], examples of the calculation of $R(t)$ by the proposed algorithm are given. They show that with a proper choice of N, the error is some fractions of a percent, and it can be reduced by increasing N.

A more complicated case of TM polarization is considered in Rothwell [22]. Transient reflection coefficients are efficiently used in the study of current distributions over thin-wire antennas placed close to a lossy half-space [23]. In the studies of Pao et al. [24, 25], an approach is used which is based on an analytic conversion of the inverse Laplace transform. Using this approach, a solution is expressed in terms of incomplete Lipschitz–Hankel integrals for which both well-converging and asymptotic expansions are obtained. Rather efficient methods for numerical inversion of the Laplace transform have also been developed [26, 27] that show good agreement of the calculation results with the results obtained by other methods.

3.2.1.2 Propagation of a Pulsed Field Generated by a Line Source

Let a conducting medium occupy the half-space $z < 0$ (Figure 3.10). An isolated line source of infinite dimensions is coincident with the y-axis and is excited by a current pulse $I(t)$.

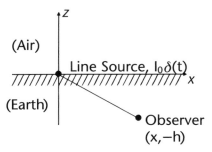

Figure 3.10 A conducting half-space with a line source on its surface (the observation point lies beneath the air–medium interface)

The problem is to derive equations for calculating the pulsed field at an arbitrary point of the conducting medium. In this case, it is convenient to proceed from the well-known solution of the problem for a line source with a uniform harmonic current: $I = I(i\omega)\exp(i\omega t)$. Assume that the medium conductivity σ^e and permeability μ_0 are constant. It is well known that a conducting medium is a low-pass filter. Hence, only fairly low frequencies are essential at the observation point located in the nonstationary field. Therefore, the displacement currents in the medium can be neglected.

The problem formulation contains only one nonzero electric field component, E_y. For $z = -h$, it can be expressed as [28]

$$E_y(i\omega) = -\frac{i\omega\mu_0 I(i\omega)}{\pi}\int_0^\infty \frac{\cos\xi x}{u + \xi}\exp(-uh)\,d\xi \qquad (3.100)$$

where $u = (\xi^2 + i\sigma^e\mu_0\omega)^{1/2}$.

Suppose that since the time $t = 0$, a current $I(t)$ flows in the line source that varies with time synchronously at all points of the source. The Laplace transform of the function $I(t)$ is given by

$$I(s) = \int_0^\infty I(t)\exp(-st)\,dt \qquad (3.101)$$

The transform $E_y(s)$ of the nonstationary electric field $E_y(t)$ generated by the source at the point $z = -h$ is found in a similar manner by merely changing $i\omega$ in (3.100) by a complex parameter s:

$$
\begin{aligned}
E_y(s) &= -\frac{\mu_0 s I(s)}{\pi}\int_0^\infty \cos\xi x\,\frac{\exp\left[-\left(\xi^2 + \sigma^e\mu_0 s\right)^{1/2} h\right]}{\left[\left(\xi^2 + \sigma^e\mu_0 s\right)^{1/2} + \xi\right]}\,d\xi \\
&= \frac{I(s)}{\pi\sigma^e}\int_0^\infty \cos\xi x\left(\frac{\partial}{\partial h} + \xi\right)\exp\left[-\left(\xi^2 + \sigma^e\mu_0 s\right)^{1/2} h\right]\,d\xi
\end{aligned}
\qquad (3.102)
$$

The desired field $E_y(t)$ is found using the inverse Laplace transform

$$E_y(t) = \int_C E_y(s) \exp(st)\, ds \tag{3.103}$$

The integration path C in (3.103) is a straight line parallel to the imaginary axis of the s-plane, so that all the singular points of the integrand are on the left of the path. The integral representation of $E_y(t)$ (3.103) can be, in principle, transformed into its representation in terms of the residues of the integrand at its poles and of the integrals along the sides of the cuts connecting the branch points. However, this requires, first, finding singular points for each specified current $I(t)$ and, second, laborious and cumbersome calculations. In this sense, direct computer integration of (3.37) using the available codes for computing the inverse Laplace transform is preferred.

The calculations can be simplified significantly for the current given by $I(t) = I_0\delta(t)$. For this case, we have $I(s) = I_0$. Therefore, substituting (3.102) in (3.103) and changing the order of integration, we obtain

$$E_y(t) = \frac{I_0}{\pi\sigma^e} \int_0^\infty \cos\xi x \left(\frac{\partial}{\partial h} + \xi\right) \int_C \exp\left[-(\sigma^e\mu_0)^{1/2} h\left(s + \frac{\xi^2}{\sigma^e\mu_0}\right)^{1/2}\right] ds\, d\xi \tag{3.104}$$

The inner integral in (3.104) can be evaluated analytically. As a result, we find

$$E_y(t) = \frac{I_0}{\pi\sigma^e} \int_0^\infty \cos\xi x \left(\frac{\partial}{\partial h} + \xi\right) \frac{(\sigma^e\mu_0)^{1/2} h}{2(\pi t^3)^{1/2}} \exp\left[-\frac{\xi^2 t}{\sigma^e\mu_0} - \frac{\sigma^e\mu_0 h^2}{4t}\right] d\xi \tag{3.105}$$

In turn, (3.105) admits further simplification by using the well-known equations

$$\int_0^\infty \cos\xi x \exp(-\alpha\xi^2)\, d\xi = \frac{1}{2}\left(\frac{\pi}{\alpha}\right)^{1/2} \exp\left[-\frac{x^2}{4\alpha}\right] \tag{3.106}$$

$$\int_0^\infty \xi \cos\xi x \exp(-\alpha\xi^2)\, d\xi = \frac{1}{2\alpha} - \frac{x}{4\alpha^{3/2}} \sum_{k=0}^\infty \frac{(-1)^k k!}{(2k+1)!} \left(x\alpha^{-1/2}\right)^{2k+1} \tag{3.107}$$

where $\alpha = t/\sigma^e\mu_0$.

Substituting (3.106) and (3.107) in (3.105) and performing differentiation with respect to h, we obtain the following representation for the electric field [29]:

$$E_y(t) = \frac{I_0}{4\pi\sigma^e h^2 \left(\sigma^e\mu_0 h^2\right)} F(D, T) \tag{3.108}$$

where

$$F(D,T) = \exp\left(-\frac{1}{4T}\right)\left\{\left(T^{-2} - \frac{1}{2}T^{-3}\right)\exp\left(-\frac{D^2}{4T}\right) + \pi^{-1/2}T^{-5/2}\right.$$
$$\left. \times\left[1 - \frac{D}{2T^{1/2}}\sum_{k=0}^{\infty}\frac{(-1)^k k!}{(2k+1)!}(DT^{-1/2})^{2k+1}\right]\right\} \tag{3.109}$$

where $D = x/h$ and $T = t/(\sigma^e \mu_0 h^2)$.

The quantities F, D, and T in (3.109) are dimensionless, whereas I_0 has the dimension of charge.

Figure 3.11 presents the variation of the function $F(D,T)$ with T for different values of D.

It can be seen that the field pulse has a pronounced negative peak followed by a slowly decreasing tail. For the observation point located beneath the source ($x = 0$), it can immediately be seen from (3.109) that the field decreases with deeper imbedding of the observation point into the medium according to an algebraic law as h^{-4}. This distinguishes the behavior of a nonstationary field in a conducting medium from the behavior of a time-harmonic field, which decays exponentially. Finally, it should be noted that for a current arbitrarily depending on time, the field can be represented as a integral of convolution of this current and the function (3.108) having the meaning of the impulse response of the medium represented as a conducting half-space:

$$E_y(t) = \frac{1}{4\pi\sigma^e h^2 \left(\sigma^e \mu_0 h^2\right)}\int_0^t I(\tau)F\left(D, \frac{(t-\tau)}{\sigma^e \mu_0 h^2}\right)d\tau \tag{3.110}$$

3.2.2 Propagation of Pulses Generated by a Point Source in a Multilayered Medium

3.2.2.1 Formulation of the Problem: The Frequency-domain Solution

The geometry of the problem is sketched in Figure 3.12. A horizontal electric dipole is located at an altitude h above the boundary of a multilayered dielectric medium

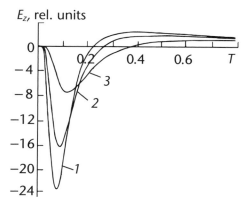

Figure 3.11 Waveforms of the normalized electric field at different observation points: $D = 0$ (curve 1), 0.5 (curve 2), and 1 (curve 3). (With permission from IEEE, Inc.)

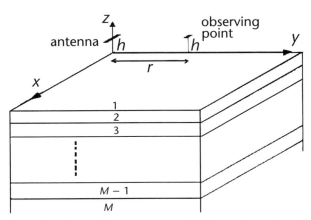

Figure 3.12   Sketch illustrating the problem geometry. (With permission from IEEE, Inc.)

and is oriented along the x-axis. Each layer is characterized by its permittivity and thickness. The permeability of all the layers is assumed to be equal to the permeability μ_1 of the half-space $z > 0$; the conductivity of the layers is zero, that is, the medium is nondispersive. The task is to find the total electromagnetic field at an arbitrary point above the multilayered medium.

The formulation of the problem is similar to that of the classical problem of the dipole field above a homogeneous conducting half-space solved by Sommerfeld. An electric dipole located above a multilayered medium generates simultaneously a transverse electric (TE) field and a transverse magnetic (TM) field with respect to the z-axis. The TM field can be completely described by an electric Hertz vector Π^e_z having only a z component, and the TE field can be completely described by a magnetic Hertz vector Π^m_x having only an x component. These vectors can be represented as [30]

$$\Pi^m_x = \frac{I(\omega)L}{i4\pi\omega\varepsilon_1} \int_0^\infty \Big[\exp\big(-u_1|z - h|\big) + E_1(\xi)\exp\big(-u_1(z + h)\big)\Big]\frac{\xi}{u_1} J_0\big(\xi\sqrt{x^2 + y^2}\big)d\xi \qquad (3.111)$$

$$\Pi^e_z = -\frac{I(\omega)Lx}{i4\pi\omega\varepsilon_1\sqrt{x^2 + y^2}} \int_0^\infty \big[E_1(\xi) + D_1(\xi)\big]J_1\big(\xi\sqrt{x^2 + y^2}\big)\exp\big(-u_1(z + h)\big)d\xi \quad (3.112)$$

where $I(\omega)$ is the current in the dipole, L is the dipole length, ε_1 is the permittivity of the half-space $z > 0$, J_0 and J_1 are Bessel functions, ξ is the horizontal spatial wave number, and u_1 is the vertical spatial wave number. The time dependence is described by $\exp(i\omega t)$.

Expressions (3.111) and (3.112) can be treated [31] as describing an infinite continuous spectrum waves propagating along the z-axis on both sides of the plane $z = h$. This spectrum includes both propagating and exponentially dumped waves. The amplitudes of the waves are complicated functions of the horizontal spatial wave number ξ.

The functions $E_1(\xi)$ and $D_1(\xi)$ represent the coefficients of reflection of the TE and the TM plane waves, respectively, from the multilayered medium. For the geometry shown in Figure 3.12, they can be found using the recurrence equations [32, 33]

$$E_1(\xi) = \frac{u_1 - u_2 E_2}{u_1 + u_2 E_2},$$

$$D_1(\xi) = \frac{n_2^2 u_1 - n_1^2 u_2 D_2}{n_2^2 u_1 + n_1^2 u_2 D_2}$$

$$E_l(\xi) = \frac{u_l th(d_l u_l) + u_{l+1} E_{l+1}}{u_l + u_{l+1} E_{l+1} th(d_l u_l)}$$

$$D_l(\xi) = \frac{n_{l+1}^2 u_l th(d_l u_l) + n_l^2 u_{l+1} D_{l+1}}{n_{l+1}^2 u_l + n_l^2 u_{l+1} D_{l+1} th(d_l u_l)} \qquad (3.113)$$

$$E_M = D_M = 1$$

where n_l is the refractive index, $u_l = \sqrt{\xi^2 - k_l^2}$ is the vertical spatial wavenumber, $k_l = \omega \sqrt{\varepsilon_l \mu}$ is the wavenumber, and d_l is the thickness of the lth layer of the medium. The one-valued branch of the two-valued function u_l is chosen so that the inequality $\mathrm{Re}(u_1, u_M) > 0$ be satisfied to ensure that the radiation condition at infinity be fulfilled.

The main goal of the subsequent consideration is to reveal the structural features of the generated fields. This can be done by examining any one of the field components, for example, E_x. In terms of subsurface probing applications, it is of interest to analyze the behavior of this component in the upper half-space in relation to the geometric parameters and permittivities of individual layers of the medium. The expression for this component is obtained using the representations (3.111) and (3.112) for the Hertz vectors:

$$\begin{aligned}
E_x &= k_1^2 \Pi_x^m + \frac{\partial^2 \Pi_x^m}{\partial x^2} + \frac{\partial^2 \Pi_z^e}{\partial x \partial z} \\
&= \frac{I(\omega) L}{i 4 \pi \omega \varepsilon_1} \int_0^\infty \left\{ \frac{\xi}{u_1} \left[\exp(-u_1 |z - h|) + E_1(\xi) \exp(-u_1(z + h)) \right] \right. \\
&\quad \times \left[\left(k_1^2 - \xi^2 x^2 r^{-2} \right) J_0(\xi r) + \xi(x^2 - y^2) r^{-3} J_1(\xi r) \right] \\
&\quad \left. + \frac{u_1}{r^3}(y^2 - x^2) \left[E_1(\xi) + D_1(\xi) \right] \exp(-u_1(z + h)) J_1(\xi r) \right\} d\xi
\end{aligned} \qquad (3.114)$$

where $r = \sqrt{x^2 + y^2}$.

Formula (3.114) is substantially simplified if the field is sought at a point with coordinates $x = 0$, $z = h$, and $y = r$:

$$\begin{aligned}
E_x \equiv E(\omega) &= \frac{I(\omega) L}{i 4 \pi \omega \varepsilon_1} \int_0^\infty \left\{ \frac{\xi}{u_1} \left[1 + E_1(\xi) \exp(-2 h u_1) \right] \left[k_1^2 J_0(\xi r) - \xi r^{-1} J_1(\xi r) \right] \right. \\
&\quad \left. + u_1 r^{-1} \left[E_1(\xi) + D_1(\xi) \right] \exp(-2 h u_1) J_1(\xi r) \right\} d\xi
\end{aligned} \qquad (3.115)$$

The ordinary way of seeking a time-domain solution is to use the inverse Fourier transform. As applied to (3.115), this yields the following representation for $E(t)$:

$$E(t) = \frac{1}{\pi}\text{Re}\left\{\int\limits_0^\infty E(\omega)\exp(i\omega t)\,d\omega\right\} + 2\pi i\left\{\frac{1}{2}\text{Re}\,sE(\omega)\Big|_{\omega=0}\right\}$$

(3.116)

$$= \frac{1}{\pi}\text{Re}\left\{\int\limits_0^\infty E(\omega)\exp(i\omega t)\,d\omega\right\} + \frac{I(0)L}{4\pi\varepsilon_1 r}\int\limits_0^\infty D_0(\xi)\xi J_1(\xi r)\exp(-2h\xi)\,d\xi$$

where the second term describes the constant component of $E(t)$, which is obtained by the integration along a semicircle of vanishingly small radius centered at the point $\omega = 0$ of the complex ω-plane. The expression for $D_0(\xi)$ is obtained from that for $D_1(\xi)$ by substituting ξ for u_1 and u_2.

As noted earlier, the calculation of the field by formulas (3.115) and (3.116) with arbitrary parameters is a very complicated and laborious problem. Its solution is somewhat easier when the altitude of the dipole above the boundary of the medium, h, is large enough and the distance to the observation point, r, is relatively small. In this case, computer calculations can be performed directly by formulas (3.115) and (3.116) without their preliminary transformation. As an example, Figure 3.13(a) presents the results of the calculation for $r = h = 0.5$ m and the exciting current pulse parameters specified by the formulas [30]

$$I(t)L = g(t) = 8\pi\exp\left(-\zeta^2 t^2\right)$$

(3.117)

$$I(\omega)L = g(\omega) = \frac{8\pi^{3/2}\exp\left(\dfrac{-\omega^2}{4\zeta^2}\right)}{\zeta}, \quad \zeta = 10^9\,s^{-1}$$

(3.118)

The electric and geometric parameters of the medium are indicated in Figure 3.13(b).

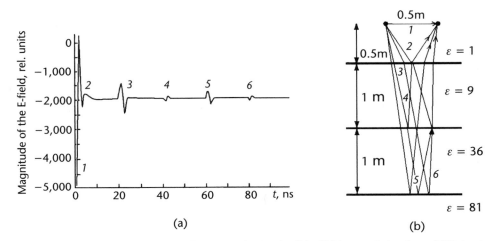

(a)

(b)

Figure 3.13 Pulse waveforms (a) and propagation paths (b). (With permission from IEEE, Inc.)

In Figure 3.13(a), pulse 1 represents the direct wave. Pulse 2 arrives at the observation point after reflection from the first interface between the media, which is the nearest to the dipole. Pulses 3 and 5 are reflected only from the second and the third interface, respectively. Each of pulses 4 and 6 undergoes three reflections, respectively, from the first and second and from the second and third interfaces. A geometric interpretation of the paths of the pulses is shown in Figure 3.13(b). The nonzero static field observed at large times is due to the close proximity of the observation point to the dipole.

3.2.2.2 The Time-domain Solution of the Problem

We restrict our consideration to a three-layered medium with two interfaces. In addition, we assume that $h = 0$, i.e., that the dipole and the observation point are located on the surface of the medium. Using formulas (3.113), it is easy to see that the functions $E_1(\xi)$ and $D_1(\xi)$ are determined by the expressions

$$E_1 = \frac{R_{12} + R_{23} \cdot \exp(-2du_2)}{1 + R_{12} \cdot R_{23} \cdot \exp(-2du_2)}, \quad D_1 = \frac{T_{12} + T_{23} \cdot \exp(-2du_2)}{1 + T_{12} \cdot T_{23} \cdot \exp(-2du_2)} \quad (3.119)$$

where d is the layer thickness,

$$R_{12} = \frac{u_1 - u_2}{u_1 + u_2}, \quad R_{23} = \frac{u_2 - u_3}{u_2 + u_3}, \quad T_{12} = \frac{n_2^2 u_1 - n_1^2 u_2}{n_2^2 u_1 + n_1^2 u_2}, \quad T_{23} = \frac{n_3^2 u_2 - n_2^2 u_3}{n_3^2 u_2 + n_2^2 u_3} \quad (3.120)$$

Expanding each of the functions (3.119) in an infinite series by using the formulas for the sum of terms of an infinite geometric progression:

$$E_1 = R_{12} + \left(R_{12} - \frac{1}{R_{12}} \right) \sum_{m=1}^{\infty} (-R_{12}R_{23})^m \exp(-2mdu_2)$$

$$D_1 = T_{12} + \left(T_{12} - \frac{1}{T_{12}} \right) \sum_{m=1}^{\infty} (-T_{12}T_{23})^m \exp(-2mdu_2) \quad (3.121)$$

and substituting (3.121), in view of (3.118), in (3.115), we can represent the total field as the sum of three terms that describe, respectively, the field of the direct wave, $E_d(\omega)$, the field of the wave reflected from the upper boundary of the layered medium, $E_b(\omega)$, and the field of the waves experienced a different number of reflections from the medium boundaries, $E_r(\omega)$:

$$E_d(\omega) = \frac{2G(\omega)}{i\omega\varepsilon_1} \int_0^{\infty} \frac{\xi}{u_1} \left[k_1^2 J_0(\xi r) - \xi r^{-1} J_1(\xi r) \right] d\xi$$

$$= -2 \frac{G(\omega)c\mu_1}{n_1} \frac{\exp(-ik_1 r)}{r} \left[\frac{1}{r} + ik_1 + \frac{i}{k_1 r^2} \right] \quad (3.122)$$

$$
E_h(\omega) = \frac{G(\omega)\mu_1 c}{ik_1 n_1}
$$
$$
\times \int_{-\infty}^{\infty} \left\{ \frac{\xi}{u_1} R_{12} \left[k_1^2 H_0^{(2)}(\xi r) - \xi r^{-1} H_1^{(2)}(\xi r) \right] + u_1 r^{-1} \left[T_{12} + R_{12} \right] H_1^{(2)}(\xi r) \right\} d\xi
$$
(3.123)

$$
E_r(\omega) = \frac{G(\omega)\mu_1 c}{ik_1 n_1}
$$
$$
\times \int_{-\infty}^{\infty} \left\{ \frac{\xi}{u_1} \left(R_{12} - \frac{1}{R_{12}} \right) \sum_{m=1}^{\infty} \left(R_{12} R_{23} \right)^m \left[k_1^2 H_0^{(2)}(\xi r) - \xi r^{-1} H_1^{(2)}(\xi r) \right] \exp(-2mdu_2) \right.
$$
$$
\left. + \sum_{m=1}^{\infty} \left[\left(R_{12} - \frac{1}{R_{12}} \right) \left(-R_{12} R_{23} \right)^m + \left(T_{12} - \frac{1}{T_{12}} \right) \left(-T_{12} T_{23} \right)^m \right] \frac{u_1}{r} H_1^{(2)}(\xi r) \exp(-2mdu_2) \right\} d\xi
$$
(3.124)

In (3.122)–(3.124), c is the velocity of light in the free space. Using well-known equations, the Bessel functions in (3.123) and (3.124) are replaced by Hankel functions. The integration is extended over the entire ξ-axis. For sufficiently large r, this transformation allows using the steepest descent method to evaluate the integral in (3.123). The result is a good approximation [30] for the sum of the fields determined by (3.122) and (3.123):

$$
E_d(\omega) + E_h(\omega) \approx \frac{2G(\omega)\mu_1 c}{ik_1 n_1} \int_0^{\infty} \frac{\xi k_1^2}{u_1} \left[1 + E_1 \right] J_0(\xi r)\, d\xi
$$
$$
= \frac{4G(\omega)c\mu_1}{\left(n_2^2 - n_1^2 \right) r^2} \left[\frac{n_2}{n_1} \exp(-ik_2 r) - \exp(-ik_1 r) \right]
$$
(3.125)

whence, using the well-known property of Fourier transforms, we immediately obtain a time-domain analog to (3.125):

$$
E_d(t) + E_h(t) \approx \frac{4c\mu_1}{\left(n_2^2 - n_1^2 \right) r^2} \left[\frac{n_2}{n_1} g\left(t - \frac{n_2 r}{c} \right) - g\left(t - \frac{n_1 r}{c} \right) \right]
$$
(3.126)

Examining (3.126), we note an interesting feature. It turns out that two direct pulses propagating along the upper interface arrive at the observation point. Their propagation velocities, which are determined by the refractive indices of the dielectric media adjacent to the interface, are different. This feature is realized in a number of practical applications. For instance, if the upper medium is saline, its permittivity can be promptly measured provided that an individual pulse propagating along the opposite side of the interface can be selected during the measuring procedure.

Next, we turn to a discussion of the main features of the method [30] used for finding a time-domain analog to (3.124). The expression for the mth term of the infinite series in (3.124) can be written as

$$E_m(\omega) = \frac{G(\omega)\mu_1 c}{i k_1 n_1} \int_{-\infty}^{\infty} \left[\frac{\xi k_1^2 R_m}{u_1} H_0^{(2)}(\xi r) + \left(T_m u_1 - \frac{k_1^2 R_m}{u_1} \right) \frac{1}{r} H_1^{(2)}(\xi r) \right] e^{-2mdu_2} \, d\xi \quad (3.127)$$

The function $E_m(\omega)$ is the Fourier transform of the function $E_m(t)$. Perform its analytic continuation for pure imaginary values of ω. As a result, the variable ω appears in the parameter of the Laplace transform, $s = i\omega$, which is subsequently considered to be positive. Changing the integration variable in (3.127) by setting $\xi = -is\chi$, we arrive at the following representation for $E_m(s)$:

$$E_m(s) = -\frac{4\mu}{\pi} G(s) \left\{ s^2 \operatorname{Im}\left[\int_0^{i\infty} \frac{\chi R_m}{v_1} K_0(s\chi r) \exp(-2mds v_2) \, d\chi \right] \right.$$
$$\left. + s \operatorname{Im}\left[\int_0^{i\infty} \left(\frac{R_m}{v_1} + \frac{c^2}{n_1^2} T_m v_1 \right) \frac{K_1(s\chi r)}{r} \exp(-2mds v_2) \, d\chi \right] \right\} \quad (3.128)$$

where K_0 and K_1 are Macdonald functions of the zero and the first order. The coefficients R_m and T_m are determined by the formulas

$$R_m = \left(R_{12} - \frac{1}{R_{12}} \right) (-R_{12} R_{23})^m \quad (3.129)$$

$$T_m = \left(T_{12} - \frac{1}{T_{12}} \right) (-T_{12} T_{23})^m \quad (3.130)$$

where

$$R_{12} = \frac{v_1 - v_2}{v_1 + v_2}, \quad R_{23} = \frac{v_2 - v_3}{v_2 + v_3}, \quad T_{12} = \frac{n_2^2 v_1 - n_1^2 v_2}{n_2^2 v_1 + n_1^2 v_2}, \quad T_{23} = \frac{n_3^2 v_2 - n_2^2 v_3}{n_3^2 v_2 + n_2^2 v_3} \quad (3.131)$$

and the one-valued branch of the function $v_l = \sqrt{(n_l/c)^2 - \chi^2}$ is chosen so that we have $\operatorname{Re}(v_l) > 0$.

The above transformations allow us to derive expressions for $E_m(t)$ directly from (3.128), not invoking (3.127) and using the inverse Fourier transform in ω. Indeed, in view of the integral representation of K_0,

$$K_0(\chi rs) = \int_0^{\infty} \exp(-s\chi rch\theta) \, d\theta = \int_{\chi r}^{\infty} \frac{\exp(-st)}{\sqrt{t^2 - (\chi r)^2}} \, dt \quad (3.132)$$

we may conclude that the function $K_0(\chi rs)$ is the direct integral Laplace transform of the function $U(t - \chi r)/\sqrt{t^2 - (\chi r)^2}$. Here, like in the original study [30], $U(t)$ denotes the Heaviside function. Now, based on the well-known properties of Laplace transforms, we can readily find the functions-originals for the integrals entering into (3.128):

$$I_1(t) = \mathrm{Im}\left\{\int\limits_{\chi(\tau)} \frac{\chi R_m}{\upsilon_1} \frac{U(t - \chi r - 2md\upsilon_2)}{\sqrt{(t - 2md\upsilon_2)^2 - (\chi r)^2}} d\chi\right\} \qquad (3.133)$$

$$I_2(t) = \mathrm{Im}\left\{\int\limits_{\chi(\tau)} \left(\frac{R_m}{\upsilon_1} + \frac{c^2}{n_1^2} T_m \upsilon_1\right) \frac{(t - 2md\upsilon_2)}{\chi r^2} \frac{U(t - \chi r - 2md\upsilon_2)}{\sqrt{(t - 2md\upsilon_2)^2 - (\chi r)^2}} d\chi\right\} \qquad (3.134)$$

Finally, we use the relationship between the Laplace image of the function $g(t)$ and its first- and second-order derivatives and the well-known relationship between the product of images and the convolution of functions-originals to obtain

$$E_m(t) = -\frac{4\mu_1}{\pi}\left[\frac{d^2 g(t)}{dt^2} \otimes I_1(t) + \frac{dg(t)}{dt} \otimes I_2(t)\right] \qquad (3.135)$$

where the symbol $\otimes$ implies the operation of convolution.

The integration path $\chi(\tau)$ in (3.133) and (3.134) should pass in the complex χ-plane so that the quantity τ, given by $\tau = \chi r + 2md\upsilon_2$, would take real and positive values. This requirement can be equivalently rewritten as

$$\chi = \frac{\left(r\tau + 2mdi\sqrt{\tau^2 - t_0^2}\right)}{R^2} \qquad (3.136)$$

where $R = \sqrt{r^2 + 4(md)^2}$ and $t_0 = n_2 R/c$. The required integration path is depicted in Figure 3.14.

As can be seen from Figure 3.14, the integration path $\chi(\tau)$ passes through a portion of the positive axis and a portion of the hyperbola lying in the first quadrant of the χ-plane. This path is obtained by deforming the original integration path passing along the positive part of the imaginary axis. This deformation does not change the values of the integrals because, first, the integral over the arc of a circle

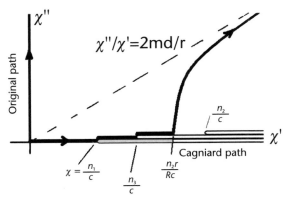

Figure 3.14 Integration paths and cuts in the complex χ-plane. (With permission from IEEE, Inc.)

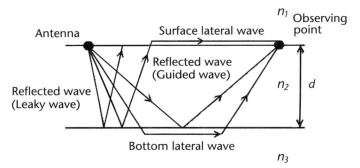

Figure 3.15   Lines of propagation of different waves. (With permission from IEEE, Inc.)

of infinite radius that connects the ends of the original and deformed paths is zero. Second, the integrands have no singularities in the region enclosed by the paths.

Noteworthy is the clear relationship between the positions of branch points in the χ-plane and the characteristic times of propagation of the pulsed field. Thus, the time $\tau = t_1 = (rn_1 + 2md\sqrt{n_2^2 - n_1^2})/c$ corresponding to the branch point $\chi = n_1/c$ is the time it takes for the lateral wave to propagate along the upper interface from the dipole to the observation point. The time $\tau = t_3 = (rn_3 + 2md\sqrt{n_2^2 - n_3^2})/c$ corresponding to the branch point $\chi = n_3/c$ is the time it takes for the lateral wave propagating along the lower interface to cover the distance from the dipole to the observation point. To the branch point $\chi = n_2r/Rc$, there corresponds the propagation time of the reflected wave, $\tau = t_0 = n_2r/c$. Figure 3.15 gives a geometric interpretation of the aforesaid for the case $m = 1$. Note that the contributions to the integrals due to the integration over the path section from the point $\chi = 0$ to the point $\chi = n_1/c$ of the real axis are zero, because the integrands take real values on this section. The integral over the segment of the real axis from the point $\chi = n_1/c$ to the point $\chi = n_3/c$ describes the contribution to the field made by the lateral wave propagating along the upper interface. A similar integration between the points $\chi = n_3/c$ and $\chi = n_2r/Rc$ makes the contribution of the lateral wave propagating along the low interface. The integration over the hyperbolic part of the integration path describes the contributions of the waves experienced different numbers of reflections from the interfaces.

The situation sketched in Figures 3.14 and 3.15 corresponds to $n_1 < n_3 < n_2$. If $n_3 > n_2$ and $n_1 > n_2$, the branch points $\chi = n_1/c$ and $\chi = n_3/c$ are located on the right of the point $\chi = n_2r/Rc$, and, therefore, the integration path $\chi(\tau)$ contains no sections passing along the cut sides connecting these points. As a consequence, lateral waves are absent.

If the point $\chi = n_2r/Rc$ lies on the left of the branch point $\chi = n_1/c$ (it is assumed that $n_2 > n_1$ and $n_3 > n_1$, that is, r is less than the critical distance $2n_1md/\sqrt{n_2^2 - n_1^2}$), the integration path not contains no sections passing along the cut sides. Therefore, in this case, also no lateral wave is generated. If $r < 2n_1md/\sqrt{n_2^2 - n_1^2}$, the reflected wave is a leaky wave, as follows from the value of the function v_1. If the point $\chi = n_2r/Rc$ lies on the right of the points of branching of the functions v_1 and v_3 (this corresponds to the case $r > 2n_3md/\sqrt{n_2^2 - n_3^2}$), the reflected wave is the guided wave judging from the values of v_1 and v_3.

3.2.2.3 Numerical Calculations

Before performing calculations, it is important to have a clear idea of the pulse waveforms of the direct, reflected, and lateral waves. Necessary information can be obtained by examining the asymptotic behavior of the solution. Thus, (3.126) shows that the waveforms of both direct waves resemble the waveform of the exciting current pulse described by (3.118). Therefore, when the dipole is excited by a current pulse of this type, the observed pulses of the direct wave fields are monopolar. Asymptotic evaluation of the integrals entering into (3.124) for large r indicates that the waveform of a reflected wave is determined by the behavior of the time derivative of the exciting current. This implies that for the pulse waveform described by (3.118), the reflected field pulses are bipolar. A more complicated and laborious asymptotic analysis shows that the field pulse of the lateral wave is monopolar and its waveform resembles that described by (3.118).

Figure 3.16 presents the results of calculations [30] for several values of r and specified geometric and electric parameters of the medium.

The numbers in Figure 3.16(a) indicate different waves whose paths are depicted in Figure 3.16(b). As can be seen from Figure 3.16(a), the direct waves (1 and 2) and lateral waves (3, 5, 7, and 9) have the waveforms of monopolar pulses of positive or negative polarity. The reflected waves (4, 6, and 8) are represented by bipolar pulses. It can clearly be seen that at $r = 2$ m the reflected waves have larger amplitudes than at $r = 0.5$ m. This is a direct consequence of the directivity effects that occur in the dipole radiation pattern in the presence of a medium. At the distance $r = 0.5$ m, there is no lateral wave, as r is less than the smallest critical distance (0.71 m in this case). The permittivities of the media adjacent to the lower interface preclude the propagation of a lateral wave along this interface. The nonzero value of the field that persists throughout the time interval of short-path observation ($r = 0.5$ m) indicates the presence of a static field due to the smallness of r. For other paths, the static field is negligible.

Figure 3.17 presents the calculation results [30] for another ratio of the permittivities of the medium regions adjacent to the interfaces.

As in the previous case, the waveforms of the lateral and direct waves are different from the waveforms of the reflected waves. Attention is drawn to the existence of a lateral wave propagating along the lower interface (numbered as 11). Physically, its presence can be explained if we take into account the ratio of the permittivities

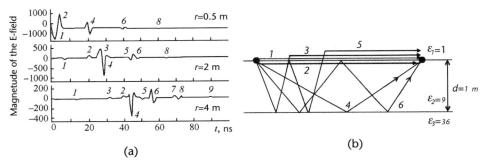

(a)   (b)

Figure 3.16 Behavior of the nonstationary field at different distances from the dipole. (With permission from IEEE, Inc.)

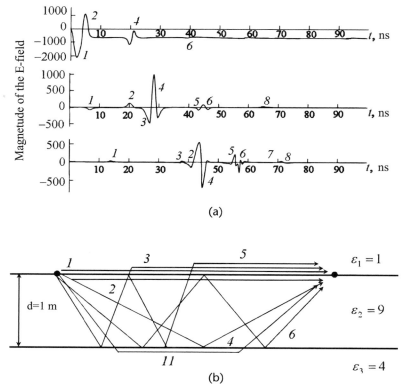

(a)

(b)

Figure 3.17 Behavior of the nonstationary field at different distances from the dipole in the case of two lateral waves. (With permission from IEEE, Inc.)

of the medium regions. An increase in the directivity of the dipole radiation can be revealed if we compare the amplitudes of the reflected waves for the first and the second paths. Finally, comparing Figures 3.16 and 3.17, we see that the reflected pulses differ in the sequence order of the lobe polarities, which can be explained by the difference in the ratio of the permittivities of the contiguous regions of the media.

Conclusion

The effect of various factors, including the dispersion of the medium and the nonlinear interaction of a high-power radiation pulse with the medium, on the propagation of nonstationary plane waves in unbounded conducting media has been investigated. The nature of the influence of the Earth's atmosphere on the electromagnetic field pulses of nanosecond and picosecond duration is described. It is pointed out that there is a fundamental limitation on the use of ultrawideband picosecond pulses in radar. The threshold intensity is given for microsecond pulses propagating in the Earth's atmosphere and an optimal proportion between the durations and amplitudes of subnanosecond pulses in a pulse sequence providing their least absorption during the propagation in the lower atmosphere.

Several problems on the reflection of ultrawideband electromagnetic pulses from medium interfaces and their transmission through the interfaces are considered. Formulas for the time-domain calculation of the reflection coefficient important for applications are given. The features of the propagation of the pulsed field generated by an infinite line source in a conducting half-space have been investigated neglecting the displacement currents in the medium.

The propagation of pulses generated by a point source in a layered medium has been investigated using the Cagniard–de Hoop method. The advantage of this method is the possibility of a sequential analysis of the entire spectrum of the waves generated in the medium. This makes it possible to find out not only a qualitative but also a quantitative contribution of these waves to the formation of the total field at various points of the medium.

Problems

3.1 The equation

$$\frac{\partial^2 u}{\partial t^2} = a^2 \frac{\partial^2 u}{\partial z^2} + b^2 u \tag{3.137}$$

where $u = u(z,t)$, and a and b are constants, is called the telegraph equation. It is used, for instance, in studying lossy transmission lines excited by nonstationary sources. Find the solution of this equation that satisfies the initial conditions

$$u(z,0) = f(z), \quad \left.\frac{\partial u(z,t)}{\partial t}\right|_{t=0} = g(z) \tag{3.138}$$

where $f(z)$ and $g(z)$ are given functions.

Solution:

$$u(z,t) = \frac{1}{2}\left[f(z + at) + f(z - at)\right] + \frac{1}{2a}\int_{-at}^{at} f(z + \beta)\frac{\partial}{\partial t} I_0\left(b\sqrt{t^2 - \left(\frac{\beta}{a}\right)^2}\right) d\beta$$

$$+ \frac{1}{2a}\int_{-at}^{at} g(z + \beta)\frac{\partial}{\partial t} I_0\left(b\sqrt{t^2 - \left(\frac{\beta}{a}\right)^2}\right) d\beta \tag{3.139}$$

where $I_0(x)$ is a modified Bessel function whose integral representation reads

$$I_0(x) = \frac{1}{\pi}\int_{-\pi/2}^{\pi/2} \cos(x\cos\xi)\,d\xi \tag{3.140}$$

3.2 Find the spectral function of the pulse described by (3.59):

$$E(t) = E_0 \left(\frac{t}{T}\right)^n \left[M^{n+1} \exp\left(-\frac{Mt}{T}\right) - \exp\left(-\frac{t}{T}\right) \right] \chi(t) \qquad (3.141)$$

Hint: Use the integral representation of the Gamma function

$$\Gamma(x) = \int_0^\infty e^{-s} s^{x-1} \, ds, \; x > 0 \qquad (3.142)$$

3.3 Calculate the initial temporal moments for the pulse described by (3.59).
   Hint: Use the integral representation of the Gamma function.

3.4 Derive formulas (3.111) and (3.112) for the magnetic and electric Hertz vectors by using the general representations for the field generated by an arbitrary set of electric and magnetic currents in cylindrical coordinates [31].

3.5 Check that the coefficients in (3.119) can be represented by infinite series as in (3.121).

References

[1] Stratton, J. A., *Electromagnetic Theory*, New York: McGraw-Hill, 1941.

[2] Il'inskii, A. S., V. V. Kravtsov, and A. G. Sveshnikov, *Mathematical Models of Electrodynamics*, Moscow: Vysshaya Shkola, 1991 (in Russian).

[3] Heaviside, O., "The General Solution of Maxwell's Electromagnetic Equations in a Homogeneous Isotropic Medium, Especially in Regard to the Derivation of Special Solutions, and Formulae for Plane Waves," *Electrical papers*. Vol. II, L.; N. Y.: Macmillan, 1892, pp. 468–485.

[4] Poincaré, H.,"Théorie Analytique de la Propagation de la Chaleur," *Comptes Rendus*, Ch. 8, 1893, p. 1030.

[5] Nikiforov, A. F., and V. B. Uvarov, *Special Functions of Mathematical Physics*, Boston: Birkhäuser, 1988.

[6] Vainshtein, L. A., "Propagation of Pulses," *Sov. Phys. Usp.*, Vol. 19, No. 2, 1976, pp. 189–205.

[7] Gutman, A. L., "Space-Time Green Function and Short Pulse Propagation in Different Media," In *Ultra-Wideband, Short-Pulse Electromagnetics 4*, pp. 301–311, E. Heyman, B. Mandelbaum, and J. Shiloh (eds.), New York: Plenum Press, 1999.

[8] Borisov, V. V., *Transient Electromagnetic Waves*, Leningrad: LSU Publishers, 1987 (in Russian).

[9] Sobolev, S. L., *Partial Differential Equations of Mathematical Physics*, London: Pergamon Press, 1964.

[10] Vakman, D. Ye., "Evolution of the Parameters of a Pulse Propagating with Dispersion and Attenuation," *Sov. J. Commun. Technol. Electron.*, Vol. 36, No. 7, 1991, pp. 100–105.

[11] Shvartsburg, A. B., "Single-Cycle Waveforms and Non-Periodic Waves in Dispersive Media (exactly solvable models)," *Phys. Usp.*, Vol. 41, 1998, pp. 77–94.

[12] Gibbins, C. J., "Propagation of Very Short Pulses Through the Absorptive and Dispersive Atmosphere," *IEE Proc.*, Vol. 137, Pt. H, No. 5, 1990, pp. 304–310.

[13]   Liebe, H. J., "An Updated Model for Millimeter Wave Propagation in Moist Air," *Radio Sci.* Vol. 20, No. 6, 1985, pp. 1069–1089.

[14]   Stadnik, A. M., and G. V. Ermakov, "The Distortion of Ultrawideband Electromagnetic Pulses in the Earth's Atmosphere," *J. Commun. Technol. Electron.*, Vol. 40, No. 10, 1995, pp. 25–32.

[15]   Stadnik, A. M., and G. V. Ermakov, "Atmospheric Distortions of Ultrashort, Ultrawideband Pulses," *Radiofiz. Radioastron.*, Vol. 5, No. 2, 2000, pp. 125–130.

[16]   Stadnik, A. M., and G. V. Ermakov, "Atmospheric Distortions of Ultra-Wideband Pulses: Method of Temporal Moments," *Proc. 13th Int. Conf. Mathematical Methods in Electrom. Theory*, Kharkov, Ukraine, September 12–15, 2000, pp. 143–145.

[17]   Golubev, A. I., et al., "Kinetic Model of the Propagation of Intense Subnanosecond Electromagnetic Pulse Through the Lower Atmosphere," *IEEE Trans. Plasma Sci.*, Vol. 28, No. 1, 2000, pp. 303–311.

[18]   Suk, J., and E. J. Rothwell, "Transient Analysis of TE Plane-Wave Reflection from a Layered Medium, *J. Electromagn. Waves Appl.*, Vol. 16, No. 2, 2002, pp. 281–297.

[19]   Suk, J., and E. J. Rothwell, "Transient Analysis of TM-Plane Wave Reflection From a Layered Medium," *J. Electromagn. Waves Appl.*, Vol. 16, No. 9, 2002, pp. 1195–1208.

[20]   Rothwell, E. J., and J. Suk, "Efficient Computation of the Time-Domain TE Plane-Wave Reflection Coefficient." *IEEE Trans. Antennas Propagat.*, Vol. 51, No. 12, 2003, pp. 3283–3285.

[21]   *Handbook of Mathematical Functions*, M. Abramovitz and I. A. Stegun (eds.), New York: Dover, 1970.

[22]   Rothwell, E. J., "Efficient Computation of the Time-Domain TM Plane-Wave Reflection Coefficient," *IEEE Trans. Antennas Propagat.*, Vol. 53, No. 10, 2005, pp. 3417–3419.

[23]   Pantoja, M. Fernández, et al., "Time Domain Analysis of Thin-Wire Antennas Over Lossy Ground Using the Reflection-Coefficient Approximation," *Radio Sci.*, Vol. 44, No. 6, 2009, pp. 1–14.

[24]   Pao, H. Y., S. L. Dvorak, and D. G. Dudley, "An Accurate and Efficient Analysis for Transient Plane Waves Obliquely Incident on a Conducting Half Space (TE case)," *IEEE Trans. Antennas Propagat.*, Vol. 44, No. 7, 1996, pp. 918–924.

[25]   Pao, H.-Y., S. L. Dvorak, and D. G. Dudley, "An Accurate and Efficient Analysis for Transient Plane Waves Obliquely Incident on a Conducting Half Space (TM case)," *IEEE Trans. Antennas Propagat.*, Vol. 44, No. 7, 1996, pp. 925–932.

[26]   Zeng, Q., and G. Y. Delisle, "Time-Domain Analysis for Electromagnetic Pulses Reflected from a Conducting Half Space," *Proc. Canadian Conf. Electrical and Computer Engineering*, Ottawa, Ont., May 7–10, 2006, pp. 92–95.

[27]   Zeng, Q., and G. Y. Delisle, "Transient Analysis of Electromagnetic Wave Reflection From a Stratified Medium," *Proc. Asia-Pacific Symp. Electromagn. Compat.*, Beijing, China, Apr. 12–16, 2010, pp. 881–884.

[28]   Hill, D. A., and J. R. Wait, "Diffusion of Electromagnetic Pulses Into the Earth from a Line Source," *IEEE Trans. Antennas Propagat.*, Vol. 22, No. 1, 1974, pp. 145–146.

[29]   Wait, J. R., "Transient Excitation of the Earth by a Line Source of Current," *Proc. IEEE*, Vol. 59, No. 8, 1971, pp. 1287–1288.

[30]   Dai, R., and C. T. Young, "Transient Fields of a Horizontal Electric Dipole on a Multilayered Medium," *IEEE Trans. Antennas Propagat.*, Vol. 45, No. 6, 1997, pp. 1023–1031.

[31]   Markov, G. T., and A. F. Chaplin, *Excitation of Electromagnetic Waves*, Moscow: Radio i Zvyaz, 1983 (in Russian).

[32]   Wait, J. R., *Wave Propagation Theory*, New York: Pergamon Press, 1981.

[33]   Brekhovskikh, L. M., *Waves in Layered Media*, New York: Academic Press, 1980.

CHAPTER 4

Scattering of Ultrawideband Electromagnetic Pulses by Conducting and Dielectric Objects

Introduction

In practical use of UWB electromagnetic radiation to examine the structure of media or objects, parts of an object or irregularities of a medium may vary in dimension, that is, be much less than, comparable to, and much greater than the wavelengths of the intense radiation spectral components. This suggests that the radiation scattering may occur simultaneously by different mechanisms: low frequency, resonance, and quasi-optical. Whether these mechanisms will be compatible or one of them will be dominant largely depends on the specifics of a particular task. Direct computer simulation is often ineffective to settle this question. As an alternative, either approximate methods [1–3] or analytic–numerical methods are used [4, 5]. The adequacy of an approximate method to a problem to be investigated depends, first of all, on the approximation tools it uses. The main feature of an efficient analytic–numerical method is the availability of a comprehensive analytic background for a problem to be considered. This allows one, in some situations, to find approximate solutions to the problem and assess their errors, and, in the general case, to develop a completely mathematically validated algorithm for its numerical investigation.

To demonstrate the capabilities and features of the methods alternative to direct numerical techniques, we consider two key problems. The first one deals with a pulsed plane electromagnetic wave scattered by a three-dimensional perfectly conducting body of arbitrary geometry. In this, quite general, formulation of the problem, the characteristic features of the scattering process can be revealed only in terms of an approximate approach. Here, we use a time-domain Kirchhoff approximation, which, however, does not cover a specific scattering process related to so-called creeping waves. The essence of this process is explained by the example of a qualitative analysis of the problem of a plane electromagnetic wave scattered by a perfectly conducting infinite circular cylinder.

Studying the second problem is aimed at revealing and interpreting the basic mechanisms of wave scattering by dielectric bodies. For an object of investigation, a dielectric sphere is used. The constancy of the radius of curvature of the surface of a sphere, its unique geometric symmetry, and the homogeneity of a dielectric make possible efficient application of a mathematical approach using wavelet transforms in frequency and time domains. The structure of the obtained solution allows a pictorial geometric interpretation of the basic scattering mechanisms and elucidation of their characteristic features.

The problems to be discussed become currently central in view of the active implementation of UWB signals in radar and sounding of natural and artificial media.

4.1   Scattering of Pulsed Electromagnetic Waves by Conducting Objects

4.1.1   Statement of the Problem. Derivation of Calculation Formulas

Let a plane nonharmonic electromagnetic wave be incident on a perfectly conducting object of arbitrary smooth shape (Figure 4.1). The wave induces currents on the object that generate a secondary (scattered) field in the surrounding space. The field can easily be found if the problem of the current distribution over the object surface has been solved. However, a rigorous solution of this problem involves considerable difficulties. Therefore, to solve it, we use an approximate method [1] referred to as the Kirchhoff method.

In this method, the portion S_0 of the object surface S illuminated by the wave is selected. The electric surface current density on S_0 is determined on the assumption that at each point $\mathbf{r}_s$ (Figure 4.1), it coincides with the current density $\mathbf{J}$ on a perfectly conducting plane tangent to the object at this point. For $\mathbf{J}$, we have the following formula:

$$\mathbf{J}(\mathbf{r}_s,t) = 2\mathbf{n} \times \mathbf{H}_i(\mathbf{r}_s,t) \qquad (4.1)$$

where $\mathbf{H}_i(\mathbf{r}_s,t)$ is the function that describes the magnetic field of the incident wave, $\mathbf{n}$ is the unit vector of the normal external to S_0 at the point $\mathbf{r}_s$, and t is time. The current density on the shaded portion of the surface S is assumed to be zero.

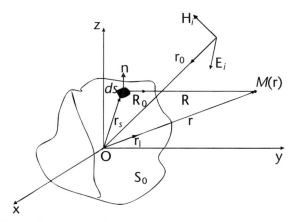

Figure 4.1   Sketch illustrating the problem geometry.

For the case that the object is irradiated by the wave in the direction specified by the unit vector $\mathbf{r}_0$ (Figure 4.1), this relation becomes

$$\mathbf{J}(\mathbf{r}_s,t) = 2\mathbf{n} \times \mathbf{H}_i\left(t - \frac{\mathbf{r}_0 \cdot \mathbf{r}_s}{c}\right) \tag{4.2}$$

where c is the wave propagation velocity in the space surrounding the object. The time is counted from the "arrival" of the wave at the origin of coordinates (point O).

As mentioned earlier, the object under consideration is rather smooth, but having an arbitrary shape. In other words, the object surface irregularities, such as corners, edges, and pointed protrusions are not fully excluded from consideration. However, it seems logical to suppose that the quality of the approximation depends on the proportion between the size of the surface patch in the vicinity of which the field is irregular because of the edge effect and the size of the entire surface of the object. The field regularity is violated near the shade boundary in the surface region whose silhouette projection width is comparable to

$$c|\mathbf{H}_i| \cdot \left|\frac{\partial \mathbf{H}_i}{\partial t}\right|^{-1} \tag{4.3}$$

Therefore, the condition for the applicability of the above assumptions can be written as [1]

$$\frac{1}{c}\left|\frac{\partial \mathbf{H}_i}{\partial t}\right| \gg \frac{|\mathbf{H}_i|}{l_c} \tag{4.4}$$

where l_c is the least linear dimension of the object silhouette projection.

Assuming the surface current density distribution over S_0 known [in our case, given by (4.2)], we can obtain a representation for the scattered field of a nonharmonic plane wave. We subdivide S_0 into surface elements ds with mutually perpendicular curved sides in such a way that, at every point of the body, the corresponding surface element is oriented by its longer side l in the direction of the vector $\mathbf{J}$ and by its smaller side a perpendicular to the vector. Each surface element can be considered as an elementary dipole of length l with a current I of amplitude $J\,da$.

The radiation field of the dipole at a point located distance R (Figure 4.1) can be determined using the earlier obtained representations (2.11) and (2.15), (2.16). These representations can be equivalently rewritten as

$$\mathbf{H} = \frac{l}{4\pi c}\left[\frac{1}{r}\frac{dI}{dt} + \frac{c}{r^2}I\right]\frac{\mathbf{l} \times \mathbf{r}}{lr} \tag{4.5}$$

$$\mathbf{E} = \frac{Z_0 l}{4\pi c}\left\{\frac{1}{r}\frac{dI}{dt}\frac{\mathbf{r} \times (\mathbf{r} \times \mathbf{l})}{lr^2} + \frac{cI}{r^2}\left[\frac{3(\mathbf{l} \cdot \mathbf{r}) \cdot \mathbf{r}}{lr^2} - \frac{1}{l}\right] + \frac{c^2}{r^3}\int I\,dt\left[\frac{3(\mathbf{l} \cdot \mathbf{r}) \cdot \mathbf{r}}{lr^2} - \frac{1}{l}\right]\right\} \tag{4.6}$$

where $\mathbf{l}$ is the vector that determines the orientation of the moment and the length of the dipole, $I = I(t - r/c)$. It should be noted that (4.5) and (4.6) determine the near field only to a constant of integration. In general, this may result to faulty estimates of the permanent field at the early and final stages of the pulse radiation process.

Substituting $\mathbf{J}ds$ (in view of the approximate relation (4.2) for $\mathbf{J}$) in (4.5) and (4.6) for the product of the current I by the vector $\mathbf{l}$, changing r by R and $\mathbf{r}/r$ by $\mathbf{R}_0$, and integrating the resulting relations over the illuminated portion S_0 of the object surface, we can formally find the scattered field at an arbitrary point of the space outside the object. However, the applicability of the obtained cumbersome calculation formulas to the intermediate radiation region, not to mention the near one, is doubtful because of the rather crude approximation used for the current distribution and of the highly irregular behavior of an actual scattered field in these regions. At the same time, the behavior of the scattered field in the far region is much more regular, and the calculation formulas for this field can be substantially simplified. Ignoring the terms of order $O(1/R^2)$ and $O(1/R^3)$ in the general relations, we obtain

$$\mathbf{E}(\mathbf{r},t) = \frac{Z_0}{2\pi c} \int_{S_0} \mathbf{R}_0 \times \left(\mathbf{R}_0 \times \left(\mathbf{n} \times \frac{\partial \mathbf{H}_i(t^*)}{\partial t} \right) \right) \frac{ds}{R} \tag{4.7}$$

$$\mathbf{H}(\mathbf{r},t) = \frac{Z_0}{2\pi c} \int_{S_0} \left(\mathbf{R}_0 \times \left(\mathbf{n} \times \frac{\partial \mathbf{H}_i(t^*)}{\partial t} \right) \right) \frac{ds}{R} \tag{4.8}$$

where $t^* = t - (\mathbf{r}_0 \cdot \mathbf{r}_s)/c - R/c$.

For large distances from the illuminated body, we have $R \gg D$, where D is the largest linear dimension of the body. As $r_s < D$, the quantity R can be expanded in a power series of the ratio r_s/r and truncate the series to the first two terms. Actually, we have

$$R^2 = (\mathbf{r} - \mathbf{r}_s) \cdot (\mathbf{r} - \mathbf{r}_s) = r^2 - 2\mathbf{r} \cdot \mathbf{r}_s + r_s^2 \tag{4.9}$$

hence,

$$R = \left(r^2 - 2\mathbf{r} \cdot \mathbf{r}_s + r_s^2 \right)^{1/2} = r\left(1 - \frac{2\mathbf{r} \cdot \mathbf{r}_s}{r^2} + \frac{r_s^2}{r^2} \right)^{1/2} \approx r\left(1 - \frac{2\mathbf{r} \cdot \mathbf{r}_s}{r^2} \right)^{1/2} \approx r\left(1 - \frac{\mathbf{r} \cdot \mathbf{r}_s}{r^2} \right) \tag{4.10}$$

and, thus,

$$R = r - (\mathbf{r}_1 \cdot \mathbf{r}_s) \tag{4.11}$$

where $\mathbf{r}_1$ is the unit vector directed along $\mathbf{r}$.

In view of the last relation, we can select in the expression for t^* a parameter $\hat{t} = t - r/c$ independent of the coordinates over which the integration is performed:

$$t^* = t - \frac{\mathbf{r}_0 \cdot \mathbf{r}_s}{c} - \frac{R}{c} = t - \frac{\mathbf{r}_0 \cdot \mathbf{r}_s}{c} - \frac{r}{c} + \frac{\mathbf{r}_1 \cdot \mathbf{r}_s}{c} = \hat{t} + \frac{(\mathbf{r}_1 - \mathbf{r}_0) \cdot \mathbf{r}_s}{c} \qquad (4.12)$$

Now, the quantity $1/R$ in (4.7) and (4.8) can be put equal to the constant $1/r$. Neglecting the variation of the angle between the unit vectors $\mathbf{R}_0$ and $\mathbf{r}_1$, we can put $\mathbf{R}_0 = \mathbf{r}_1$. With these simplifications, formulas (4.7) and (4.8) become

$$\mathbf{H}(\mathbf{r},t) = -\frac{1}{2\pi cr} \mathbf{r}_1 \times \int_{S_0} \left(\mathbf{n} \times \frac{\partial}{\partial t} \mathbf{H}_i \left(\hat{t} + \frac{(\mathbf{r}_1 - \mathbf{r}_0) \cdot \mathbf{r}_s}{c} \right) \right) ds \qquad (4.13)$$

$$\mathbf{E}(\mathbf{r},t) = -Z_0 \left(\mathbf{r}_1 \times \mathbf{H}(\mathbf{r},t) \right) \qquad (4.14)$$

For the direction of backscattering, we have $\mathbf{r}_1 = -\mathbf{r}_0$, and formula (4.13) becomes [1]

$$\mathbf{H}(\mathbf{r},t) = -\frac{1}{2\pi cr} \int_{S_0} \frac{\partial}{\partial t} \mathbf{H}_i \left(\hat{t} + \frac{2(\mathbf{r}_1 \cdot \mathbf{r}_s)}{c} \right) \cos(\mathbf{n} \cdot \mathbf{r}_1) ds \qquad (4.15)$$

Once the magnetic field $\mathbf{H}(\mathbf{r},t)$ is found, we can use formula (4.14) to find the electric field.

4.1.2   Wave Scattering by a Perfectly Conducting Rectangular Plate

Let a perfectly conducting plate with sides a and b located in the plane xOy be illuminated by a nonharmonic plane wave along a direction whose vector lies in the plane xOz and makes an angle θ with the z-axis (Figure 4.2).

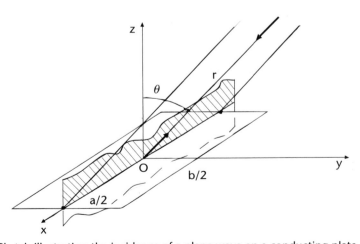

Figure 4.2   Sketch illustrating the incidence of a plane wave on a conducting plate.

The field scattered by the plate is determined by the relations [1]

$$\mathbf{H}(\mathbf{r},t) = \mathbf{H}_{i0}\frac{ab\cos\theta}{2\pi cr}\frac{1}{2\tau_a}\left\{H_i\left(\hat{t}+\tau_a\right) - H_i\left(\hat{t}-\tau_a\right)\right\} \tag{4.16}$$

$$\mathbf{E}(\mathbf{r},t) = -Z_0\left(\mathbf{r}_1 \times \mathbf{H}(\mathbf{r},t)\right) \tag{4.17}$$

Here, $\mathbf{H}_{i0}$ is the unit vector directed along the vector of the incident wave magnetic field, $\mathbf{H}_i$; $\tau_a = a\sin\theta/c$ is the parameter that characterizes the delay of the waves reflected from the plate edges normal to the plane xOz relative to the wave reflected from the plate center. Numerically, τ_a equals the projection of the electrical dimension of the plate, a/c, on the illumination direction.

In relations (4.16), the scattered field at a space point is represented as a superposition of two waves originated from the plate edges. These waves can be associated with two independent secondary sources (point scatterers) located at the plate edges (heavy points in Figure 4.2). The secondary sources generate spherical waves coinciding in shape with the incident wave, but having different retarded times. A superposition of the source fields in space forms a resulting interference field, which depends not only on the illumination direction, but also on time.

The time dependence of the field at every point in space defines the waveform of the reflected signal. The waveform is largely determined by the proportion between the incident pulse duration and the parameter τ_a.

The dependence of the resulting scattered wave on the angle θ at a fixed time characterizes the scattering properties of the plate for different directions. The case under consideration differs from the case of diffraction of a harmonic wave by that the backscattering pattern varies with time and never vanishes, whatever the illumination angle.

In particular, from (4.16), it can be seen that as $\theta \to 0$ (normal incidence), τ_a also tends to zero, whereas

$$\frac{1}{2\tau_a}\left\{H_i\left(\hat{t}+\tau_a\right) - H_i\left(\hat{t}-\tau_a\right)\right\} \to \frac{\partial H_i(\hat{t})}{\partial t} \tag{4.18}$$

Thus, for a plane nonharmonic wave incident normal to a plate, the scattered field in the far region is proportional to the time derivative of the function describing the incident field.

For the general case, when the direction of incidence of a plane nonharmonic wave is characterized by arbitrary angles θ and φ, where φ is the azimuth angle measured in the xOy plane from the positive direction of the x-axis, (4.15) becomes

$$\mathbf{H}(\mathbf{r},t) = \mathbf{H}_{i0}\frac{ab\cos\theta}{2\pi cr}\frac{1}{4\tau_a\tau_b}\int_{-\tau_b}^{\tau_b}\left\{H_i\left(\hat{t}+\tau_a+\tau\right) - H_i\left(\hat{t}-\tau_a+\tau\right)\right\}d\tau \tag{4.19}$$

where $\tau_a = a\sin\theta\cos\varphi/c$ and $\tau_b = b\sin\theta\sin\varphi/c$ are the dimensions of the projections of the respective plate sides on the illumination direction.

Considering the antiderivative $F(t)$ of the function $H_i(t)$ $(\int H_i(t)dt = F(t) + C)$, we can rewrite (4.19) as

$$\mathbf{H}(\mathbf{r},t) = \mathbf{H}_{i0}\frac{ab\cos\theta}{2\pi cr}\frac{1}{4\tau_a\tau_b}\Big\{F\big(\hat{t} + \tau_a + \tau_b\big) - F\big(\hat{t} - \tau_a + \tau_b\big) - F\big(\hat{t} + \tau_a - \tau_b\big)$$
$$+ F\big(\hat{t} - \tau_a - \tau_b\big)\Big\}$$

(4.20)

Analysis of this expression shows that the sources of the scattered field are the scattering centers coinciding with the edge points of the plate ($x = \pm a/2$, $y = \pm b/2$). In this case, the reflected signal can be more long, and if the duration of the illumination pulse is less than τ_a and $\tau_b - \tau_a$, the reflected signal can be split into four parts.

4.1.3   Wave Scattering by a Perfectly Conducting Ellipsoid or Sphere

For a triaxial ellipsoid (Figure 4.3) illuminated by a wave propagating in the negative z direction, the general (4.15) for the scattered field becomes [1]

$$\mathbf{H}(\mathbf{r},t) = \frac{ab}{2dr}\left\{\mathbf{H}_i\big(\hat{t} + \tau_0\big) - \frac{1}{\tau_0}\int_0^{\tau_0}\mathbf{H}_i\big(\hat{t} + \tau\big)d\tau\right\}$$

(4.21)

Here, a, b, and d are the ellipsoid semiaxes aligned, respectively, with the coordinate axes x, y, and z (Figure 4.3), and $\tau_0 = 2d/c$ is the electrical length of the ellipsoid along the z-axis.

Denoting the antiderivative of the function $\mathbf{H}_i(t)$ by $\mathbf{F}(t)$, we rewrite (4.21) as

$$\mathbf{H}(\mathbf{r},t) = \frac{ab}{2dr}\left\{\mathbf{H}_i\big(\hat{t} + \tau_0\big) - \frac{1}{\tau_0}\Big[\mathbf{F}\big(\hat{t} + \tau_0\big) - \mathbf{F}(\hat{t})\Big]\right\}$$

(4.22)

For the particular case of $a = b = d$, (4.21) and (4.22) describe the scattered field for a perfectly conducting sphere. It can be seen that the scattered field can be

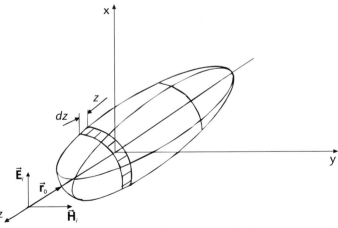

Figure 4.3   Sketch of a perfectly conducting ellipsoid in the field of a plane nonharmonic wave

treated as if it has been generated by two secondary sources (local scattering centers). The first source, located at the ellipsoid vertex, generates two waves arriving at the observation point: a reflected wave, which coincides in waveform with the incident one and differs from it in amplitude and time delay, and a diffraction wave different in waveform from the incident one. The second source is located in the region of the ellipsoid geometric shadow; it generates a wave similar to the diffraction wave produced by the first source.

4.1.4   Wave Scattering by a Perfectly Conducting Finite Circular Cone

Let a cone of this type be illuminated on the vertex side with a wave propagating along its symmetry axis (Figure 4.4).

For the scattered field, we have the expression [1]

$$\mathbf{H}(\mathbf{r},t) = -\frac{a}{2r} tg\gamma \left\{ \mathbf{H}_i(\hat{t} - \tau_0) - \frac{1}{\tau_0} \int_0^{\tau_0} \mathbf{H}_i(\hat{t} - \tau) d\tau \right\} \tag{4.23}$$

Here, a is the base radius of the cone, h is its height, 2γ is its vertex angle, and $\tau_0 = 2h/c$. As in the previous subsection, (4.23) can be transformed to

$$\mathbf{H}(\mathbf{r},t) = -\frac{a}{2r} tg\gamma \left\{ \mathbf{H}_i(\hat{t} - \tau_0) - \frac{1}{\tau_0} \left[\mathbf{F}(\hat{t} - \tau_0) - \mathbf{F}(\hat{t}) \right] \right\} \tag{4.24}$$

Comparing (4.24) and (4.22), we see their perfect analogy. The differences are only in amplitude factor and in retardation parameter. The scattered field can be treated as a field generated by two secondary sources. The first source is at the vertex of the cone; the second one, located at the cone base, generates two waves one

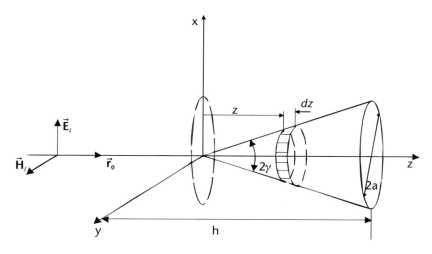

Figure 4.4   Sketch illustrating the illumination of a finite circular cone with a plane nonharmonic electromagnetic wave.

of which is similar to the wave reflected from an ellipsoid and the second one, as well as the wave generated by the first source, is similar to the wave diffracted by an ellipsoid. For $h/\tau_0 \gg 1$, the contribution of the diffraction waves is relatively small.

Note that when the length of the cone generator is fixed and $h \to 0$ ($\gamma \to \pi/2$), the cone turns into a disk. The field scattered by a disk is described, as can be seen from (4.23), by the expression

$$\mathbf{H}(\mathbf{r},t) = -\frac{a^2}{rc}\frac{\partial \mathbf{H}_i(\hat{t})}{\partial t} \qquad (4.25)$$

Thus, the field scattered by a disk illuminated with a normally incident plane nonharmonic electromagnetic wave is proportional to the time derivative of the incident field.

4.1.5 Creeping Waves

The scattering of UWB electromagnetic pulses by finite smooth convex objects has some features that distinguish it from the usual mirror reflection from extended flat objects. In this case, pulses following complicated paths arrive at the observation point in addition to mirrored pulses. Possible propagation paths of a plane electromagnetic wave normally incident on an infinite perfectly conducting cylinder [6] are illustrated in Figure 4.5.

As can be seen in the top-left picture, the pulse path consists of two fragments. The first one is the cylinder arc of radius a between point A, at which the incident wave front is tangent to the cylinder, and point B, the origin of the half-tangent to the cylinder passing through the observation point P. The second fragment of the path is the half-tangent segment BP. The path is depicted by a line of decreasing thickness to emphasize that the pulse is damping as it travels along the shady side of the cylinder.

The top-right picture shows another path of a pulse that makes almost a full tracking of the cylinder before reaching the observation point. From physical considerations, it is natural to expect that in this case, the pulse damping will be much more pronounced.

The following two pictures illustrate the paths of pulses for another location of the observation point P.

Finally, the last two pictures correspond to the cases where either the observation point is in the shadow zone of the cylinder or the direction to it coincides with the direction reverse to that of the incident pulse.

As can be seen from the Figure 4.5, there are two pulses in the field scattered by the cylinder. One of them results from the reflection of the incident pulse from the cylinder, and the other is the incident pulse having rounded the cylinder. The geometric path difference between the two pulses is equal to $a(2 + \pi)$.

The damping of the incident pulse during its propagation over the shady side of the cylinder is difficult to characterize quantitatively. However, it can be analyzed in frequency domain for a harmonic incident plane wave:

$$E_0(r,\theta) = \exp(-ikr\cos\theta) \qquad (4.26)$$

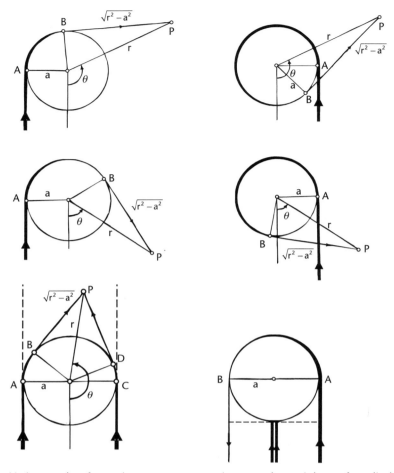

Figure 4.5 Various paths of creeping waves propagating over the periphery of a cylinder.

For this case, an asymptotic representation has been obtained for the field back-scattered by a cylinder of radius $ka \gg 1$ [6]:

$$E_s(r,\theta) = \sqrt{\frac{a}{2r}}\exp[ik(r-2a)]\left\{1 \mp 2(ka)^{-1/6}\sum_s C_s \frac{\exp[i(2ka - \pi v_s + \pi/12)]}{1 - \exp(2\pi v_s)}\right\} \quad (4.27)$$

In this formula, the minus and the plus sign correspond to the incident wave polarized, respectively, along and normal to the cylinder axis. The coefficients C_s are real and positive. The values of the first five coefficients are given in [6]. The values of v_s are determined, in accordance with the polarization, by the relations

$$v_s = ka + \frac{1+i\sqrt{3}}{2}\left(\frac{ka}{6}\right)^{1/3}\left(\frac{q_s}{q_s'}\right) \quad (4.28)$$

where q_s is the sth zero of the Airy function and q'_s is the sth zero of its derivative.

In view of (4.28), the argument of the exponential function in (4.27) can be written as

$$-\frac{\sqrt{3}\pi}{2}\left(\frac{ka}{6}\right)^{1/3} q_s + i\left[ka(\pi + 2) + \frac{\pi}{2}\left(\frac{ka}{6}\right)^{1/3} q_s + \frac{\pi}{12}\right] \qquad (4.29)$$

Each term of the series in (4.27) describes a so-called creeping wave. The value of the summation index s is the number of paths traveled by the creeping wave round the cylinder. As the first term in (4.29) is negative and the zeros q_s increase in value with s, it can be concluded that each wave is decaying during its propagation along the cylinder, and the larger the number s, the quicker the decay. The sum of the second and third terms in square brackets determines the additional phase shift relative to that determined from geometrical considerations. This implies that the creeping waves propagate along the cylinder surface with a decreasing phase velocity.

4.2   Scattering of Pulsed Plane Electromagnetic Waves by Dielectric Objects

4.2.1   Wavelet Analysis of the Wave Scattering by a Dielectric Sphere

When considering the scattering of a UWB electromagnetic pulse by a dielectric body of finite size, one can distinguish several scattering mechanisms, each making a certain contribution to the total scattered field. These mechanisms are usually investigated in either frequency or time domain. With separate consideration, it is not simple to identify a scattering mechanism and evaluate its relative contribution to a complex scattered field by analyzing data on the field structure. A much more efficient and fruitful approach is to examine simultaneously the time and frequency characteristics of the scattered field. In the framework of this approach, information about local time changes of the scattered field and about its local frequency content is extracted from the available data. This may provide a deeper insight into a particular scattering mechanism.

For this approach, a special term "time–frequency analysis" is currently used in the literature. The well-known examples of time–frequency analysis methods are the short-time Fourier transform, the Wigner–Ville distribution, and the wavelet transform [7–14]. The wavelet transform [15–18] seems to be the most powerful tool to study scattered fields on different scales.

Let our aim be to study the mechanisms of the wave scattering by a dielectric body of the simplest shape, namely by a sphere. For this purpose, it is expedient to interpret data obtained in both the time and the frequency domain by using a wavelet transform.

The continuous wavelet transform of a signal $s(t)$ is given by [15, 16]

$$W_t(a,b) = \frac{1}{\sqrt{a}} \int\limits_{-\infty}^{\infty} s(t)\psi^*\left(\frac{t-b}{a}\right) dt \tag{4.30}$$

where $\psi(t) \in L^2(\mathbf{R})$ is the prototype wavelet, which is conventionally termed the mother wavelet, that satisfies all requirements placed by the wavelet theory on mathematical objects of this type. Representation (4.30) indicates that $W_t(a,b)$ is an integral transform with the kernel $\psi((t-b)/a)$ being the mother wavelet subject to time translation and frequency scaling. Physically, the scaling parameter a and the translation parameter b correspond, respectively, to reciprocal frequency $1/\omega$ and time t.

For the mother wavelet, we consider the Gaussian wavelet [11]:

$$\psi(t) = \frac{1}{\sqrt{2\pi}\sigma_t} \exp\left(-\frac{t^2}{2\sigma_t^2}\right) \exp(-i\omega_0 t) \tag{4.31}$$

where σ_t and ω_0 are constants that specify the wavelet width and central frequency, respectively. Substituting (4.31) in (4.30) and using a direct Fourier transform, we find

$$
\begin{aligned}
W_t(T,\Omega) &= \frac{\Omega}{\sqrt{2\pi\sigma_t^2}} \int\limits_{-\infty}^{\infty} s(t)\exp\left(-\frac{\Omega^2}{2\sigma_t^2\omega_0^2}(t-T)^2\right)\exp\left[i\Omega(t-T)\right] dt \\
&= \frac{1}{2\pi}\sqrt{\frac{\omega_0}{\Omega}} \int\limits_{-\infty}^{\infty} S(\omega)\exp\left(-\frac{\omega_0^2\sigma_t^2}{2}\left(\frac{\omega}{\Omega}+1\right)^2\right)\exp(i\omega t) d\omega
\end{aligned}
\tag{4.32}
$$

where $S(\omega)$ is the Fourier transform of $s(t)$ that represents the frequency-domain data of the signal. In (4.32), the parameters a and b are replaced by $\omega_0/a = \Omega$ and $t = T$, respectively. Recall that these parameters correspond, respectively, to reciprocal frequency and time. It can clearly be seen that W_t is expressed in terms of the short-time Fourier transform $S(\omega)$ with the Gaussian window of varied width that is applied to frequency-domain data. The window "slides" along the ω-axis, providing the extraction of local information from frequency-domain data. As the window varies in width with frequency Ω, W_t is well time resolved in the high frequency range and well frequency-resolved in the low frequency range. This characteristic is very convenient to determine fast time variations of signal components.

The continuous wavelet transform of the signal spectrum $S(\omega)$, that is, of its frequency-domain data, is given by [15, 16]

$$W_f(a,b) = \frac{1}{\sqrt{a}} \int\limits_{-\infty}^{\infty} S(\omega)\xi^*\left(\frac{\omega-b}{a}\right) d\omega \tag{4.33}$$

where $\xi(\omega) \in L^2(\mathbf{R})$ is the mother wavelet.

For the mother wavelet, we again consider the Gaussian wavelet [11]:

$$\xi(\omega) = \frac{1}{\sqrt{2\pi}\sigma_f} \exp\left(-\frac{\omega^2}{2\sigma_f^2}\right) \exp(-i\omega t_0) \tag{4.34}$$

where σ_f and t_0 are constants that determine the time and frequency resolution and the wavelet center, respectively. Substituting (4.34) in (4.33) and using an inverse Fourier transform, we find

$$
\begin{aligned}
W_f(T,\Omega) &= \sqrt{\frac{T}{2\pi\sigma_f^2 t_0}} \int_{-\infty}^{\infty} S(\omega)\exp\left(-\frac{T^2}{2\sigma_f^2 t_0^2}(\omega-\Omega)^2\right)\exp[-iT(\omega-\Omega)]d\omega \\
&= \frac{1}{\sqrt{T}} \int_{-\infty}^{\infty} s(t)\exp\left(-\frac{t_0^2\sigma_f^2}{2}\left(\frac{t}{T}-1\right)^2\right)\exp(-i\Omega t)dt
\end{aligned}
\tag{4.35}
$$

where the parameters a and b are replaced by t_0/T and Ω, respectively. Expression (4.35) indicates that W_f is obtained by applying a short-time Fourier transform, $s(t)$, with the Gaussian window of varied width to frequency-domain data. As the window width varies with time T, W_f is well time resolved at the early stage of the signal variation and well frequency resolved at the late stage. This characteristic is very convenient to separate multi-scale (multifrequency) signal components on the frequency axis.

4.2.2   Numerical Results and Discussion

The wavelet transforms (4.32) and (4.35) were used to analyze the structure of the pulsed electromagnetic field scattered by a dielectric sphere of radius a, and various scattering mechanisms were investigated [11] (the same problems are discussed elsewhere [12, 13]). The incident pulsed field was represented by a plane wave. The time dependence of the field was specified by the time derivative of the Gaussian. To make the interpretation of calculation results convenient, the time variable was normalized to the time it took for the pulse to travel a distance equal to the sphere radius [$\tau = t/(a/c)$], where c is the velocity of light in free space.

Data on the scattered field were obtained using a rigorous frequency-domain solution of the problem of the diffraction of a plane wave by a dielectric sphere. The refractive index of the sphere material was chosen to be three. Analysis of the calculation results made it possible to distinguish several scattering mechanisms. These mechanisms are schematically illustrated by the ray paths of wave propagation in Figure 4.6. The mirror scattering [Figure 4.6(a)] and various variants of the reflection of the radiation having penetrated into the sphere from its surface [Figures 4.6(c–e)] are presented. The creeping wave is sketched in Figure 4.6(b). A combination of the mechanism of internal reflections and the creeping wave mechanism is presented in Figures 4.6(f–h).

In the study of Nishimoto and Ikuno, [11] the delay times for the radiation propagating along different paths calculated by the formulas of geometrical optics are compared with those calculated based on a rigorous analysis. There is good agreement between the results, which indirectly confirms their authenticity. Moreover, the comparison of the real parts of the natural frequencies found as a result of the analysis [11] has shown their good agreement with the exact values of the real parts of some natural resonant frequencies. It should be emphasized that a dielectric sphere has a number of resonant frequencies with small imaginary values, which makes them difficult to detect and precisely resolve.

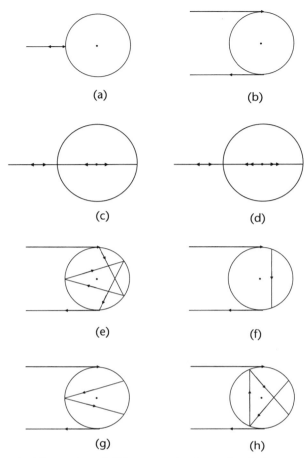

Figure 4.6 Geometrical illustration of different mechanisms of radiation scattering by a dielectric sphere.

Conclusion

The above formulation and solution of the problem of the backscattering of a pulsed plane electromagnetic wave by a conducting object allow the following conclusions:

1. The characteristics of a pulsed electromagnetic wave reflected from a scattering surface depend in a complicated manner on the parameters of the illumination pulse and on the surface shape and dimensions.
2. The main contribution to the formation of the scattered wave is made only by some regions or points of the scattering surface (so-called local scattering centers) rather than its all continuously distributed elements.
3. The scattering centers correspond to the parts where the geometric uniformity of the scattering object is violated. Their properties and features are determined by the nature of the object surface irregularity. Depending on

the type of irregularity, some scattering centers (mirror and edge points of a rectangular plate, shadow-forming points of a finite cone) generate reflected pulses of the same waveform as that of the illuminating pulse, whereas the other (points of an ellipsoid, sphere, or cone) cause changes in the waveform of the reflected pulse, endowing it with certain features.

4. The characteristics of the reflected pulse substantially depend on the proportion between the duration of the illumination pulse and the difference between the retarded times of the pulses generated by individual scattering centers. When the pulse duration is greater than the difference in retarded times of the pulses generated by these centers, they are superposed, and the resulting "mixed" reflected pulse is extended. If the difference in retarded times is greater than the duration of the illumination pulse, the reflected pulse is a sequence of individual pulses varied in intensity and, generally, in shape and duration. The number of pulses is determined by the number of scattering centers.

It should also be noted that approximate methods, having been used for a long time, are now not only of academic interest [2, 3]. Thus, the scalar problem of a pulsed plane wave passing through a rectangular or circular aperture in an infinite screen has been solved [3]. Mikhailov and Golovinskii [3] used a nonstationary version of the Kirchhoff method combined with expansion of the pulsed field in terms of wavelets. Interest in this kind of problems arose due to recently obtained interesting experimental results in the field of femtosecond optics and UWB radar.

The treatment of the radiation scattering problem for a dielectric sphere presented in this section has shown that for this case, the mechanisms of the backscattering of pulsed radiation can be clearly identified and resolved by wavelet analysis. In addition, some natural resonant frequencies of the sphere can also be detected.

Problems

4.1 Derive formula (4.16) for the scattered magnetic field proceeding from formula (4.15).

Solution: The general formula (4.15) reads

$$\mathbf{H}(\mathbf{r},t) = \frac{1}{2\pi c r}\int_{S_0}\frac{\partial}{\partial t}\mathbf{H}_i\left(\hat{t} + \frac{2(\mathbf{r}_1 \cdot \mathbf{r}_s)}{c}\right)\cos(\mathbf{n}\cdot\mathbf{r}_1)\,ds \qquad (4.36)$$

If the scattering object is a plate, in view of the problem geometry (Figure 4.2), we have

$$ds = dx\,dy, \quad \cos(\mathbf{n}\cdot\mathbf{r}_1) = \cos\theta, \quad \mathbf{r}_1\cdot\mathbf{r}_s = x\cos\left(\frac{\pi}{2}-\theta\right) = x\sin\theta \qquad (4.37)$$

and, hence,

$$\mathbf{H}(\mathbf{r},t) = \frac{\cos\theta}{2\pi cr}\int_{S_0}\frac{\partial}{\partial t}\mathbf{H}_i\left(\hat{t} + \frac{2\sin\theta}{c}x\right)dxdy$$

$$= \mathbf{H}_{i0}\frac{\cos\theta}{2\pi cr}\int_{-a/2}^{a/2}\int_{-b/2}^{b/2}\frac{\partial}{\partial t}H_i\left(\hat{t} + \frac{2\sin\theta}{c}x\right)dxdy \qquad (4.38)$$

$$= \mathbf{H}_{i0}\frac{b\cos\theta}{2\pi cr}\int_{-a/2}^{a/2}\frac{\partial}{\partial t}H_i\left(\hat{t} + \frac{2\sin\theta}{c}x\right)dx$$

Change the integration variable by putting

$$\tau = \frac{2\sin\theta}{c}x, \quad dx = \frac{c}{2\sin\theta}d\tau \qquad (4.39)$$

For $x = a/2$ and $x = -a/2$, the new integration variable becomes, respectively, $\tau_a = a\sin\theta/c$ and $-\tau_a = -a\sin\theta/c$.

This yields

$$\mathbf{H}(\mathbf{r},t) = \mathbf{H}_{i0}\frac{b\cos\theta}{2\pi cr}\frac{c}{2\sin\theta}\int_{-\tau_a}^{\tau_a}\frac{\partial}{\partial t}H_i(\hat{t} + \tau)d\tau$$

$$= \mathbf{H}_{i0}\frac{ab\cos\theta}{2\pi cr}\frac{1}{2\tau_a}\int_{-\tau_a}^{\tau_a}\frac{\partial}{\partial \tau}H_i(\hat{t} + \tau)d\tau \qquad (4.40)$$

$$= \mathbf{H}_{i0}\frac{ab\cos\theta}{2\pi cr}\frac{1}{2\tau_a}\left\{H_i(\hat{t} + \tau_a) - H_i(\hat{t} - \tau_a)\right\}$$

and, finally,

$$\mathbf{H}(\mathbf{r},t) = \mathbf{H}_{i0}\frac{ab\cos\theta}{2\pi cr}\frac{1}{2\tau_a}\left\{H_i(\hat{t} + \tau_a) - H_i(\hat{t} - \tau_a)\right\} \qquad (4.41)$$

Note that this representation of the scattered field offers a more pictorial way to analyze the case of an electromagnetic wave normally incident on a plate.

4.2 Derive formula (4.21) for the field scattered by an ellipsoid.

4.3 Obtain formulas (4.23) and (4.24) for the field scattered by a finite cone.

Solution:

In view of the problem geometry (Figure 4.4), the integral in the general formula (4.15) is convenient to evaluate using spherical coordinates. For the case under consideration, we have

$$ds = r_s \sin\gamma \, dr_s d\varphi, \quad \cos(\mathbf{n}\cdot\mathbf{r}_1) = \sin\gamma, \quad \mathbf{r}_1\cdot\mathbf{r}_s = -r_s\cos\gamma$$

and then

$$
\begin{aligned}
\mathbf{H}(\mathbf{r}, t) &= \frac{1}{2\pi cr} \int\limits_{0}^{h/\cos\gamma} \int\limits_{0}^{2\pi} \frac{\partial}{\partial t} \mathbf{H}_i\left(\hat{t} - \frac{2r_s \cos\gamma}{c}\right) \sin\gamma\, r_s \sin\gamma\, dr_s d\varphi \\
&= \frac{\sin^2\gamma}{cr} \int\limits_{0}^{h/\cos\gamma} \frac{\partial}{\partial t} \mathbf{H}_i\left(\hat{t} - \frac{2r_s \cos\gamma}{c}\right) r_s\, dr_s
\end{aligned}
\tag{4.42}
$$

Change the integration variable by putting

$$
\tau = \frac{2r_s \cos\gamma}{c}, \quad r_s = \frac{c}{2\cos\gamma}\tau, \quad dr_s = \frac{c}{2\cos\gamma} d\tau
\tag{4.43}
$$

For $r_s = h/\cos\gamma$, the new integration variable is $\tau_0 = 2h/c$, and for $r_s = 0$, we have $\tau_0 = 0$.

As a consequence, we obtain

$$
\begin{aligned}
\mathbf{H}(\mathbf{r}, t) &= \frac{c}{4r}\tan^2\gamma \int\limits_{0}^{\tau_0} \frac{\partial}{\partial t}\mathbf{H}_i(\hat{t} - \tau)\tau d\tau = -\frac{c}{4r}\tan^2\gamma \int\limits_{0}^{\tau_0} \frac{\partial}{\partial \tau}\mathbf{H}_i(\hat{t} - \tau)\tau d\tau \\
&= -\frac{c}{4r}\tan^2\gamma\left[\tau_0 \mathbf{H}_i(\hat{t} - \tau_0) - \int\limits_{0}^{\tau_0} \mathbf{H}_i(\hat{t} - \tau)\,d\tau\right] \\
&= -\frac{a}{2r}\tan\gamma\left[\mathbf{H}_i(\hat{t} - \tau_0) - \frac{1}{\tau_0}\int\limits_{0}^{\tau_0}\mathbf{H}_i(\hat{t} - \tau)\,d\tau\right]
\end{aligned}
\tag{4.44}
$$

and, finally,

$$
\mathbf{H}(\mathbf{r}, t) = -\frac{a}{2r}\tan\gamma\left[\mathbf{H}_i(\hat{t} - \tau_0) - \frac{1}{\tau_0}\int\limits_{0}^{\tau_0}\mathbf{H}_i(\hat{t} - \tau)\,d\tau\right]
\tag{4.45}
$$

Given the antiderivative $\mathbf{F}_i(\hat{t} - \tau)$, this formula can be rewritten as

$$
\mathbf{H}(\mathbf{r}, t) = -\frac{a}{2r}\tan\gamma\left[\mathbf{H}_i(\hat{t} - \tau_0) - \frac{1}{\tau_0}\left[\mathbf{F}_i(\hat{t} - \tau) - \mathbf{F}_i(\hat{t})\right]\right]
\tag{4.46}
$$

References

[1] Chernousov, V. S., "Scattering of Nonharmonic Electromagnetic Waves by Finite-Size Perfectly Conducting Bodies," *Radiotekh. Elektron.*, Vol. 10, No. 1, 1965, pp. 31–39.

[2] Gutman, A. L., "The Kirchhoff Method of Calculating Pulsed Fields," *J. Commun. Technol. Electron.*, Vol. 42, No. 3, 1997, pp. 247–252.

[3] Mikhailov, E. M., and P. A. Golovinskii, "Description of Diffraction and Focusing of Ultrashort Pulses on the Basis on a Nonstationary Kirchhoff–Sommerfeld Method," *J. Exp. Theor. Phys.*, Vol. 90, No. 2, 2000, pp. 240–249.

[4]   Doroshenko, V. A., V. F. Kravchenko, and V. I. Pustovoit, "Meler-Fock Transformations in Problems of Wave Diffraction on Unclosed Structures in the Time Region," *Doklady Physics*, Vol. 50, No. 11, 2005, pp. 560–564.

[5]   Doroshenko, V. A., and V. F. Kravchenko, *Diffraction of Electromagnetic Waves by Unclosed Conical Structures*, Moscow: Fizmatlit, 2009 (in Russian).

[6]   Hoenl, H. A., A. W. Maue, and K. Westpfahl, *Theorie der Beugung*, Berlin: Springer-Verlag, 1961.

[7]   Moghaddar, A., and E. K. Walton, "Time-Frequency Distribution Analysis of Scattering from Waveguide Cavities," *IEEE Trans. Antennas Propagat.*, Vol. 41, No. 5, 1993, pp. 677–680.

[8]   Trintinalia, L. C., and H. Ling, "Interpretation of Scattering Phenomenology in Slotted Waveguide Structures via Time-Frequency Processing," *IEEE Trans. Antennas Propagat.*, Vol. 43, No 11, 1995, pp. 1253–1261.

[9]   Kim, H., and H. Ling, "Wavelet Analysis of Radar Echo from Finite-Size Targets," *IEEE Trans. Antennas Propagat.*, Vol. 41, No. 2, 1993, pp. 200–207.

[10]   Nishimoto, M., and H. Ikuno, "Time-Frequency Analysis of Scattering Data Using Wavelet Transform," *IEICE Trans. Electron.*, Vol. E80-C, No. 11, 1997, pp. 1440–1447.

[11]   Nishimoto, M., and H. Ikuno, "Time-Frequency Analysis of Scattering Responses from a Dielectric Sphere," *Proc. Int. Symp. Antennas and Propagation*, Vol. 1, Fukuoka, Japan, 2000, pp. 417–420.

[12]   Nishimoto, M., and H. Ikuno, "Time-Frequency Processing of Scattering Responses from a Dielectric Sphere," *IEICE Trans. Electron.*, Vol. 84, No. 9, 2001, pp. 1256–1259.

[13]   Chen, D., and Y. Q. Jin, "Time-Frequency Analysis of Electromagnetic Pulse Response from a Spherical Target," *Chinese Phys. Lett.*, Vol. 20, No. 5, 2003, pp. 660–663.

[14]   Ling, H., et al., "Time-Frequency Analysis of Backscattering Data from a Coated Strip with a Gap," *IEEE Trans. Antennas Propagat.*, Vol. 41, No. 8, 1993, pp. 1147–1150.

[15]   Daubechies, I., *Ten Lectures on Wavelets*, SIAM, 1992.

[16]   Astaf'eva, N. M., "Wavelet Analysis: Basic Theory and Some of Applications," *Phys. Usp.*, Vol. 39, 1996, pp. 1085–1108.

[17]   Vorobjov, V. I., and V. G. Gribunin, *Theory and Practice of Wavelet Transforms*, St.-Petersburg: VUS Publishers, 1999 (in Russian).

[18]   Stark, H. G., *Wavelets and Signal Processing: An Application-Based Introduction*, New York: Springer, 2005.

CHAPTER 5

Impulse Responses of Objects and Propagation Channels

Introduction

In the context of linear processes and phenomena, which occur in a variety of devices and systems, the response of a device or system to an action $x(t)$ is described by a function $y(t)$. Mathematically, this can be formulated by a relation of the form $y(t) = Ax(t)$, where A is an operator whose structure is determined by the properties of the device or system. From physical considerations, it is clear that for such an operator, the ranges of allowed actions $x(t)$ and the responses $y(t)$ should be specified. Furthermore, in many cases, the property of linearity of the operator A can be complemented with the property of conservation of time shifts. In other words, the properties of a device or system are invariable in time and their responses to actions $x(t)$ and $x(t - t_0)$ differ only by a time shift t_0.

In applications, two important problems are generally to be solved. The first one aims at finding the response $y(t)$ of a system to an allowed action $x(t)$. The second, more complicated problem is to give, as far as possible, an adequate description of the system behavior during the exposure to an action $x(t)$ proceeding from the waveform of the response $y(t)$. It is noteworthy that to solve the first problem, it suffices to know the response $h(t)$ of the system to a test action specified by a Dirac delta function, $x(t) = \delta(t)$. Then, the solution of the problem is given by a convolution-type integral:

$$y(t) = \int_{-\infty}^{\infty} x(\tau)h(t - \tau) \, d\tau \qquad (5.1)$$

The function $h(t)$ is differently termed in different fields of science: an apparatus function in optics, a unit step function in electrical engineering, and an impulse response in radar. The scalar representation of the impulse response in (5.1) suffices to solve the problems discussed later. Generally, it is a vector that depends not only on time but also on the angle of incidence of the field on the object and on the field polarization.

The solution of the second problem is reduced to solving the integral of the first kind (5.1). This is an ill-posed problem and, hence, it is difficult to solve.

5.1   The Impulse Response: Models of Signals and Their Spectral Characteristics

5.1.1   Forms and Properties of the Impulse Response

In UWB radar, the most complete information about an object can be obtained only by properly processing the signal reflected from the object. This involves some operations on the received signal and subsequent computer processing. The parameters of the probing and reflected signals are in a certain relationship. If a time-dependent probing signal is described by a function $s(t)$, the signal reflected from the object can be represented as the convolution of the function $s(t)$ with a function $h(t)$. The latter, from the physical point of view, represents the signal reflected from the object as if it was irradiated with a probing pulse being a Dirac delta function $\delta(t)$.

The function $h(t)$ is termed the IR of an object and considered a principal radar characteristic of the object. The commonly practiced transient, amplitude-frequency, and phase-frequency characteristics of an object can be derived from the IR using well-known techniques. In principle, any object can be fully described by these characteristics found for various combinations of irradiation and receiving directions and types of polarization.

A function $h(t)$ that describes the signal at the output of a propagation channel as its convolution with the input signal $s(t)$ is termed the impulse response of the channel. In the physical sense, this function is similar to that mentioned earlier: it describes the signal at the output of a propagation channel excited at the input by a signal being a Dirac delta function $\delta(t)$.

In practice, only an approximate IR corresponding to a chosen approximation of $\delta(t)$ can be obtained. The possible approximations

$$\delta(t) = \frac{\alpha}{\sqrt{\pi}}\exp(-\alpha^2 t^2), \ \delta(t) = \frac{\alpha}{\pi(1 + \alpha^2 t^2)}, \ \delta(t) = \frac{\sin \alpha t}{\pi t} \qquad (5.2)$$

are the better, the larger the parameter α. It is advisable to choose α so that the spectrum of the approximating signal would be distributed over the frequency domain that is intended to be used for solving the radar problem. A direct IR calculation is quite a challenge. Therefore, the studies aimed at developing efficient methods for evaluating IRs of objects and propagation channels still remain relevant.

5.1.2   The Envelope, Instantaneous Phase, and Instantaneous Frequency of a Signal: The Analytic Signal

In quite many cases, a signal $s(t)$ can be described by the following general relation:

$$s(t) = A(t)\cos\psi(t) = A(t)\cos\left[\omega_0 t + \varphi(t) + \varphi_0\right] \qquad (5.3)$$

At first glance, it seems that $A(t)$ and $\psi(t)$ could be naturally interpreted as the signal envelope and instantaneous phase, respectively. However, these quantities are not independent. Indeed, if one of them is set arbitrarily, the other should be selected so that (5.3) would describe precisely the signal $s(t)$. Mathematically, this immediately makes evident the ambiguity in the definitions of the envelope and instantaneous phase of a signal, and physically, one has to deal with situations in which the concepts of amplitude and phase are meaningless. To eliminate the ambiguity, a function $s_1(t)$ conjugate to $s(t)$ is also considered, and the envelope $A(t)$, the instantaneous phase $\psi(t)$, and the instantaneous frequency $\omega(t)$ are defined as

$$A(t) = \sqrt{s^2(t) + s_1^2(t)}, \ \psi(t) = \arctan\frac{s_1(t)}{s(t)}, \ \omega(t) = \frac{d\omega(t)}{dt} \tag{5.4}$$

It was shown that these quantities can be determined consistently only if the functions $s(t)$ and $s_1(t)$ are related with one another by the Hilbert transforms

$$s_1(t) = -\frac{1}{\pi}\int_{-\infty}^{\infty}\frac{s(\tau)}{t-\tau}d\tau, \ s(t) = \frac{1}{\pi}\int_{-\infty}^{\infty}\frac{s_1(\tau)}{t-\tau}d\tau \tag{5.5}$$

The conjugate function $s_1(t)$ is also helpful in that it can be used to introduce the notion of the complex amplitude of a nonharmonic signal. To do this, an actual $s(t)$ is put in correspondence with an analytic complex signal $z(t) = s(t) + is_1(t)$, which can be represented in exponential form as

$$z(t) = A(t)\exp[i\psi(t)] = A(t)\exp\left[i\left(\omega_0 t + \varphi(t) + \varphi_0\right)\right] = \dot{A}(t)\exp\left(i\omega_0 t\right) \tag{5.6}$$

where

$$\dot{A}(t) = A(t)\exp\left[i\left(\varphi(t) + \varphi_0\right)\right] \tag{5.7}$$

is the complex envelope of the signal.

5.1.3   Kramers–Kronig-Type Relations

Mathematically, Kramers–Kronig-type relations are consequences of the fundamental properties of functions of complex variable. Let us summarize the most important of these consequences [1]. Suppose that a function of a complex variable $p = \sigma + i\omega$,

$$W(p) = U(\sigma,\omega) + iV(\sigma,\omega) \tag{5.8}$$

1. has no singularities in the half-plane $\operatorname{Re} p \geq 0$ and
2. tends to zero, as $|p| \to \infty$, in the half-plane $\operatorname{Re} p \geq 0$.

We introduce special designations for the real and imaginary parts of this function corresponding to the case where the complex variable p takes purely imaginary values:

$$U(0,\omega) = A(\omega), \ V(0,\omega) = B(\omega) \tag{5.9}$$

It was shown that if the above conditions are fulfilled, the functions $A(\omega)$ and $B(\omega)$ are not independent, and they are related by Hilbert transforms:

$$A(\omega) = -\frac{1}{\pi} \int\limits_{-\infty}^{\infty} \frac{B(\omega')}{\omega' - \omega} \, d\omega', \; B(\omega) = \frac{1}{\pi} \int\limits_{-\infty}^{\infty} \frac{A(\omega')}{\omega' - \omega} \, d\omega' \qquad (5.10)$$

Here, the integrals are understood in the sense of the Cauchy principal value. In applications, these relations are often used in equivalent form:

$$A(\omega) = -\frac{2}{\pi} \int\limits_{0}^{\infty} \frac{\omega' B(\omega')}{\omega'^2 - \omega^2} \, d\omega', \; B(\omega) = \frac{2\omega}{\pi} \int\limits_{0}^{\infty} \frac{A(\omega')}{\omega'^2 - \omega^2} \, d\omega' \qquad (5.11)$$

The Kramers–Kronig relations describe the relationships between the real part $\varepsilon'(\omega)$ and imaginary part $\varepsilon''(\omega)$ of the complex permittivity in integral form:

$$\varepsilon''(\omega) = -\frac{1}{\pi} \int\limits_{-\infty}^{\infty} \frac{\varepsilon'(\omega') - 1}{\omega' - \omega} \, d\omega', \; \varepsilon'(\omega) - 1 = \frac{1}{\pi} \int\limits_{-\infty}^{\infty} \frac{\varepsilon''(\omega')}{\omega' - \omega} \, d\omega' \qquad (5.12)$$

Physically, they are the consequence of the causality principle that states that the response of a system to an external action cannot occur ahead of the action. On the other hand, comparing these relations with (5.10), we see that this is a particular example of functions related by Hilbert transforms.

The domain of applicability of (5.10) and (5.11) is restricted by condition (2). However, it appears that, with appropriate modification of the formulas, this condition can be substantially weakened. In particular, the formulas

$$A(\omega) - A(0) = -\frac{2\omega^2}{\pi} \int\limits_{0}^{\infty} \frac{B(\omega')}{\omega'(\omega'^2 - \omega^2)} \, d\omega', \; B(\omega) = \frac{2\omega}{\pi} \int\limits_{0}^{\infty} \frac{A(\omega') - A(0)}{\omega'^2 - \omega^2} \, d\omega' \qquad (5.13)$$

are valid if the functions $A(\omega)$ and $B(\omega)$ are, respectively, even and odd, and $|W(p)|$ is limited or even increases (however, slower than $|p|$) with $|p| \to \infty$.

In addition, formulas were derived to express the argument (phase) of the function $W(p)$ in terms of its modulus and vice versa.

From (5.11), it is clear that to determine one component, $A(\omega)$ or $B(\omega)$, from the other requires knowledge of the latter throughout the frequency range from zero to infinity. Performing this operation on a limited data set needs to extrapolate the data to both smaller and larger values of ω and then approximate them by a function of certain form. The approximating function is put to some limitations. Thus, the function $A(\omega)$ and $B(\omega)$ should be absolutely integrable on the semiaxis $\omega \in [0,\infty)$.

5.1.4   A Pole Model of Exponentially Decaying Signals

For a signal decaying exponentially with time t, which is described as

$$s(t) = A\exp(\gamma t)\cos(\omega_0 t + \varphi) = \frac{A}{2}\exp(\gamma t)\left[\exp(i(\omega_0 t + \varphi)) + \exp(-i(\omega_0 t + \varphi))\right]$$

$$= \frac{A}{2}\exp(i\varphi)\exp(i(\omega_0 - i\gamma)t) + \frac{A}{2}\exp(-i\varphi)\exp(-i(\omega_0 + i\gamma)t)$$

$$= B\exp(-iqt) + B^*\exp(iq^*t),$$

$$t \in (0,\infty), \; \gamma < 0, \; q = \omega_0 + i\gamma, \; B = \frac{A}{2}\exp(-i\varphi)$$

(5.14)

where the symbol "*" implies complex conjugation, there exists a direct exponential Fourier transform having the sense of the spectral characteristic (spectral density) of the signal $s(t)$:

$$Q(\omega,C,q) = \int_{-\infty}^{\infty} s(t)\exp(i\omega t)\,dt = B\int_0^{\infty}\exp[i(\omega - q)t] + B^*\int_0^{\infty}\exp[i(\omega + q^*)t]\,dt$$

$$= \frac{C}{\omega - q} - \frac{C^*}{\omega - q^*}, \quad C = iB$$

(5.15)

The spectral characteristic

$$Q(\omega,C,q) = \frac{C}{\omega - q} - \frac{C^*}{\omega - q^*}$$

(5.16)

allows unique reconstruction of the signal $s(t)$ by using the inverse Fourier transform

$$s(t) = \frac{1}{2\pi}\int_{-\infty}^{\infty} Q(\omega,C,q)\exp[-i\omega t]\,d\omega$$

(5.17)

Apart from the physical meaning of the frequency ω as a real quantity, it turns out that a direct Fourier transform (i.e., a function $Q(\omega,C,q)$) exists not only for real values of ω, but also for its complex values belonging to the half-plane $\mathrm{Im}\,\omega > \gamma$ of the complex ω plane. Within this half-plane, the function $Q(\omega,C,q)$ is analytical, and its analytic continuation to the whole ω plane is a meromorphic function with two simple poles: $\omega = q$ and $\omega = -q^*$.

These considerations allow us hereinafter to refer to the function $Q(\omega,C,q)$ as the pole model of a signal $s(t)$ (as accepted in the scientific literature) along with its traditional name, a complex spectrum (CS). The informative capacity of this term is quite understandable, as when we have information about the locations of only two complex poles, $\omega = q$ and $\omega = -q^*$, in the complex ω plane and know the amplitude and initial phase of a signal, we, in principle, can reconstruct the signal at any time, $0 < t < \infty$.

For a signal of general form,

$$s(t) = \left(a_0 + a_1 t + a_2 t^2 \ldots + a_m t^m\right)e^{\gamma t}\cos(\omega_0 t + \varphi)$$

(5.18)

the structure of the signal pole model is much more complicated. Besides the first-order poles, it contains poles of order $2,3...m$.

The structure of the pole model of a signal $s(t)$ occurring not at $t = 0$, but at a later time $t = \tau$ is somewhat different. Actually, for this case, we have

$$Q(\omega,C,q,\tau) = \int_{-\infty}^{\infty} s(t - \tau)\exp(i\omega t)\, dt = B\int_{\tau}^{\infty} \exp\left[-iq(t - \tau)\right]\exp(i\omega t)\, dt$$

$$+ B^*\int_{\tau}^{\infty} \exp\left[iq^*(t - \tau)\right]\exp(i\omega t)\, dt = B\int_{0}^{\infty} \exp(-iqt')\exp\left[i\omega(t' + \tau)\right] dt' \qquad (5.19)$$

$$+ B^*\int_{0}^{\infty} \exp(iq^*t')\exp\left[i\omega(t' + \tau)\right] dt' = \exp(i\omega\tau)\left[\frac{C}{\omega - q} - \frac{C^*}{\omega + q^*}\right]$$

We can also note the following:

1. The exponential decay of an oscillatory signal gives promise that in calculating its CS, the contribution from the interval $t_k < t < \infty$ (t_k is the end time of the signal under approximation) will be small.
2. On the other hand, in constructing a pole model of a signal by selecting the values of ω_0 and q, we should be sure that
   (i)   our approximation is good enough for $0 < t < t_k$, and
   (ii) the contribution to the CS from the interval $t_k < t < \infty$ is small enough for the CS to be as close in structure as possible to the actual spectrum of the signal under approximation.

5.1.5 The Singular Value Decomposition Method in Problems of Impulse Response Estimation and Reconstruction

Suppose that an unknown signal $x(t)$ arrives at the input of a linear system. At the system output, a signal $y(t)$ is observed, and we have an n-component vector $\mathbf{y}$ of measured signal samples $y_1,...,y_n$ distorted by noise. The problem is to find an m-component vector $\mathbf{x}$ of received signal samples $x_1,...,x_m$. The problem formulation can be reduced to a matrix equation of the form

$$\mathbf{y} = \mathbf{A}\mathbf{x} \qquad (5.20)$$

where $\mathbf{A}$ is an $n \times m$ matrix that characterizes the system properties. It was shown that in this formulation, the problem is ill-posed [2], and to solve it, regularizing algorithms should be invoked. To inverse the matrix $\mathbf{A}$, the so-called singular value decomposition method [3, 4] can be used. The essence of the method is difficult to understand without at least a brief explanation of its idea. Therefore, we first turn to some important definitions and statements of the theory of matrices [5].

For an arbitrary rectangular matrix $\mathbf{A}$ of size $m \times n$, there exists a universal decomposition $\mathbf{A} = \mathbf{U}\mathbf{\Lambda}\mathbf{V}$, where $\mathbf{U}$ and $\mathbf{V}$ are unitary matrices and $\mathbf{\Lambda}$ is a rectangular diagonal matrix of size $m \times n$ with nonincreasing non-negative diagonal elements. This decomposition is referred to as the singular decomposition of the matrix $\mathbf{A}$.

The set of elements a_{ij} of the matrix $\mathbf{A}$ for which $i = j$ is called the main diagonal and the corresponding elements are called diagonal elements. All other elements are called off-diagonal. A matrix $\mathbf{A}$ is called diagonal if all its off-diagonal elements are equal to zero. A diagonal matrix with elements $a_{11}, a_{22}, \ldots, a_{mm}$ is denoted as diag $(a_{11}, a_{22}, \ldots, a_{mm})$.

The transpose of a matrix $\mathbf{A}$ all elements of which are complex conjugate is the conjugate (Hermitian conjugate) of the matrix (denoted as $\mathbf{A}^*$).

A number λ is called the eigenvalue of a matrix $\mathbf{A}$ if there exists a nonzero vector $\mathbf{x}$, such that $\mathbf{A}\mathbf{x} = \lambda\mathbf{x}$. The nonzero eigenvalues of the matrices $\mathbf{A}^*\mathbf{A}$ and $\mathbf{A}\mathbf{A}^*$ are always the same.

The arithmetic square roots of the common eigenvalues of the matrices $\mathbf{A}^*\mathbf{A}$ and $\mathbf{A}\mathbf{A}^*$ are called singular (principal) values of the matrix $\mathbf{A}$. The singular values of a matrix are not changed upon its premultiplication or postmultiplication by any unitary matrix.

A complex matrix $\mathbf{U}$ is termed unitary if its conjugate matrix $\mathbf{U}^*$ coincides with its inverse matrix $\mathbf{U}^{-1}$; that is, if $\mathbf{U}\mathbf{U}^* = \mathbf{U}^*\mathbf{U} = \mathbf{E}$.

If a system $\mathbf{A}\mathbf{x} = \mathbf{y}$ is consistent (i.e., it has at least one solution), it is equivalent to the system $\mathbf{A}^*\mathbf{A}\mathbf{x} = \mathbf{A}^*\mathbf{y}$. The latter system, in turn, is consistent for any matrix $\mathbf{A}$ and any right-hand side $\mathbf{y}$, and its solution is a pseudosolution or a generalized solution.

From the preceding, an important conclusion can be made. If the solvability of the system $\mathbf{A}\mathbf{x} = \mathbf{y}$ cannot be guaranteed, one can solve instead the system $\mathbf{A}^*\mathbf{A}\mathbf{x} = \mathbf{A}^*\mathbf{y}$. This ensures minimization of the residual $\mathbf{A}\mathbf{x} - \mathbf{y}$.

Of great importance is the matrix denoted by the symbol $\mathbf{A}^+$. This $m \times n$ matrix is called the pseudoinverse or generalized inverse of an $n \times m$ matrix $\mathbf{A}$ if

$$\mathbf{A}\mathbf{A}^+\mathbf{A} = \mathbf{A}, \quad \mathbf{A}^+ = \mathbf{U}\mathbf{A}^* = \mathbf{A}^*\mathbf{V} \tag{5.21}$$

where $\mathbf{U}$ and $\mathbf{V}$ are some matrices.

The pseudoinverse matrix is uniquely determined by the so-called Penrose equations

$$\mathbf{A}\mathbf{A}^+\mathbf{A} = \mathbf{A}, \quad \mathbf{A}^+\mathbf{A}\mathbf{A}^+ = \mathbf{A}^+, \quad (\mathbf{A}^+\mathbf{A})^* = \mathbf{A}^+\mathbf{A}, \quad (\mathbf{A}\mathbf{A}^+)^* = \mathbf{A}\mathbf{A}^+ \tag{5.22}$$

5.2   Use of Regularization and a Kramers-Kronig-Type Relation for Estimating Transfer Functions and Impulse Responses

5.2.1   General Relations

The variations of a signal propagating in a linear system are completely determined by the properties of the system IR. Therefore, the IR estimation is of importance. There are two well-known conventional methods for doing this. The first one is directly solving a system of linear equations corresponding to the convolution of the impulse response $h(t)$ with the input signal, $x(t)$:

$$y(t) = \int_0^\infty h(\tau)x(t - \tau)\,d\tau \tag{5.23}$$

The second method uses the well-known spectral relation corresponding to the convolution operation:

$$Y(\omega) = H(\omega)X(\omega) \qquad (5.24)$$

where $H(\omega)$ is the transfer function of the system.

In view of (5.24), we have

$$h(t) = \frac{1}{2\pi} \int\limits_{-\infty}^{\infty} \frac{Y(\omega)}{X(\omega)} \exp(-i\omega t)\,d\omega \qquad (5.25)$$

The difficulties encountered in evaluating an IR using any of these approaches are due to the same reasons: the presence of noise in the output signal $y(t)$ and the lack of information about the input and output signals. Indeed, the convolution operation using (5.23) or (5.24) assumes that the values of a continuous signal or its spectrum are known throughout the time or the frequency domain. However, this condition cannot be fulfilled in an actual physical experiment.

Analysis of the spectral method of IR estimation shows that the presence of noise in the output signal causes errors in the evaluated transfer function $H(\omega)$. The errors are especially large in the neighborhood of the zeros of the input signal CS. To reduce the errors, various regularization techniques are used [6, 7].

The limited amount of input information is also the reason for errors in the estimated $Y(\omega)$ and $X(\omega)$ complex spectra. The problem of estimating a continuous CS based on a limited set of data is an ill-posed. The ill-posedness can be eliminated by using an additional criterion for selecting a single evaluated estimated CS from the set of admissible ones.

The finiteness of the sampling frequency of a signal restricts the frequency range of its spectrum analysis. The finiteness of the frequency range, in turn, can violate the principle of causality. The violation shows up in the appearance of precursors in IRs reconstructed from transfer functions in a limited frequency range. This section presents an approach that reduces the errors in estimated IRs caused by the finiteness of the sampling frequency.

For an estimated IR to contain no precursor, the CS corresponding to its transfer function $H(\omega)$ must be approximated by an analytic function whose real and imaginary parts are related by a Kramers–Kronig-type dispersion relation [1, 8, 9]. Let us employ a relation of this type [10] to estimate a transfer function $H(\omega)$. First, we write this function in exponential form:

$$H(\omega) = \exp\left[-\left(\beta(\omega) + i\varphi(\omega)\right)\right] \qquad (5.26)$$

According to the Kramers–Kronig-type relation, the phase in (5.26) can be determined as

$$\varphi(\omega) = \omega\tau + \frac{2\omega}{\pi} \int\limits_{0}^{\infty} \frac{\beta(\omega')}{\omega'^2 - \omega^2}\,d\omega' \qquad (5.27)$$

where $\beta(\omega)$ is the extinction coefficient determined by the transfer function modulus as $\beta(\omega) = -\ln(|Y(\omega)|/|X(\omega)|)$, where τ is the delay of the output pulse $y(t)$ relative to $x(t)$.

Thus, to completely evaluate the complex transfer function $H(\omega)$, it is necessary to previously estimate the extinction coefficient for the whole real frequency axis. According to the sampling theorem, the actual signal spectrum can be estimated only in a frequency band whose width is equal to half the signal sampling frequency. In addition, it should be taken into account that the presence of zeros in the input pulse spectrum causes large errors in the transfer function $H(\omega)$ estimated by directly dividing the complex spectrum $Y(\omega)$ by the spectrum $X(\omega)$. The use of regularization reduces these errors. The expression for $H(\omega)$ then becomes [6]

$$H(\omega) = \frac{Y(\omega)X^*(\omega)}{X(\omega)X^*(\omega) + \alpha}$$

(5.28)

where $X^*(\omega)$ is the complex conjugate spectrum and α is the regularization parameter. The regularization parameter is chosen based on the a priori information about the level of noise of the pulse spectra $X(\omega)$ and $Y(\omega)$ or, in the absence of this information, by using iterative methods. This approach gives good estimates for the spectrum $H(\omega)$ in the frequency ranges for which the inequality $|Y(\omega)|/|X(\omega)| < 1$ is fulfilled. For other frequency ranges, such that the values of the spectra $X(\omega)$ and $Y(\omega)$ are comparable to the measurement noise level, the use of regularization will give $H(\omega)$ values close to zero. Thus, if the frequency band of the input pulse $X(\omega)$ does not overlap the frequency band of a passive system under investigation, the application of a regularization technique will give an IR abounding with unnecessary details, such as time fluctuations, whose waveform may differ significantly from that of the actual IR of the system.

Another approach is based on an extrapolation of the amplitude spectrum $|H(\omega)|$ to frequency ranges not overlapped by the spectrum of the input pulse. The form of the extrapolating function is chosen based on a priori information about the system under investigation. When choosing the form of this function, one should bear in mind that the operation of convolution (5.23) admits a unique solution if the modulus of the transfer function $|H(\omega)|$ decreases at infinity as ω^{-2} or more rapidly [7]. The parameters of the extrapolating function are chosen to minimize the quadratic residual between the measured samples of the output signal, y_j, and the model samples $<y_j>$ obtained by applying an inverse Fourier transform to the transfer function:

$$\Phi = \sum_{j=1}^{N} \left| y_j - <y_j> \right|$$

(5.29)

5.2.2   Reconstruction of Transfer Functions and Impulse Responses using Regularization and Kramers–Kronig-type Relations

To test the procedure of impulse response estimation, special experiments were performed [10]. A UWB pulse generator was used to produce a monopolar or a bipolar voltage pulse, $x(t)$, of duration 1 or 2 ns, respectively (Figure 5.1). The pulses were

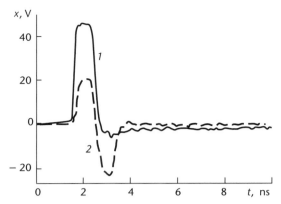

Figure 5.1   A monopolar (1) and a bipolar pulse (2) at the generator output. (With permission from Springer.)

applied to a piece of cable served as the test system. Two different cables of imped-ance 50 Ω were used: cable 1 (PK50-4-11, 8 m long) and cable 2 (PK50-2-11, 20 m long). At the cable output, the output pulses $y(t)$ were recorded.

The spectra of the $x(t)$ and $y(t)$ pulses were calculated by two independent meth-ods. The first one was based on a conventional approach using fast Fourier trans-formation (FFT). The second approach, based on the maximum entropy method and ensuring high resolution, gave similar results for the frequency range where most of the input pulse energy was concentrated. Thus, it was demonstrated that the signals had no fine spectral structure, and, therefore, to estimate their spectra, it sufficed to use the conventional FFT procedure. The plots of the spectrum moduli ($|X(\omega)|$ at the cable input; $|Y1(\omega)|$ and $|Y2(\omega)|$ at the outputs of cable 1 and cable 2, respectively) are given in Figure 5.2. The spectral function of a bipolar pulse is substantially different from that of a monopolar pulse only in the low-frequency range, where it takes values close to zero.

If no regularization is used, the unity level of the IR amplitude spectra can be exceeded at the points where the levels of $|X(\omega)|$ and $|Y(\omega)|$ are comparable to the

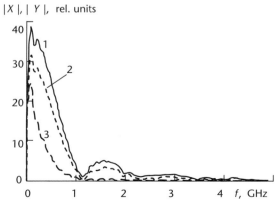

Figure 5.2   Spectra of a monopolar pulse at the generator output (1) and after the passage through cable 1 (2) and cable 2 (3). (With permission from Springer.)

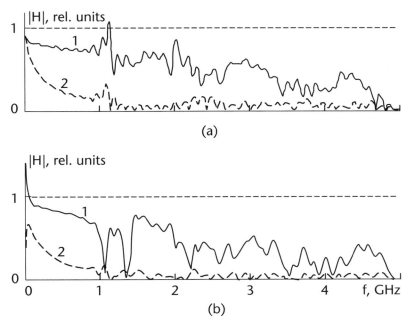

Figure 5.3 Impulse response spectra of cable 1 (1) and cable 2 (2) evaluated using regularization for a monopolar (*a*) and a bipolar pulse (*b*). (With permission from Springer.)

measurement noise level. In this case, the reconstructed IR has a heavily jugged main lobe and a level of spurious fluctuations comparable to the base level. Figure 5.3 presents the IR amplitude spectra calculated using regularization ($\alpha = 0.1$).

As can be seen from Figure 5.3, regularization, in most cases, moderates the "nonphysical" behavior of IR amplitude spectra, i.e., the excess of the unity level.

At the same time, regularization gives rise to heavily damped $|H(\omega)|$ windows. The reconstructed IRs of the two cables obtained using monopolar pulses are shown in Figure 5.4. For bipolar pulses, the IR amplitude spectra are difficult to estimate in the low-frequency range, where the level of the pulse amplitude spectra is comparable to that of the measurement noise. This error has a significant effect on the waveform of the reconstructed IR.

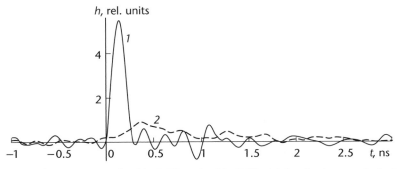

Figure 5.4 Impulse responses of cable 1 (1) and cable 2 (2) obtained using regularization. (With permission from Springer.)

It is beyond reason to think that the strong fluctuations in the IR amplitude spectra are associated with something other than instrumental errors. These fluctuations were accounted for by the significant effect of noise in the frequency range where the amplitude spectrum level of the input pulse was low. To eliminate this effect, the IR amplitude spectrum was estimated using a model function given by [10]

$$|H(\omega)| = \frac{|Y(\omega)|}{|X(\omega)|} = \frac{d}{\omega^2 + g} \tag{5.30}$$

The parameters d and g were estimated by minimizing the quadratic residual between model (5.30) and the experimental values of the ratio $|Y(\omega)|/|X(\omega)|$ for the frequency ranges where the level of the amplitude spectrum was higher than that of the measurement noise. For the cables used in the experiment, this model was well matched to their certified damping constants [11].

In calculating the phase $H(\omega)$ with the use of (5.27), the data on IR amplitude spectra in the frequency range from zero to 5 GHz obtained using regularization (α = 0.1) and calculated by the model relation (5.30) were utilized. In the latter case, the use of the model was restricted to the frequency ranges where the levels of the amplitude spectra $|X(\omega)|$ and $|Y(\omega)|$ were comparable to the measurement noise level. For other frequency ranges, the values of $|H(\omega)|$ were estimated from experimental data. The reconstructed IRs of cable 1 for monopolar pulses are shown in Figure 5.5.

The IRs evaluated using regularization contain many unnecessary details, such as spurious fluctuations. There are several reasons for their occurrence. The dominant reason is the small width of the input pulse spectrum compared to that of the IR spectrum. In addition, the regularization parameter cannot correct the phase distortions caused by the measurement noise. The extrapolation of $|H(\omega)|$ by the model function (5.30) along with the calculation of the phase by a Kramers–Kronig-type relation allow an artificial extension of the frequency range for the probe and the received pulses. Thus, for a monopolar pulse, as can be seen from Figure 5.2, most of its energy is concentrated in the frequency range up to 1 GHz, and this is the least noise-affected range. For a bipolar pulse, the exception is the low-frequency range ($f < 0.1$ GHz), where the level of the amplitude spectra $|X(\omega)|$ and $|Y(\omega)|$ is comparable to the noise level. Extrapolation allows one to estimate the damping

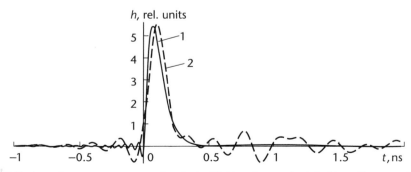

Figure 5.5 Impulse responses reconstructed with the phase calculated by a Kramers–Kronig-type relation using an extrapolation of the amplitude spectrum (1) and regularization (2). (With permission from Springer.)

factor for monopolar and bipolar pulses in the frequency range $f > 1$ GHz and for a bipolar pulse in the low-frequency region as well. In this case, the level of the precursors in the IR significantly decreases, the leading edge becomes steeper, and the spurious fluctuations disappear.

As can be seen from Figure 5.5, the use of a Kramers–Kronig-type relation does not eliminate precursors in this case. The reason is that a relation of this type proposes that the amplitude spectrum $|H(\omega)|$ is analytic. In the case under consideration, the function $|H(\omega)|$ was not analytic, as it was set identically zero for a frequency of 5 GHz and greater. The analyticity of the amplitude spectrum can be restored by continuously extending the measured or calculated values of $|H(\omega)|$ to the range of higher frequencies.

The impulse responses $h1$ and $h2$ obtained for a monopolar and a bipolar pulse, respectively, are closely similar to each other. Figure 5.6 shows the data for cable 2. The main differences are caused by an error in the estimated damping factor of $|H(\omega)|$ at low frequencies in extrapolating the spectrum $h2$ and by the effect of the measurement noise in estimating the damping factor of $|H(\omega)|$ in the neighborhood of the first side lobe. For both the monopolar and the bipolar pulse, the error of the estimation of $y(t)$ by $x(t)$ with the use of $h1$ and $h2$ was not over 1%.

These data show that in probing a system by UWB pulses different in waveform and spectral composition, one can obtain IRs similar in waveform. In addition, these IRs allow one to estimate a signal $y(t)$ whose spectral band is wider than that of the pulse used to obtain the IR.

5.2.3 Comparison of the Impulse Responses Estimated Using Two Phase Spectrum Models

Let us compare the estimated IRs with the results presented by Glebovich and Kovalev [11], where the algorithms for calculating the IRs of coaxial cables by the

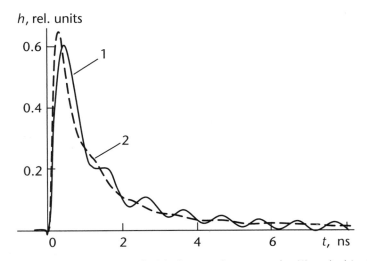

Figure 5.6 Impulse responses obtained with the use of a monopolar (1) and a bipolar pulse (2). (With permission from Springer.)

values of their electromagnetic characteristics are described in detail. The spectrum $H(\omega)$ is determined by (5.26), where

$$\beta(\omega) = b\omega^{1/2} + \frac{a\omega^{3/2}}{1 + m\omega} \tag{5.31}$$

and the phase is calculated as

$$\varphi(\omega) = \omega\tau + b\omega^{1/2} + \frac{b\omega^{1/2} + bm\omega^{3/2} - a\omega^{3/2}}{(1 + m\omega)a\omega k} \tag{5.32}$$

The values of the coefficients a, b, and m are determined by the electromagnetic characteristics of the cable. In the experiment [10], the signal propagation channel incorporated, besides the cable, attenuators and adapters. For channels of this type, the available reference books contain no data on the electromagnetic parameters in the frequency range of several gigahertz. Therefore, the values of a and b were found numerically by minimizing the quadratic residual between the damping factor measurements and the theoretical predictions obtained using relations (5.24) and (5.26) with $m = 2 \cdot 10^{-11}$ s/rad (related to the dielectric losses in the cable) and $k = 1/3 \cdot 10^9$.

Figure 5.7 shows the damping factor (curve 1) and the phase (curve 2) calculated as functions of frequency by relations (5.31) and (5.32) for cable 2. Herein, the phase spectrum is given (curve 3) that was evaluated by relation (5.27). In both cases, $\varphi(\omega)$ was calculated for $\tau = 0$. The IRs corresponding to the obtained complex spectra are given in Figure 5.8. The IRs presented in Figure 5.8 indicate that the calculation of the phase by (5.32) violates the principle of causality that shows up in the appearance of a precursor in the IRs.

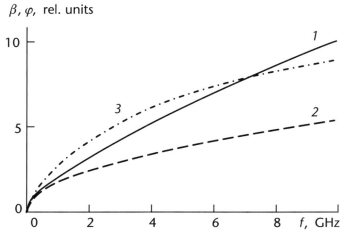

Figure 5.7 The damping factor (curve 1) and phase (curve 2) calculated as functions of frequency by (5.31) and (5.32), and the phase spectrum (curve 3) calculated by (5.27). (With permission from Springer.)

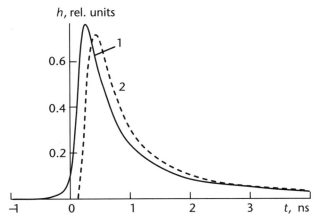

Figure 5.8 Impulse responses calculated for the same damping factor $\beta(\omega)$ using the phase spectrum given by (5.32) (curve 1) and by (5.27) (curve 2). (With permission from Springer.)

Summarizing the above results, we can state the following. To obtain the impulse response of a linear system, it is advisable to use experimental measurements only in estimating the amplitude spectrum. The phase spectrum should be calculated from the amplitude spectrum using a Kramers–Kronig-type relation. This procedure provides an impulse response free of precursors and spurious fluctuations caused by the errors in the estimated phase spectrum.

If the frequency band of the probe pulse is narrower than that of the channel under investigation, the impulse response will contain precursors and have a flatter leading edge. Extrapolation of the amplitude spectrum of the impulse response based on a priori information about the system under investigation lowers the level of spurious time lobes.

5.3   A Pole Model of the Signal in the Problem of Estimating the Impulse Response of a Propagation Channel

5.3.1   Signal Representation and Impulse Response Estimation using Pole Functions

The complex spectra of signals starting at different times can be adequately approximated by the expression [12]

$$S(\omega) = \sum_{m=1}^{M} \int_{-\infty}^{\infty} P_m(\tau)Q_m(\omega)\exp(i\omega\tau)\,d\tau \tag{5.33}$$

$$Q_m(\omega) = \frac{C_m^*}{\omega + q_m^*} - \frac{C_m}{\omega - q_m}, \quad q_m = \omega_m + i\gamma_m \tag{5.34}$$

Here, $P_m(\tau)$ is the density distribution function of the delays of pole functions $Q_m(\omega)$. Each mth pole function with a complex amplitude C_m contains a pole q_m

with corresponding values of frequency ω_m and decrement γ_m. The symbol "*"
denotes complex conjugation. This representation reduces the number of parameters
in the signal description and increases the stability of IR estimation. The positions of
the IR poles in the complex plane and the form of the function $P_m(\tau)$ are identification attributes of the object under investigation and can be used for its recognition.

In what follows we will use three forms of representation of a signal s: time
representation $s(t)$, spectral representation $S(\omega)$, and pole representation $S(q)$. A pole
representation supposes that a set of real density distribution functions $\{P_m(\tau)\}$ and a
set of complex quantities $\{q_m, C_m\}$, $m = 1 \ldots M$, are specified. This form of signal representation admits a unique transformation into a time or a frequency representation.

In the literature, the so-called generalized Prony method [3] is well known. To
this method, there corresponds a pole model for which all $P_m(\tau)$ are Dirac delta functions: $P_m(\tau) = \delta(\tau)$. The functions $P_m(\tau)$ are to be estimated for damped oscillations
starting at different times. A possible way of estimating $P_m(\tau)$ is to approximate
them by trigonometric polynomials. However, this way calls for the development of
complicated algorithms requiring a substantiated choice of the number of poles and
the order of the trigonometric polynomial. In some cases, the functions $P_m(\tau)$ may
be chosen to have a rather simple form. For example, one may use an equiprobable
distribution of the onset times of damped oscillations $P_m(\tau)$ in a limited time interval:

$$
\begin{aligned}
P_m(\tau) &= 1/T_m, \quad \tau \in \left[\tau_m, T_m + \tau_m\right] \\
P_m(\tau) &= 0, \quad\quad \tau \notin \left[\tau_m, T_m + \tau_m\right]
\end{aligned}
\tag{5.35}
$$

According to (5.33), the CS of a signal corresponding to this distribution will
be given by

$$
S(\omega) = \sum_{m=1}^{M} Q_m(\omega)\exp\left(i\omega\tau_m\right)\left(\exp\left(i\omega T_m\right) - 1\right)/i\omega T_m
\tag{5.36}
$$

In estimating the IRs of UWB systems, we will assume that a probe pulse $x(t)$
is characterized by a complex spectrum $X(\omega)$ whose expression is similar to (5.36).
The complex spectrum $\tilde{H}(\omega)$ of the impulse response is sought in a similar form.

The procedure of IR estimation consists of the following steps:

1. For a known time sequence of the probe pulse samples $(x(t_i), i = 1, N)$, a pole
   function $\tilde{X}(q)$ is found that minimizes the quadratic residual

$$
\Phi_x = \frac{\displaystyle\sum_{i=1}^{N}\left|x\left(t_i\right) - \tilde{x}\left(t_i\right)\right|^2}{\displaystyle\sum_{i=1}^{N}\left|x\left(t_i\right)\right|^2}
\tag{5.37}
$$

where $\tilde{x}(t_i)$ is the time representation of the pole function $\tilde{X}(q)$. The
values of the $\tilde{X}(q)$ parameters $\{q_m, T_m, \tau_m\}$ providing the minimum of the
quadratic residual Φ_x are found by the coordinate descent method. The

complex amplitudes C_m corresponding to the poles are calculated by solving the system of linear algebraic equations that follows from the condition

$$\frac{\delta \Phi_x}{\delta C_m} = 0 \tag{5.38}$$

2. For a known time sequence of the output signal samples $(y(t_i), i = 1,M)$, a pole function $\tilde{H}(q)$ is found that minimizes the quadratic residual

$$\Phi_y = \frac{\sum\limits_{i=1}^{M}\left|y(t_i) - \tilde{y}(t_i)\right|^2}{\sum\limits_{i=1}^{M}\left|y(t_i)\right|^2} \tag{5.39}$$

where $\tilde{y}(t_i)$ is the time representation of the output signal obtained by evaluating the Fourier integral for the complex spectrum $\tilde{Y}(\omega) = \tilde{H}(\omega)\tilde{X}(\omega)$ based on the residue theorem.

The search for the model parameters for $\tilde{H}(q)$ is similar to that for $\tilde{X}(q)$ with a little difference: in the expression for the complex spectrum $\tilde{Y}(\omega)$, the parameters of $\tilde{X}(\omega)$ are fixed and only the parameters of $\tilde{H}(\omega)$ are to be estimated.

As mentioned earlier, the change from the frequency domain (5.33) to the time domain is performed using an integral Fourier transformation based on the residue theorem. The expression for the model functions $\tilde{x}(t)$ and $\tilde{h}(t)$ is the following:

$$s(t) = \sum_{m=1}^{M}\int_0^t P_m(t - \tau)\left[C_m\exp\left(-iq_m\tau\right) - C_m^*\exp\left(iq_m^*\tau\right)\right]d\tau \tag{5.40}$$

Accordingly, the model function $\tilde{y}(t)$ is obtained from an analytic relation as a result of the convolution

$$\tilde{y}(t) = \tilde{h}(t) \otimes \tilde{x}(t) \tag{5.41}$$

The condition for the residual of the functional (5.39) to be a minimum is used as an accuracy criterion for the IR estimation. To describe the model, the minimum number of poles is sought that yields the residual threshold determined by the noise level in the signal. If the noise level is not known a priori, the number of poles, N_p, is determined from the condition $[\Phi_y(N_p) - \Phi_y(N_p + 1)] \ll [\Phi_y(N_p - 1) - \Phi_y(N_p)]$ implying that increasing the number of poles over N_p insignificantly decreases the residual.

The quality of the IR evaluation was judged by the shift in the estimates of the poles and by the confidence intervals. These quantities were estimated by statistical modeling.

An additional criterion for the quality of the IR estimation was obtained by solving the inverse problem of reconstruction of the probe pulse using the evaluated IR and the output signal.

5.3.2 Estimation of the Impulse Response of a Coaxial Cable Transmission Line

To test the efficiency of the approach described, the experimental data on the passage of UWB pulses through coaxial cables (cable 1: PK50-4-11, 8 m long; cable 2: and PK50-2-11, 20 m long) [12, 13] were used.

Figure 5.9 presents monopolar pulses produced by the generator, $x(t)$ (curve 1), passed through cable 1, $y1$ (curve 2), and passed through cable 2, $y2$ (curve 3). In addition, a bipolar pulse was used in the experiment, which was composed of two monopolar pulses of opposite polarity of duration 1 ns time shifted from one another by 1 ns. The choice of two probe pulses different in waveform was dictated by the aim to demonstrate that the proposed approach to IR reconstruction is insensitive to the probe pulse waveform.

The search for the poles of the impulse response $\tilde{h}(t)$ was carried out by the coordinate descent method using relations (5.39) and (5.41). The generator pulses were approximated with high accuracy by a set of five pole functions whose parameters were also sought by the coordinate descent method. In this case, the density distribution functions of the times of invoking of the pole functions $P_m^x(\tau)$ were characterized by the following parameter values: $\tau_m^x = 0$ ($m = 1...5$), $T_1^x = 0.91$ ns, and $T_2^x, T_3^x, T_4^x, T_5^x = \Delta t = 0.016$ ns, where Δt is the sampling step of the signal $x(t)$.

Figure 5.10 shows the IRs of cable 1 calculated using three pole functions. Curve 1 and curve 2 correspond to the IRs calculated, respectively, for a monopolar and a bipolar pulse passed through the cable. Similarly, the IRs of cable 2 were calculated using two pole functions. In both cases, the parameters of $P_k^p(\tau)$ were set as $\tau_k^p = 0$ and $T_k^p = \Delta t$.

It can be seen that the IR calculated using a monopolar pulse has the higher amplitude and decays within a shorter time. This can be due to that the spectrum of a bipolar pulse is narrower than that of a monopolar pulse.

To test the stability of the algorithm to external noise, a numerical experiment was performed [12] in which the IR of cable 2 was calculated for 50 realizations of

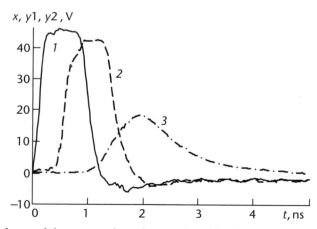

Figure 5.9 Waveforms of the monopolar pulses produced by the generator (1) and passed through cable 1 (curve 2) and cable 2 (curve 3). (With permission from Springer.)

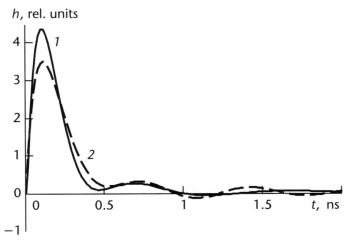

Figure 5.10 Impulse responses of cable 1 reconstructed using a monopolar (curve 1) and a bipolar pulse (curve 2). (With permission from Springer.)

$y(t)$. Each realization was a noise imposed on a test signal with a variance of 5% of the signal maximum. For each realization, poles with fixed $P_k^p(\tau)$ were calculated.

The numerical experiment has shown that the average positions of the poles, $\bar{q}_1^b = 0.025 - 1.003i$ and $\bar{q}_2^b = 0.586 - 2.59i$ were shifted relative to their positions in the absence of noise, $q_1^b = 0.015 - 0.781i$ and $q_2^b = 0.34 - 3.06i$, by a value twice that of the error in the estimated averages. Despite this, the IR waveform was stable to external noise.

As noted earlier, the proposed pole model allows one to reconstruct a probe pulse from a known output signal and an IR modeled by a pole function by using the same approach as that used for IR reconstruction. This operation serves to additionally validate the IR reconstruction procedure, and it is of independent significance in evaluating the input pulse waveform when the IR of the system is known. Figure 5.11 shows the results of reconstruction of a monopolar pulse from the output signals of cable 1 and cable 2 and their previously calculated impulse responses. Curve 1 corresponds to the generator pulse $x(t)$, curve 2 to the reconstructed signal $\tilde{x}1(t)$ for cable 1, and curve 3 to the reconstructed signal $\tilde{x}2(t)$ for cable 2. In reconstructing the pulse $\tilde{x}1(t)$, a set of five pole functions and $P_m^x(\tau)$, whose parameters were close to the parameters of the pole functions, and $P_m^x(\tau)$ approximating $x(t)$ were obtained. The residual calculated by (5.37) was 0.2%. To reconstruct the pulse $\tilde{x}2(t)$, it sufficed to find four pole functions with their $P_m^x(\tau)$. The fifth pole function, which was present in the approximation of the initial pulse $x(t)$ and was associated with a high-frequency pole, was absent in the approximation of $\tilde{x}2(t)$. This was due to a decrease in the high-frequency amplitude of the pulse CS during the passage of the pulse through the cable. The physical consequence of this was smoothed edges of the reconstructed pulse $\tilde{x}2(t)$. The residual calculated by (5.37) was 0.8%.

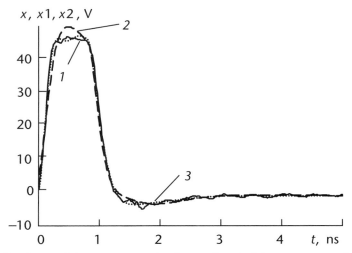

Figure 5.11 Waveforms of the generator pulse (curve 1) and of the reconstructed pulses passed through cable 1 (curve 2) and cable 2 (curve 3). (With permission from Springer.)

The experiments have revealed the following basic features of the cable UWB transmission systems. To reconstruct the IR of a system of this type, it is advisable to use monopolar pulses whose spectra contain low frequencies. In addition, to estimate correctly the low-frequency poles in an IR pole models, it is necessary that the pulse observation window be as large as possible.

The obtained IR of a test cable can be used to estimate the damping constant of the cable in the frequency band of the probe pulse.

5.3.3   Stability of the Reconstruction of Impulse Responses to the Probe Pulse Waveform and Measurement Noise

Of great interest is to investigate the stability of the estimates of IR model parameters to the probe pulse waveform and measurement noise. For doing this, a numerical simulation was performed to check whether the parameters of the model given by (5.35) are stable to white noise added to an output signal $y(t)$. To model the IRs of cables 1 and 2, monopolar signals $y1(t)$ and $y2(t)$ were used whose normalized waveforms were different in amplitude by no more than 7%. White noise was set to make 5% of the maximum amplitude of the signal $y(t)$.

It turned out that the model (5.35) yielded stable estimation of the cable IRs with only one IR pole taken into account. The variations in the pole position found using the model (5.35) for cable 1 and cable 2 are shown in Figure 5.12 with respective confidence ellipses. The spreads in other model parameters, T and τ, for a 5% noise variance are shown in Figure 5.13. It can be seen that there is no overlap for all model parameters. The small number of parameters required to approximate the signal and the lack of overlap between the parameter values make the model given by (5.35) promising for solving object recognition problems.

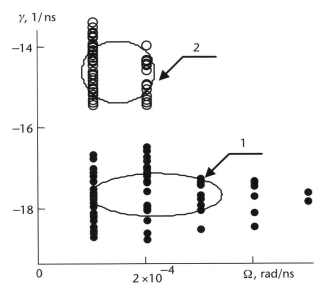

Figure 5.12 Variations in pole positions for the impulse responses of cables 1 and 2 determined using the model (5.35). (With permission from Springer.)

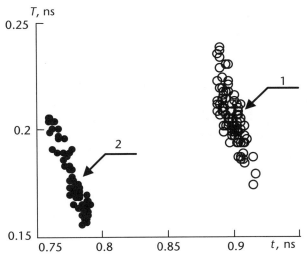

Figure 5.13 Values of the parameters T and τ of the model (5.35) for cables 1 and 2. (With permission from Springer.)

5.4   A Pole Model of a Signal in Estimating the Impulse Responses of a Conducting Sphere and Cylinder

Another application of pole models [12, 13] is the calculation of the IRs of objects probed by UWB pulses. We used the experimentally recorded waveform of a UWB pulse scattered by a sphere of radius $R = 10$ cm [14]. In addition, using a pole model, we calculated the IR of the sphere based on the data reported in the study

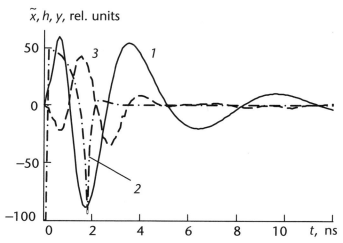

Figure 5.14 The reconstructed probe pulse waveform (curve 1), the impulse response of a sphere of radius $R = 10$ cm (curve 2), and the measured pulse scattered by the sphere (curve 3). (With permission from Springer.)

of Kennaugh and Moffatt [15]. The information obtained sufficed to reconstruct the waveform of the probe pulse $\tilde{x}(t)$ (Figure 5.14, curve 1). It should be emphasized that we took into account the distortion introduced by the receive path and used only five polar pole functions, each specified by $P_m^x(\tau)$ with $\tau_m^x = 0$ and $T_m^x = 0.09$ ns ($m = 1...5$).

The reconstructed probe pulse was used to calculate the IR of a sphere of radius $R = 4$ cm (Figure 5.15, curve 1) using five polar pole functions and $P_k^p(\tau)$ with $\tau_k^p = 0$ and $T_k^p = 0.09$ ns ($k = 1...5$). For comparison, we give (Figure 5.15, curve 2) the IR calculated for the same sphere with the use of the experimental data [15]. The criterion for the IR reconstruction accuracy served the residual between the measured signal scattered by the sphere (Figure 5.16, curve 1) and the signal $\tilde{y}(t)$

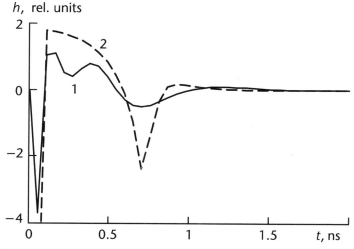

Figure 5.15 The reconstructed IR (curve 1) and the IR calculated using the data of [15] (2) for a sphere of radius $R = 4$ cm. (With permission from Springer.)

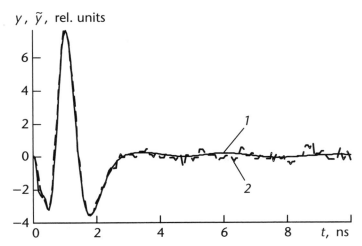

Figure 5.16 The scattered signal calculated from the reconstructed impulse response of a sphere of radius $R = 4$ cm (curve 1) and the measured scattered signal (2). (With permission from Springer.)

obtained by the convolution of the probe pulse $\tilde{x}(t)$ with the reconstructed IR of the sphere (Figure 5.16, curve 2). The residual was calculated by (5.39) and made 2.8%.

The distortion of the reconstructed IR waveform (Figure 5.15) is due to the errors having accumulated during successive calculations of the probe pulse and of the sphere IR and to the absence of high frequencies in the CS of the probe pulse.

It should be noted that the position of the second minimum in the IR of the sphere characterizes the sphere radius and is related to the effect of its rounding by the so-called creeping wave. In addition, for a given spatial length of the probe pulse, as the sphere radius is increased, the portion of energy going into the creeping wave decreases. We performed additional simulations to elucidate the behavior of the error η in determining the sphere radius R by the position of the second global minimum in the reconstructed IR for different sphere sizes and a fixed spatial length of the probe pulse, $\tau_p c$, where c is the velocity of light and τ_p is the pulse duration. The IR was approximated by five pole functions with fixed $P_k^p(\tau)$ having the parameters $\tau_k^p = 0$ and $T_k^p = 0.09$ ns ($k = 1...5$).

Figure 5.17 shows the simulation results for no noise in the reflected signal (curve *1*) and for a noise variance of 5% (curve 2) and 10% (curve 3). The variance was determined relative to the signal maximum in the receiving system. The error in determining the radius of the sphere was a minimum of 11.5% in the absence of noise for the ratio of the diameter to the pulse spatial length equal to 1.5 and increased with the level of noise. The increase in the error with $2R/\tau_p c$ with respect to its optimum value was associated with a decrease in the proportion of the creeping wave in the reflected signal. The increase in the error on decreasing $2R/\tau_p c$ with respect to its optimum value was due to a decrease in the resolution of the probe pulse $\tau_p c/2$.

The probe pulse (Figure 5.14, curve 1) was also used to reconstruct the IR of a metal cylinder of length 62 cm and diameter 25 cm. The incident wave propagated along the cylinder axis. The cylinder was represented as a complex object with a

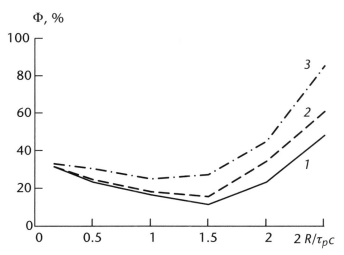

Figure 5.17 The error in determining the radius of a sphere versus the ratio of its diameter to the spatial length of the probe pulse with no noise (curve 1) and with a noise variance of 5% (curve 2) and 10% (curve 3). (With permission from Springer.)

certain $P_k^p(\tau)$ containing two pole function invoking times τ_k^p and $T_k^p = 0.09$ ns. The first invoking time was set to correspond to the position of the first end of the cylinder and the second one was chosen based on the minimum residual between the scattered measured and model signals. To each invoking time, there corresponded three pole functions whose pole parameters were estimated by the coordinate descent method. The dependence of the residual on the second pole function invoking time had a pronounced minimum corresponding to a distance of 63 cm, which was close to the length of the cylinder. Thus, using a pole model, one can determine the length of a cylinder from the measurement data for the probe and scattered UWB pulses. Figure 5.18 shows the calculated IR of the cylinder with the found $P_k^p(\tau)$

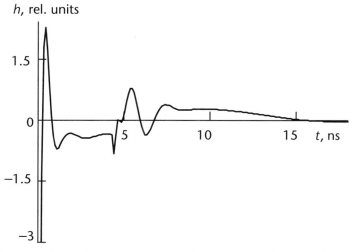

Figure 5.18 The reconstructed impulse response of a cylinder. (With permission from Springer.)

containing two pole function invoking times. For the given IR, the residual calculated by (5.39) was 2.6%.

Thus, the above approach provides stable estimation of the impulse responses of objects and propagation channels for UWB electromagnetic pulses. The approximation of signals and impulse responses using a pole model with a rectangular density distribution of time delays of the pole functions, $P(\tau)$, significantly reduces the number of model parameters compared with conventional methods. The possibility to determine the size of an object using its evaluated impulse response has been demonstrated for a metal sphere and cylinder.

5.5   Reconstruction of Ultrawideband Pulses Passed Through Channels with Linear Distortions

We already emphasized that the transformation of signals by linear systems, including objects and UWB pulse propagation channels, is completely determined by the object and channel IRs. Knowledge of the IRs is necessary to recognize a probed object and to obtain information on the signal distortion produced by the system connecting, for instance, a voltage pulse generator and a transmitter or a receiving antenna and a recorder. Therefore, the search for signal processing methods for evaluating the IR of a probed object from discrete sequences of noisy measurements of UWB pulses is a basic UWB radar problem.

In Sections 5.3 and 5.4, we described an approach to evaluating the IRs of UWB systems and reconstructing the waveforms of radar pulses, which is based on the time-domain approximation of signals by exponentially damped oscillations described by pole functions. Unlike the generalized Prony method [3], in which all functions are invoked at the same time, the invoking times for oscillations of this type could be different. Thus, the IR waveform was described by a few pole functions. The disadvantage of this approach is the difficulty of finding the poles from noisy measurements, as this operation becomes unstable with increasing number of pole functions used.

As the input signal and the IR are equally significant components of the convolution equation, a method of estimating the IR of a system by known signals at the system input and output should be applicable to the estimation of an input signal using a known signal at the output of the system and the system impulse response. This problem often arises in performing experiments with high-power UWB radiation. In this section, we aim at reconstructing the waveform of a probe pulse from a known IR of the system.

5.5.1   Solution of the Pulse Reconstruction Problem

As discussed in Section 5.1.5, the problem of evaluating the sample vector $\mathbf{x} = (x_1,...x_M)$ by the vector of the observed signal samples $\mathbf{y} = (y_1,...y_N)$ taking into account the measurement noise is reduced to solving the matrix equation

$$\mathbf{y} = \mathbf{W}\mathbf{x} \tag{5.42}$$

where $\mathbf{W}$ is an $N \times M$ matrix describing the measuring system. This problem is ill-posed, and to solve it requires using regularizing algorithms.

The matrix $\mathbf{W}$ can be inverted by using the singular value decomposition method [3]. The singular value decomposition theorem states that for a matrix $\mathbf{W}$ there exist positive real numbers $\sigma_1 \geq \sigma_2 \ldots \geq \sigma_k > 0$ (so-called singular values of the matrix), an unitary $m \times m$ matrix $\mathbf{U} = (\mathbf{u}_1, \ldots, \mathbf{u}_m)$, and an unitary $n \times n$ matrix $\mathbf{V} = (\mathbf{v}_1, \ldots, \mathbf{v}_n)$, such that the matrix $\mathbf{W}$ can be represented as

$$\mathbf{W} = \mathbf{U}\mathbf{D}\mathbf{V}^H, \quad \mathbf{D} = \mathrm{diag}(\sigma_1, \ldots, \sigma_k) \tag{5.43}$$

The symbol "H" implies Hermitean conjugation. The pseudoinverse Moore–Penrose matrix $\mathbf{W}^+$ corresponding to an $m \times n$ matrix $\mathbf{W}$ of rank k is uniquely determined by the components of a singular value decomposition as a matrix of the form

$$\mathbf{W}^+ = \mathbf{V}\mathbf{D}^{-1}\mathbf{U}^H \tag{5.44}$$

A singular value decomposition of a matrix $\mathbf{W}$ helps one to determine the rank of the matrix by the number of its significant singular values. It was shown [16] that the evaluation of an IR can be improved by reducing the number of nonzero singular values. However, reducing the number of singular values is equivalent to reducing the rank of a matrix, which in turn can lead to loss of the fine structure of the solution. This approach is similar to the regularization methods discussed above. At the same time, the number of singular values to be zeroed is a priori unknown, and therefore the proposed approach cannot be fully computerized. To eliminate this phenomenon, a regularization of singular values was proposed [17] which is implemented as

$$\mathbf{D}_\alpha^{-1} = (\tilde{\sigma}_1^{-1}, \ldots, \tilde{\sigma}_k^{-1}), \quad \tilde{\sigma}_k^{-1} = \frac{\sigma_k}{\sigma_k^2 + \alpha} \tag{5.45}$$

where α is the regularization parameter. In this case, the input signal can be found from the relation

$$\mathbf{x}_\alpha = \mathbf{V}\mathbf{D}_\alpha^{-1}\mathbf{U}^H\mathbf{y} \tag{5.46}$$

To obtain the optimum value of the regularization parameter, the maximum entropy criterion [18] can be applied to the residual signal as

$$\Phi(\alpha = \alpha_{opt}) = \sum_i \Delta Y_i \ln(\Delta Y_i) = \max, \quad \Delta \mathbf{Y} = F(\mathbf{y} - \mathbf{y}_\alpha), \quad \mathbf{y}_\alpha = \mathbf{W}\mathbf{x}_\alpha \tag{5.47}$$

where F denotes a Fourier transform. The maximum of the functional implies that the useful information in the residual signal is minimal and it contains only the noise component. To find the functional maximum, the coordinate descent method was used.

In a numerical experiment [19], it was found that the use of this regularization method yields a decreased amplitude of the reconstructed input signal compared to the original one. The simulation has shown that the original-to-reconstructed signal ratio at their maxima depends on the regularization parameter and on the waveform of the input signal. This dependence was obtained in the numerical experiment as the relation

$$\frac{\max(\mathbf{x})}{\max(\mathbf{x}_\alpha)} = 1 + \beta(\alpha) \tag{5.48}$$

where $\beta(\alpha) = 3.4\alpha^{0.9}$ for a three-lobe pulse, $\beta(\alpha) = 1.8\alpha^{0.9}$ for a bipolar pulse, and $\beta(\alpha) = 0.7\alpha^{0.7}$ for a monopolar pulse.

The proposed approach was compared with a conventional one based on Wiener filtering for which the frequency domain evaluation of the input signal is performed using the relation [6]

$$X_\alpha(\omega) = \frac{W^*(\omega)Y(\omega)}{W(\omega)W^*(\omega) + \alpha} \tag{5.49}$$

where $W(\omega)$ and $Y(\omega)$ are calculated by applying a discrete Fourier transform to the vectors $\mathbf{W}$ and $\mathbf{y}$, respectively. Once the vector $\mathbf{X}_\alpha$ has been estimated, the estimate of $\mathbf{x}_\alpha$ is obtained using the inverse Fourier transform. In this algorithm, the entropy maximum criterion was also used to find the optimum regularization parameter.

5.5.2 Numerical Simulation

In the numerical experiment [19], an additive noise with a normally distributed amplitude was added to the signal. Calculations were carried out to reconstruct the waveform of a monopolar probe pulse at various γ depending on the signal/noise ratio q_p (1.10). Here, γ is the bandwidth ratio of the transfer characteristic to the probe pulse. The bandwidth was determined from the amplitude spectrum at 0.1 of the spectrum peak. The waveform reconstruction error Δ was calculated as the residual between the actual and the reconstructed signal:

$$\Delta = \frac{\sqrt{\sum_i (x_i - x_i^\alpha)^2}}{\sqrt{\sum_i x_i^2}} 100\% \tag{5.50}$$

Figure 5.19(a) presents the numerical simulation results that demonstrate the dependence of Δ on q_p for $\gamma = 1.5$ (curve 1) and 2 (curve 2) for the algorithm based on singular value decomposition. Similar calculations were performed using a conventional frequency-domain reconstruction algorithm (curves 3 and 4, respectively). In some cases, the researchers are interested not in the waveform of the reconstructed signal, but in its peak value. The dependence of the amplitude reconstruction error Δ_a on the signal/noise ratio q_p is shown in Figure 5.19(b). The error Δ_a was calculated by the formula

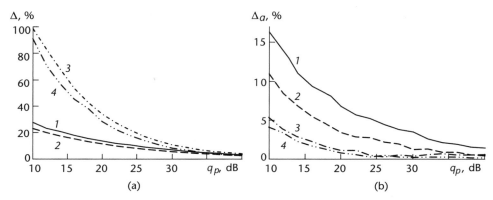

Figure 5.19 The pulse waveform reconstruction error (a) and the pulse amplitude estimation error (b) as functions of the signal/noise ratio. (With permission from Science & Technology Publishing House.)

$$\Delta_a = \left| \frac{\max(\mathbf{x}_\alpha)}{\max(\mathbf{x})} - 1 \right| \cdot 100\% \tag{5.51}$$

The numbers of curves in Figure 5.19(b) correspond to those in Figure 5.19(a). As can be seen from the figures, the algorithm based on singular value decomposition gives smaller waveform reconstruction errors compared with the spectral algorithm. At the same time, the amplitude estimation error Δ_a is smaller for the spectral algorithm. For q_p greater than 35 dB, the algorithms give similar results. Comparison of the errors shows that increasing γ reduces the waveform reconstruction and amplitude estimation errors. It should be noted that for γ less than unity, the information about the input signal is lost and the error increases dramatically.

Figure 5.20 shows the waveform reconstruction error for a monopolar (curve 1), a bipolar (curve 2), and a three-lobe pulse (curve 3) as a function of the signal/noise ratio q_p. Similar results for the amplitude estimation error are given in Figure 5.20(b). The figure shows that the smallest waveform reconstruction and amplitude estimation errors arise when an object is probed with a monopolar pulse.

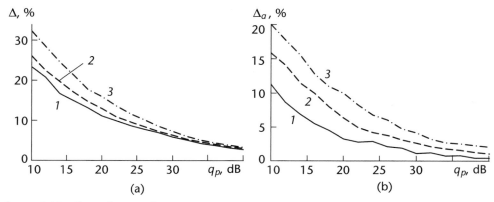

Figure 5.20 The pulse waveform reconstruction error (a) and the pulse amplitude estimation error (b) as functions of the signal/noise ratio for different probe signal waveforms. (With permission from Science & Technology Publishing House.)

Thus, based on the above simulation, we may formulate the following recommendation to perform an experiment aimed at successfully solving the input signal reconstruction problem:

1. The signal/noise ratio should be no less than 30 dB.
2. The bandwidth of the impulse response should be greater than that of the input pulse by a factor of 1.5 or more.
3. The probe pulse should be monopolar.

Implementation of these recommendations can significantly reduce the pulse waveform reconstruction error when the reconstruction is performed using the proposed algorithm.

5.5.3   Experimental Verification of the UWB Pulse Reconstruction Method

In some cases, such as recording UWB radiation with high electromagnetic field strength, the oscilloscope must be located in a screened room to be kept from direct exposure. In this case, the cable between the receiving antenna and the oscilloscope can be as long as 10–15 m. When propagating through a cable of this length, a UWB pulse is damped and distorted.

The proposed method allows one to reconstruct a pulse passed through a long cable and to estimate the signal characteristics at the output of the receiving antenna [19].

First, we need to evaluate the IR of the cable. For this purpose, a low-voltage generator is used. The generator pulse (the pulse at the input of the cable) and the pulse passed through the cable are recorded (Figure 5.21). These records are used to calculate the cable IR (Figure 5.22) by the method presented in Section 5.2. The bandwidth of the generator pulse should be greater than that of the measured signal.

To test the pulse reconstruction procedure, a known low-voltage pulse is used, which is passed through the cable and then recorded. Thereafter, the reconstruction

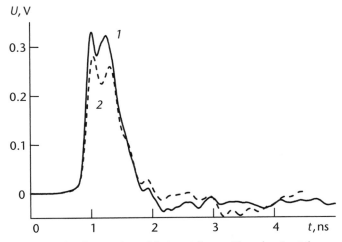

Figure 5.21   Waveforms of pulses at the cable input (curve 1) and output (curve 2). (With permission from Science & Technology Publishing House.)

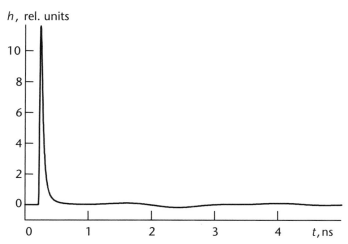

Figure 5.22 The impulse response of the cable. (With permission from Science & Technology Publishing House.)

procedure is performed using the previously obtained IR. The reconstructed signal is compared with the original one. Figure 5.23 shows the waveforms of the input pulse, of the pulse passed through the cable, and of the reconstructed pulse. The standard deviation of the reconstructed pulse waveform from the original one is not over 13%, and the amplitude difference is no more than 6%.

The above method was used to reconstruct high-power UWB radiation pulses passed through a long cable [20, 21]. Figure 5.24 shows the waveforms of the recorded and reconstructed voltage pulses at the output of the receiving TEM antenna. It can be seen that the amplitude of the reconstructed pulse is greater than that of the recorded one by a factor of 1.3.

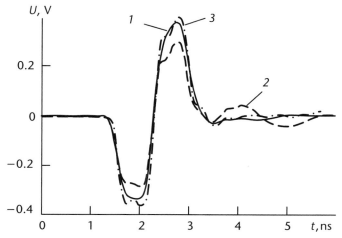

Figure 5.23 Waveforms of the pulses at the cable input (curve 1) and output (curve 2) and of the reconstructed pulse (curve 3). (With permission from Science & Technology Publishing House.)

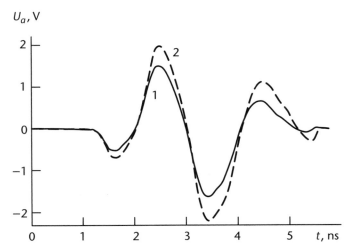

U_a, V

Figure 5.24 Waveforms of the recorded (curve 1) and reconstructed UWB pulses (curve 2). (With permission from Science & Technology Publishing House.)

Thus, the approach described in this section allows one to reconstruct the waveform and estimate the amplitude of pulses passed through transmission channels with linear distortions.

Conclusion

The field of application of methods for estimating the transfer functions and impulse responses of ultrawideband systems considered in this chapter is extremely wide. It includes the problems of investigating the basic electromagnetic characteristics of antennas and antenna arrays, scattering objects, specialized transmission lines, propagation channels, etc. The use of these methods is of particular importance as applied to problems of probing objects located in free space or under an interface between two media. The solution of these problems in real-life situations is strongly hindered by the presence of unavoidable intense noise. Therefore, the development of estimation methods stable to noise is an urgent task. Several methods that satisfy this criterion have been discussed above. They might subsequently be used as a base for the development of more precise methods.

Problems

5.1 Show that the multiplier $A(t)$ in relation (5.2) has the properties of an envelope, namely
  (i) $A(t) \geq s(t)$, i.e., the signal waveform does not intersect the envelope;
  (ii) $A(t) = s(t)$ at the points of contact between the envelope and the signal waveform, i.e., the conjugate signal turns to zero, and
  (iii) at the points of contact between the envelope and the signal waveform, we have not only $A(t) = s(t)$, but also

$$\frac{dA(t)}{dt} = \frac{ds(t)}{dt} \qquad (5.52)$$

that is, the signal and its envelope vary with the same rate.

5.2 Find the Hilbert transform for the signals given by [22]

(a) $s(t) = 1, |t| \le T; s(t) = 0, |t| > T$

(b) $s(t) = 1/(1 + t^2), -\infty < t \le \infty$

(c) $s(t) = \cos 2\pi f_0 t, -\infty < t \le \infty, f_0 = \text{const}$

(d) $s(t) = \sin 2\pi w t / 2\pi w t, -\infty < t \le \infty, w = \text{const}$

5.3 Let a pulsed radio signal $s(t)$ be transmitted through a coaxial cable and the signal $y(t)$ at the cable output be defined by the integral relation

$$\int_0^t s(\tau) h(t - \tau) \, d\tau = y(t) \qquad (5.53)$$

where $h(t)$ is the impulse response of the cable, given by [2]

$$h(t) = \chi(t) \frac{\beta}{\sqrt{4\pi t}} \exp\left(\frac{-\beta^2}{4t}\right) \qquad (5.54)$$

where β is a constant characterizing the cable type and length.

Assuming that the output signal $y(t)$ is known, describe the main steps of the algorithm for estimating the input signal $s(t)$. What measures must be taken to make the estimation stable to noise?

5.4 What is meant by the pole model of a signal?

5.5 What are the advantages of using pole functions invoked at different times?

References

[1]   Kontorovich, M. I., *Operational Calculus and Processes in Electrical Circuits*, Moscow: Sov. Radio, 1975 (in Russian).

[2]   Tikhonov, A. N., and V. Y. Arsenin, *Solution of Ill-Posed Problems*, Washington: Winston & Sons, 1977.

[3]   Marple, S. L., Jr., *Digital Spectral Analysis with Applications*, Englewood Cliffs, NJ: Prentice-Hall, 1987.

[4]   Forsythe, J. E., M. A. Malcolm, and C. B. Moler, *Computer Methods for Mathematical Computations*, Englewood Cliffs, NJ: Prentice-Hall, 1977.

[5]   Voyevodin, V. V., and Yu. A. Kuznetsov, *Matrices and Calculations*, Moscow: Nauka, 1984 (in Russian).

[6]   Tikhonov, A. N., et al., *Numerical Methods for the Solution of Ill-Posed Problems*, New York: Kluwer Academic Publishers, 1995.

[7]   Verlan, A. F., and V. S. Sizikov, *Integral Equations: Methods, Algorithms, and Codes. Reference Book*, Kiev: Naukova Dumka, 1986 (in Russian).

[8]   Nussenzveig, H. M., *Causality and Dispersion Relations*, New York: Academic Press, 1972.

[9] Landau, L. D., and E. M. Lifshitz, *Electrodynamics of Continua (revised edition)*, Oxford: Pergamon Press, 1984.

[10] Koshelev, V. I., V. T. Sarychev, and S. E. Shipilov, "Using the Kramers–Kronig Relation for Estimation of the Impulse Responses of Extra-Wide-Band Systems," *Radiophysics and Quantum Electronics*, Vol. 43, No. 5, 2000, pp. 390–395.

[11] Glebovich, G. V., and I. P. Kovalev, *Wideband Pulsed Signal Transmission Lines*, Moscow: Sov. Radio, 1973 (in Russian).

[12] Koshelev, V. I., V. T. Sarychev, and S. E. Shipilov, "Using the Pole Models of Signals for Estimation of the Impulse Responses of Ultra-Wideband Systems," *Radiophysics and Quantum Electronics*, Vol. 45, No. 1, 2002, pp. 41–47.

[13] Koshelev, V. I., V. T. Sarychev, and S. E. Shipilov, "Object Impulse Response Evaluation for Ultrawideband Application." In *Ultra-Wideband, Short-Pulse Electromagnetics 6*, pp. 63–73, E. L. Mokole, M. Kragalott, and K. R. Gerlach (eds.), New York: Plenum Press, 2003.

[14] Le Goff, M., et al., "UWB Short Pulse Sensor for Target Electromagnetic Backscattering Characterization." In *Ultra-Wideband, Short-Pulse Electromagnetics 4*, pp. 195–202, E Heyman, B. Mandelbaum, and J. Shiloh (eds.), New York: Plenum Press, 1999.

[15] Kennaugh, E. M., and D. L. Moffatt, "Transient and Impulse Response Approximations," *Proc. IEEE.*, Vol. 53, No. 8, 1965, pp. 893–901.

[16] Rahman, J., and T. K. Sarkar, "Deconvolution and Total Least Squares in Finding the Impulse Response of an Electromagnetic System from Measured Data," *IEEE Trans. Antennas Propagat.*, Vol. 43, No. 4, 1995, pp. 416–421.

[17] Yakubov, V. P., *Doppler Ultra-Large-Base Interferometry*, Tomsk: TSU Publishers, 1997 (in Russian).

[18] Sarychev, V. T., "Some Problems of Spectral Estimation," *Radiophysics and Quantum Electronics*, Vol. 40, No. 7, 1997, pp. 618–621.

[19] Shipilov, S. E., et al., "Reconstruction of Ultrawideband Pulses After Their Passage Through Channels with Linear Distortions," *Izv. Vyssh. Uchebn. Zaved., Fiz.*, Vol. 53, No. 9/2, 2010, pp. 78–82.

[20] Gubanov, V. P., et al., "Sources of High-Power Ultrawideband Radiation Pulses with a Single Antenna and a Multielement Array," *Instrum. Exp. Tech.*, Vol. 48, No. 3, 2005, pp. 312–320.

[21] Efremov, A. M., et al., "Generation and Radiation of High-Power Ultrawideband Nanosecond Pulses," *J. Commun. Technol. Electron.*, Vol. 52, No. 7, 2007, pp. 756–764.

[22] Franks, L. E., *Signal Theory*, Englewood Cliffs, NJ: Prentice-Hall, 1969.

CHAPTER 6

Receiving Antennas

Introduction

The transition of radar and radio communication systems to digital technology puts forward the problems related to developing efficient antennas capable of receiving UWB signals with minimum distortion. The degree of distortion of the received signals strongly depends on the properties of the receiving antenna. A receiving antenna is a device converting the energy of free electromagnetic waves into the energy of the currents (mobile charge carriers) occurring in the antenna load.

The efficiency of a receiving antenna is determined by the following parameters:

- the amplitude pattern $f(\theta, \varphi)$, defined as the relation between the amplitude of the load voltage or current and the direction of arrival of a plane electromagnetic wave. The pattern shape is characterized by its boresight, the main-lobe width, and the side-lobe level;
- the phase pattern $\psi(\theta, \varphi)$, defined as the relation between the signal phase and the direction of arrival of a plane electromagnetic wave. The phase pattern is characterized by the presence of a phase center and its position relative to the antenna;
- the polarization pattern $\mathbf{p}_a(\theta, \varphi)$, defined as a unit complex vector whose components depend on the direction of arrival of a plane electromagnetic wave;
- the effective antenna length l_a, defined as the ratio of the electromotive force (EMF) induced in the antenna by the electromagnetic wave arrived at boresight to the electric field strength at the receiving end;
- the antenna impedance Z_a, defined as the ratio of the EMF at the antenna terminals to the short-circuit current;
- the field sensitivity $E_{\min}$, defined as the minimum field strength of the signal that can be received by the antenna, and
- the dynamic range D_a, defined as the maximum-to-minimum amplitude ratio of the received signal for which the signal distortion is within the distortion margin.

In some cases, the following parameters can also be considered essential:

- the reradiation factor K_Σ, defined as the amplitude ratio of the field reradiated by the antenna to the field of the incident wave, and
- the overall dimensions of the antenna.

The effect of a particular parameter on the efficiency of an antenna depends on its purpose, as a UWB receiving antenna may be used as

1. an antenna receiving UWB electromagnetic pulses or UWB signals;
2. an antenna array element in radio systems using UWB signals;
3. a probing device for measuring the field structure in studying the space-time characteristics of UWB electromagnetic pulses, and
4. a detector of UWB radiation.

A receiving UWB antenna should have a high sensitivity, ensure minimum distortion of the received signal, and satisfy the condition for undistorted transmission, that is, have a constant amplitude–frequency response and a linear phase-frequency response over the frequency band of the received signal spectrum.

If a receiving antenna is used as an array element, it should ensure minimum distortion of the received signal, have high sensitivity, and weakly interact with the adjacent elements. In addition, in most cases, an array element should have small dimensions and a pattern of predetermined shape.

An antenna probe should satisfy more stringent requirements, as it must record both the waveform of the received pulse and the spatial characteristics of the electromagnetic field with minimum distortion. In some cases, for instance, in measuring the field structure near the interface between two media or near a source of UWB electromagnetic radiation, the antenna probe must be "invisible"; that is, the field scattered by the antenna must be substantially lower than the recorded field.

An antenna detector should have the greatest possible bandwidth and high field sensitivity, as the spectral power density of UWB radiation is substantially lower than that of narrow-band radiation. In addition, this antenna must also detect an electromagnetic field of any polarization. In this case, it is generally not necessary that the recorded pulse waveform be retained.

To determine the limiting and allowed values for each of the essential parameters of a receiving antenna, it is necessary to investigate the factors affecting the distortion of the spectrum and, hence, the waveform of a received pulse.

6.1 The Transfer Function of a Receiving Antenna

In the simplest case, a receiving antenna can be represented as part of a receive path whose input is exposed to an electromagnetic pulse $E(t)$ and the response is the voltage pulse across the antenna load, $U_a(t)$. In time-domain representation of a signal, the response and the action are related as

$$U_a(t) = \int_0^t E(\tau)h_a(t - \tau)\, d\tau \tag{6.1}$$

where $h_a(t)$ is the impulse response of the antenna. If $h_a(t)$ is known, $E(t)$ can be determined using the deconvolution operation. However, the above parameters of a receiving antenna are uniquely determined only in frequency domain; therefore, difficulties arise in investigating the effect of a parameter on the impulse response of the antenna. In this connection, to investigate the factors affecting the distortion of a received UWB signal, it is expedient to use its spectral representation.

6.1.1   Determination of the Transfer Function of a Receiving Antenna

If each component of the field strength of an electromagnetic pulse is represented as

$$E(t) = \frac{1}{2\pi} \int_{-\infty}^{\infty} E(\omega)\exp(i\omega t)\,d\omega, \quad E(\omega) = \int_{-\infty}^{\infty} E(t)\exp(-i\omega t)\,dt$$

where $E(\omega)$ is the pulse spectrum, the voltage across the load of a receiving antenna will be given by

$$U_a(t) = \frac{1}{2\pi} \int_{-\infty}^{\infty} E(\omega)H_a(\omega)\exp(i\omega t)\,d\omega \tag{6.2}$$

where $H_a(\omega)$, having the meaning of the antenna transfer function, can be represented as $H_a(\omega) = |H_a(\omega)|\exp[i\Phi_a(\omega)]$, where $|H_a(\omega)|$ is the amplitude–frequency response (AFR) and $\Phi_a(\omega)$ is the phase-frequency response (PFR) of the antenna. In this case, for each spectral component of the voltage pulse across the antenna load, the following condition is fulfilled:

$$U_a(\omega) = E(\omega)|H_a(\omega)|\exp\left[i\Phi_a(\omega)\right] \tag{6.3}$$

This condition implies the condition for a signal to be received undistorted (a change in signal strength and a time shift are not considered a distortion) [1]: in the frequency band of the signal spectrum, the AFR must remain constant and the PFR must be a linear function of frequency:

$$|H_a(\omega)| = \text{const}, \quad \Phi_a(\omega) = -t_a\omega, \quad \left(\omega_{\min} < \omega < \omega_{\max}\right) \tag{6.4}$$

Here, t_0 is a constant having the dimension of time; $\omega_{\min}$ and $\omega_{\max}$ are the lower-edge and the higher-edge frequency of the signal spectrum, respectively.

In what follows, we assume that all quantities are complex functions of frequency and omit the symbol (ω) for brevity.

The field strength of an incident wave is a vector quantity; therefore, U_a should be determined by the scalar product

$$U_a = \left(\mathbf{E} \cdot \mathbf{H}_a\right)\exp\left(i\Phi_a\right) \tag{6.5}$$

It follows that in the general case, the transfer function of a receiving antenna is a vector quantity whose component proportion determines the antenna polarization pattern. To account for the effect of the vector nature of the electromagnetic field and of the antenna polarization pattern on the transfer function, we may introduce a polarization field transmission efficiency $\chi = \mathbf{p}_e \cdot \mathbf{p}_a$ [2], which is the scalar product of the unit polarization vector of the incident wave, $\mathbf{p}_e$, by the unit polarization vector of the antenna, $\mathbf{p}_a$.

To calculate χ, the polarization vectors should be represented on a common basis, for instance, in a coordinate system with basis vectors $\mathbf{x}_0$, $\mathbf{y}_0$, and $\mathbf{z}_0$ related to the receiving antenna:

$$\mathbf{p}_e = \mathbf{x}_0 p_{ex} \exp(i\alpha_x) + \mathbf{y}_0 p_{ey} \exp(i\alpha_y) + \mathbf{z}_0 p_{ez} \exp(i\alpha_z), \; |\mathbf{p}_e| = 1 \qquad (6.6)$$

$$\mathbf{p}_a = \mathbf{x}_0 p_{ax} \exp(i\beta_x) + \mathbf{y}_0 p_{ay} \exp(i\beta_y) + \mathbf{z}_0 p_{az} \exp(i\beta_z), \; |\mathbf{p}_a| = 1 \qquad (6.7)$$

In view of this, (6.5) can be rewritten as

$$U_a = |\chi| E_0 |H_a| \exp\left[i(\Phi_a + \delta) \right] \qquad (6.8)$$

where E_0 is the amplitude of the incident wave field and δ is the argument of χ.

A receiving antenna, which can be treated as a source of an EMF ε_a having an impedance Z_a, is generally connected to a receiving (recording) device via a feeder whose input impedance is the impedance of the antenna load, Z_L. The equivalent circuit diagram of a receiving antenna is given in Figure 6.1. The field-induced EMF in the antenna is proportional to the electric field strength at the antenna location. The maximum proportionality factor corresponding to the case that the direction of arrival of the wave coincides with the boresight and the antenna is polarization matched to the incident wave field is termed the effective length of the antenna: $l_a = \varepsilon_{a,\max}/E_0$. As it is not feasible to measure an EMF in the UWB range, the effective antenna length is conventionally implied to be the proportionality factor between the voltage across the feeder input and the electric field strength, $l_e = U_{a,\max}/E_0$, which corresponds to the maximum of the antenna transfer function. If the antenna is matched to the feeder, we have $l_e = l_a/2$.

In the general case, according to the equivalent circuit, the voltage across the antenna load is given by

$$U_a = \frac{\varepsilon_a Z_L}{Z_a + Z_L} = I_a Z_L \qquad (6.9)$$

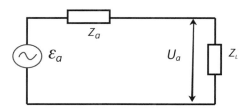

Figure 6.1   Equivalent circuit diagram of a receiving antenna.

where I_a is the load current. This relation indicates that for $Z_L \to \infty$ we have $I_a = 0$ and $U_a = \varepsilon_a = U_0$, where U_0 is the idling voltage, whereas for $Z_L \to 0$ we have $U_a = 0$ and $I_a = \varepsilon_a/Z_a = I_s$, where I_s is the short-circuit current. It follows that the impedance of a receiving antenna is given by

$$Z_a = R_a + iX_a = \frac{U_0}{\mathrm{Re}(I_s) + i\,\mathrm{Im}(I_s)} \tag{6.10}$$

which is in consistence with the Thevenin–Helmholtz theorem [3, 4].

In the general case, ε_a depends on the wave arrival direction and on the antenna polarization pattern. Given the direction of arrival and polarization of the incident wave, to determine the frequency dependence of the antenna transfer function, it suffices to know the frequency dependence of the antenna load current.

If the transfer function of an antenna is known, the received signal waveform $U_a(t)$ is found using an inverse Fourier transform as

$$U_a(t) = \frac{1}{2\pi} \int\limits_{-\infty}^{\infty} |\chi| E_0(\omega)| H_a(\omega)| \exp\left[i\left(\omega t + \Phi_a + \delta\right)\right] d\omega \tag{6.11}$$

We will consider the dependence of the transfer function of a receiving antenna on the antenna characteristics by the example of linear antennas whose parameters can readily be determined if the distribution of the current induced in the antenna by the incident wave is known.

6.1.2   The Current Distribution in the Receiving Wire of an Antenna

An efficient method to calculate the parameters of a linear receiving antenna is based on the superposition of traveling waves. As applied to the receiving wire of an antenna, it is described in the book [5]. Essentially, the method is as follows. Let a straight wire of length L and characteristic impedance ρ_a, arranged along the z-axis, be loaded at the beginning ($z = 0$) by an impedance Z_L, which we consider to be the load impedance of the receiving wire, and at the end ($z = L$) by an impedance Z_2 (Figure 6.2). A plane harmonic linearly polarized wave with an electric field amplitude E_0 is incident on the wire at an angle θ. According to the law of electromagnetic induction, the electric field component tangential to the wire, E_z,

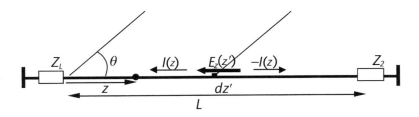

Figure 6.2   Sketch of the receiving wire of an antenna.

induces an EMF $d\varepsilon_z = E_z(z')dz'$ in an elementary section dz' of the wire. The EMF generates two current waves of amplitude $E_z/2\rho_a$: one traveling from the section under consideration to the beginning of the wire and the other traveling to its end. The direction of the current of each wave near the excited element coincides with the direction of the projection $E_z(z')$.

Denote the respective coefficients of wave reflection from the wire beginning and end by R_1 and R_2, and the current wave propagation constant by γ. The amplitude of the current wave traveling to the wire beginning is multiplied at a point z by $\exp[-i\gamma(z' - z)]$, by $R_1\exp[-i\gamma(z' + z)]$ on reflection from Z_L, by $R_1R_2\exp(-i2\gamma L)$ $\exp[-i\gamma(z' - z)]$ on reflection from Z_2, etc. The amplitude of the wave initially traveling to the wire end behaves similarly: at a point z, it is multiplied by $R_2\exp(-i2\gamma L)$ $\exp[i\gamma(z' + z)]$ on reflection from Z_2, by $R_2R_1\exp(-i2\gamma L)\exp[-i\gamma(z' - z)]$ on reflection from Z_L, etc.

The current waves multiply reflected from the wire beginning and end produce at a point z $(z < z')$ a current described by

$$dI = \frac{E_z}{2\rho_a}\Big\{\big[\exp[-i\gamma(z' - z)] + R_1\exp[-i\gamma(z' - z)] + R_1R_2\exp(-i2\gamma L)\exp[-i\gamma(z' - z)]$$

$$+ R_1^2R_2\exp(-i2\gamma L)\exp[-i\gamma(z' + z)] + \ldots\big] + \big[R_2\exp(-i2\gamma L)\exp[i\gamma(z' + z)]$$

$$+ R_2R_1\exp(-i2\gamma L)\exp[-i\gamma(z' - z)] + R_1R_2^2\exp(-i4\gamma L)\exp[i\gamma(z' + z)] \qquad (6.12)$$

$$+ R_1^2R_2^2\exp(-i4\gamma L)\exp[i\gamma(z' - z)] + \ldots\big]\Big\}dz'$$

The bracketed terms can be combined into four geometrical progressions to obtain

$$dI = \frac{E_z}{2\rho_a\big[1 - R_1R_2\exp(-i2\gamma L)\big]}$$

$$\times \Big\{\exp[-i\gamma(z' - z)] + R_1\exp[-i\gamma(z' + z)] \qquad (6.13)$$

$$+ R_2\exp(-i2\gamma L)\big[\exp(i\gamma(z' + z)) + R_1\exp[i\gamma(z' - z)]\big]\Big\}dz'$$

Similar manipulations yield an expression for the current at a point z for $z > z'$:

$$dI' = \frac{E_z}{2\rho_a\big[1 - R_1R_2\exp(-i2\gamma L)\big]}$$

$$\times \Big\{\exp[i\gamma(z' - z)] + R_1\exp[-i\gamma(z' + z)] \qquad (6.14)$$

$$+ R_2\exp(-i2\gamma L)\big[\exp(i\gamma(z' + z)) + R_1\exp[-i\gamma(z' - z)]\big]\Big\}dz'$$

In the above equations, $R_1 = (\rho_a - Z_L)/(\rho_a + Z_L)$ and $R_2 = (\rho_a - Z_2)/(\rho_a + Z_2)$ are the reflection coefficients of the loads Z_L and Z_2, respectively.

The total current at a point z is the sum of the EMF-driven elementary currents (6.13) distributed over the section $z < z' < L$ and of the EMF-driven currents (6.14) distributed over the section $z < z' < z$:

$$I(z) = \int_0^z dI' + \int_z^L dI \tag{6.15}$$

For the wire located in the polarization plane of the arriving plane wave, we have $E_z = E_0 \sin\theta \exp(ikz'\cos\theta)$, and after evaluation of the integrals in (6.15), the function of the current distribution along the wire becomes

$$I(z) = \frac{E_0 \sin\theta}{2i\rho_a \left[\exp(i\gamma L) - R_1 R_2 \exp(-i\gamma L)\right]} \left\{\left[\exp(i\gamma z) + R_1 \exp(-i\gamma z)\right]\right.$$

$$\left[\frac{\exp(i\gamma L)\left\{\exp[i(\xi - \gamma)L] - \exp[i(\xi - \gamma)z]\right\}}{\xi - \gamma} + R_2 \frac{\exp(-i\gamma L)\left\{\exp[i(\xi + \gamma)L] - \exp[i(\xi + \gamma)z]\right\}}{\xi + \gamma}\right]$$

$$\left. + \left\{\exp[i\gamma(L - z)] + R_2 \exp[-i\gamma(L - z)]\right\}\left[\frac{\left\{\exp[i(\xi + \gamma)z] - 1\right\}}{\xi + \gamma} + R_1 \frac{\left\{\exp[i(\xi - \gamma)z] - 1\right\}}{\xi - \gamma}\right]\right\} \tag{6.16}$$

where $\xi = k\cos\theta$, $k = 2\pi/\lambda$, λ is the wavelength. Formula (6.16) allows one to calculate the current distribution for an unbalanced linear receiving antenna (monopole).

For an end-matched wire, we have $R_2 \to 0$. In this case, the current distribution function becomes

$$I(z) = \frac{E_0 \sin\theta}{2i\rho_a} \left\{\left[\exp(i\gamma z) + R_1 \exp(-i\gamma z)\right]\frac{\exp[i(\xi - \gamma)L] - \exp[i(\xi - \gamma)z]}{\xi - \gamma}\right.$$

$$\left. + \exp(-i\gamma z)\left[\frac{\exp[i(\xi + \gamma)z] - 1}{\xi + \gamma}\right] + R_1 \frac{\exp[i(\xi - \gamma)z] - 1}{\xi - \gamma}\right\} \tag{6.17}$$

For an open-end wire, we have $R_2 = -1$, and the current distribution function can be represented as

$$I(z) = \frac{E_0 \gamma \sin\theta}{(Z_L + Z_a)(\gamma^2 - \xi^2)\sin\gamma L} \left\{\left[\exp(i\xi L)\cos\gamma z - \exp(i\xi z)\cos\gamma L - i\frac{\xi}{\gamma}\sin\gamma(L - z)\right]\right.$$

$$\left. + i\frac{Z_L}{\rho_a}\left[\exp(i\xi L)\sin\gamma z - \exp(i\xi z)\sin\gamma L + \sin\gamma(L - z)\right]\right\} \tag{6.18}$$

where $Z_a = -i\rho_a \cot(\gamma L)$ is the impedance of the antenna. In this case, we have $I(L) = 0$, and for a short wire ($L \le 0.5\lambda$), the current distribution pattern is like a standing wave. However, for a long wire, a decaying traveling wave is formed due to backscattering (reradiation) losses. For $L > \lambda$, the current distribution pattern is a superposition of a standing and a traveling wave. The amplitude ratio of these waves depends on the wave arrival direction and on the load impedance Z_L.

A balanced antenna (dipole) is formed by adding one more wire arranged along the negative portion of the z-axis (Figure 6.3). The current distribution in the second wire (arm) of the dipole is calculated by the same method.

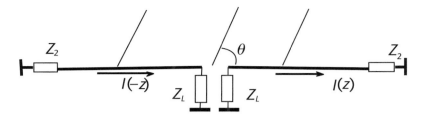

Figure 6.3   Sketch of a balanced receiving antenna.

For instance, for a balanced dipole, we can use formulas (6.13) and (6.14) where it suffices to change z' by $-z'$, take into account that $\cos(\theta + \pi) = -\cos\theta$, and perform integration from $-L$ to 0. The function of the current distribution in the second arm of the dipole is

$$I(-z) = \int_{-L}^{-z} dI + \int_{-z}^{0} dI'$$

$$= \frac{E_0 \sin\theta}{2i\rho_a[\exp(i\gamma L) - R_1 R_2 \exp(-i\gamma L)]}\left\{[\exp(i\gamma z) + R_1 \exp(-i\gamma z)]\left[\frac{\exp(i\gamma L)\{\exp[-i(\xi - \gamma)L] - \exp[-i(\xi - \gamma)]z\}}{\xi + \gamma}\right.\right.$$

$$\left. + R_2 \frac{\exp(-i\gamma L)\{\exp[-i(\xi + \gamma)L] - \exp[-i(\xi + \gamma)z]\}}{\xi - \gamma}\right] + \{\exp[i\gamma(L - z)] + R_2 \exp[-i\gamma(L - z)]\}$$

$$\left.\left[\frac{\{\exp[-i(\xi + \gamma)z] - 1\}}{\xi - \gamma} + R_1 \frac{\{\exp[-i(\xi - \gamma)z] - 1\}}{\xi + \gamma}\right]\right\}$$

$$(6.19)$$

In the general case, if the dipole arms are not collinear or different in length, the currents in the beginnings of the first and second wires, $I_1(0) = I(z)|_{z=0}$ and $I_2(0) = I(-z)|_{-z=0}$, are not equal, and we may separate out an anti-phase (I_o) or an in-phase load current (I_e):

$$I_o = 0.5[I_1(0) + I_2(0)], \quad I_e = 0.5[I_1(0) - I_2(0)] \qquad (6.20)$$

and their corresponding voltages

$$U_o = 2Z_L I_o = Z_L[I_1(0) + I_2(0)], \quad U_e = 0.5Z_L I_e = 0.25Z_L[I_1(0) - I_2(0)] \qquad (6.21)$$

Two options of connecting a load to a receiving dipole for separating out the anti-phase or the in-phase current are sketched in Figure 6.4.

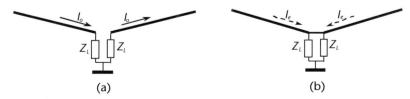

(a)   (b)

Figure 6.4   Sketch showing a receiving dipole and options of connecting a load to separate out an antiphase (a) and an in-phase current (b).

The above method can be easily extended to the case of a curvilinear receiving wire, as the progression sums in (6.13) and (6.14) do not depend on the wire shape. To do this, it suffices to substitute, in these equations, the coordinate along the curved wire for z and the component of the arriving wave electric field vector tangential to the wire, E_t, for E_z. For a ring loop of radius b located in the xOy plane (Figure 6.5), we have $L \to 2\pi b$, $z \to b\delta$, and $dz' \to bd\varphi'$. Then, for $\delta < \varphi'$, we obtain

$$
d\,I = \frac{bE_t}{2\rho_a\left[1 - R_1 R_2 \exp(-i2\gamma L)\right]}
$$
$$
\times\left(\exp\left[-i\gamma b(\varphi' - \delta)\right] + R_1 \exp\left[-i\gamma b(\varphi' + \delta)\right] + R_2 \exp(-i2\gamma L)\right. \qquad (6.22)
$$
$$
\times\left.\left\{\exp\left[i\gamma b(\varphi' - \delta)\right] + R_1 \exp\left[i\gamma b(\varphi' + \delta)\right]\right\}\right)d\varphi'
$$

If $\delta > \varphi'$, then

$$
d\,I' = \frac{bE_t}{2\rho_a\left[1 - R_1 R_2 \exp(-i2\gamma L)\right]}
$$
$$
\left(\exp\left[i\gamma b(\varphi' - \delta)\right] + R_1 \exp\left[-i\gamma b(\varphi' + \delta)\right] + R_2 \exp(-i2\gamma L)\right. \qquad (6.23)
$$
$$
\times\left.\left\{\exp\left[i\gamma b(\varphi' + \delta)\right] + R_1 \exp\left[-i\gamma b(\varphi' - \delta)\right]\right\}\right)d\varphi'
$$

Let the wave be incident along the direction characterized by a meridional angle θ and an azimuthal angle φ, and the electric field vector be parallel to the plane of the loop. For this case, we have

$$
E_t = E_\varphi = E_o \cos(\varphi - \varphi')\exp\left\{ikb\sin\theta\left[\cos(\varphi - \varphi') - \cos\varphi\right]\right\} \qquad (6.24)
$$

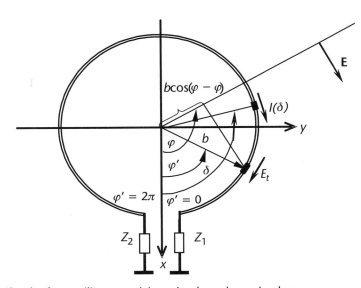

Figure 6.5   Sketch of a curvilinear receiving wire shaped as a ring loop.

and the current at a point δ is given by

$$I(\delta) = \int\limits_0^{\delta} dI' + \int\limits_{\delta}^{2\pi} dI = \frac{bE_o \exp(-ikb\cos\varphi\sin\theta)}{2\rho_a\left[\exp(i\gamma L) - R_1R_2\exp(-i\gamma L)\right]}$$

$$\left[\left\{\exp[i\gamma(L - b\delta)] + R_2\exp[-i\gamma(L - b\delta)]\right\} \times \int\limits_0^{\delta}\exp[ikb\sin\theta\cos(\varphi - \varphi')]\right.$$

$$\times\left\{\left[\exp(i\gamma b\varphi') + R_1\exp(-i\gamma b\varphi')\right]\cos(\varphi - \varphi')d\varphi' + \left[\exp(i\gamma b\delta) + R_1\exp(-i\gamma b\delta)\right]\right\}$$

$$\left.\times\int\limits_{\delta}^{2\pi}\exp[ikb\sin\theta\cos(\varphi - \varphi')]\left\{\exp[i\gamma(L - b\varphi')] + R_2\exp[-i\gamma(L - b\varphi')]\right\}\cos(\varphi - \varphi')\,d\varphi'\right]$$

$$(6.25)$$

The currents at the terminals of a receiving loop, $I(0)$ and $I(2\pi)$, are generally not equal to each other and their ratio depends on the way by which the loop is connected to the load, as shown in Figure 6.6. If the signal voltage is taken from one end of the loop with an impedance Z_1, which, in this case, is the load impedance, this corresponds to an unbalanced connection [Figure 6.6(a)]. For the load impedance Z_2, in addition to the cases $Z_2 = \rho_a$ and $Z_2 \to \infty$, the case $Z_2 = 0$ can be realized, which is difficult to realize for a straight wire.

As the loop ends are spaced closely together and the currents at the terminals are not equal to each other, we may separate out the in-phase current $I_e = 0.5[I(0) + I(2\pi)]$ and the corresponding in-phase voltage $U_e = I_e Z_e$, where $Z_e = Z_1 Z_2/(Z_1 + Z_2)$ is the load impedance, that are realized for the in-phase connection [Figure 6.6(b)]. When the antiphase current $I_o = 0.5[I(0) - I(2\pi)]$ is separated out [Figure 6.6(c)], the corresponding antiphase voltage is given by $U_o = I_o Z_0$, where $Z_0 = (Z_1 + Z_2)$ is the load impedance for the antiphase connection.

Using the traveling wave superposition method to calculate the current distribution in a linear receiving antenna requires knowledge of the antenna electromagnetic parameters, such as characteristic impedance ρ_a and propagation constant γ. The calculation accuracy depends substantially on the accuracy of determination of these parameters.

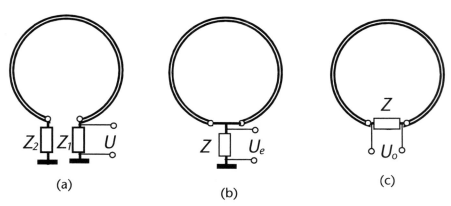

Figure 6.6 Options for connecting a loop to the receive path: unbalanced (a), in-phase (b), and antiphase connection (c)

6.1.3   Electromagnetic Parameters of a Linear Receiving Antenna

A linear receiving wire is a guiding structure for the current waves induced by the field of an arriving wave. This structure is characterized by a characteristic impedance ρ_a and a propagation constant γ. For a guiding structure with a traveling current wave, the stored magnetic and electric energies per unit length, w^m and w^e, are equal to each other. We specify these energies as $w^m = 0.5|I|^2 L'$ and $w^e = 0.5|U|^2 C'$, where L' and C' are the inductance per unit length and the capacitance per unit length, and $|I|$ and $|U|$ are the electric current and potential amplitudes on a section dz, respectively. For the no-loss case, the equality $w^m = w^e$ yields the following formula for the characteristic impedance of a linear wire:

$$\rho = \frac{U}{I} = \sqrt{\frac{L'}{C'}} \tag{6.26}$$

The propagation constant for a traveling current wave is determined from the fact that the energy density w is equal to the period-average power P transferred along the wire divided by the wave propagation velocity: $w = P/v$. In view of $w = w^m + w^e = 2w^e = 2w^m$ and $\gamma = \omega/v$, for a lossless guiding structure we have

$$\gamma = \omega \frac{2w^m}{P} = \omega \frac{|I|^2 L'}{|I|^2 \rho} = \omega \frac{L'}{\rho} = \omega\sqrt{L'C'} = \beta_0 \tag{6.27}$$

Here, β_0 is the wave number of the structure. If the wire is perfectly conducting, the velocity of the current wave is equal to the velocity of light in the medium surrounding the wire, and $\beta_0 = k = \omega\sqrt{\varepsilon\mu}$.

The linear parameters of a wire are difficult to determine accurately. Therefore, there are several formulas for estimating the characteristic impedance of a dipole of length $2L$ or of a single straight wire of length L and radius a:

1. $\rho = 120\ln(2L/a)$, obtained by solving Hallen's equation for the current distribution in a thin symmetric dipole [6]; for a monopole, $\rho = 60\ln(2L/a)$;
2. $\rho = 120[\ln(L/a) - 1]$, obtained for a quasi-static case on the assumption that the total capacitance of a symmetric dipole is uniformly distributed over its length (the Hough method) [7]; for a monopole, $\rho = 60[\ln(L/a) - 1]$;
3. $\rho = 120\ln[\cot(\psi/2)]$, obtained by Schelkunoff [8] for a biconical TEM waveguide, where ψ is the cone angle; for a cone located above a conducting surface, $\rho = 60\ln[\cot(\psi/2)]$;
4. $\rho = 120\ln(L/a) = 120[\ln(2L/a) - 0.69]$, obtained by Stratton and Chu [9] for a thin spheroidal dipole; for a monopole, $\rho = 60[\ln(2L/a) - 0.69]$, and
5. $\rho = 60[\ln(2/ka) - 0.577]$, obtained by Kessenikh [10] in a study of the energy relations for a single long wire with a traveling current wave; King [11] obtained a somewhat different relation for a single long wire: $\rho = 60[\ln(1/ka) - 0.577]$; for a dipole, the Kessenikh formula is $\rho = 120[\ln(2/ka) - 0.577]$.

The Kessenikh and King formulas are different from the other by that the characteristic impedance depends on the ratio a/λ, that is, it is a function of frequency.

Kessenikh [12] obtained an exact solution of the electromagnetic problem of finding the input resistance of an infinite cylindrical wire with a traveling current wave:

$$\rho = \cfrac{60}{\displaystyle\int_0^\infty \cfrac{k\,du}{\left(u^2 + 2ku\right)\left[K_0^2\left(a\sqrt{u^2 + 2ku}\right) + \pi^2 I_0^2\left(a\sqrt{u^2 + 2ku}\right)\right]}} \tag{6.28}$$

where $K_0^2(x)$ and $I_0^2(x)$ are modified Bessel functions. This formula is inconvenient for calculations; however, for $ka < 1.5$, it can be approximated, to within 3%, by the formula

$$\rho = 60\left[\ln(1/ka) - 0.577 - 0.8ka + (12ka)^{0.5}\right] \tag{6.29}$$

Figure 6.7 presents plots of the relation $\rho(a/L)$ for monopoles, calculated by the formulas given under items (1), (2), and (3). It can be seen that for thick wires ($a/L \sim 0.1$), the calculation results can be different even by 100%. For $a/L \sim 1$, all formulas except for (6.28) and (6.29) (the latter approximates the exact formula) become physically meaningless, as the characteristic impedance takes a negative value.

Figure 6.8 presents plots of the relation $\rho(ka)$ calculated by the Kessenikh formula (curve 1), by the King formula (curve 2), and by the approximating formula (curve 3). Herein, the experimental data taken from [10] are given (indicated by dots).

It can be seen that for thin wires ($ka < 0.01$), the best fit is between the exact calculation results and the results obtained by the King formula, whereas for $0.01 < ka < 0.1$, the Kessenikh formula gives more accurate results. In what follows, in the calculations for straight radiators, formula (6.29) will be used.

For a curvilinear conductor of arbitrary shape, no formulas are available to determine its characteristic impedance. However, a formula is known to calculate the inductance of a ring loop of radius b made of a cylindrical wire of radius a [13]:

$$L_l = \mu_0 b\left(\ln\frac{8b}{a} - 2\right) \tag{6.30}$$

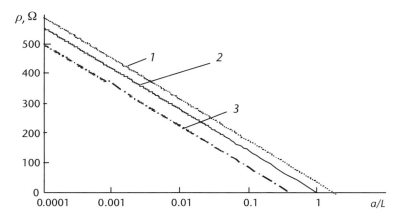

Figure 6.7 The characteristic impedance of a single wire calculated by the formulas $\rho = 60\ln(2L/a)$ (1), $\rho = 60[\ln(2L/a) - 0.69]$ (2), and $\rho = 60[\ln(L/a) - 1]$ (3).

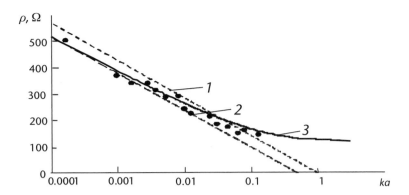

Figure 6.8   The characteristic impedance of a wire with a traveling current wave calculated by the formulas $\rho = 60[\ln(2/ka) - 0.577]$ (1), $\rho = 60[\ln(1/ka) - 0.577]$ (2), and $\rho = 60[\ln(1/ka) - 0.577 - 0.8ka + (12ka)^{0.5}]$ (3).

Assuming that the linear inductance $L' = L_l/2\pi b$ is uniformly distributed along the ring, in view of the relation $\rho = L'v$, where v is the velocity of propagation of a current wave along the wire, for a ring loop in free space, we have

$$\rho_l = 60\left(\ln\frac{8b}{a} - 2\right) \tag{6.31}$$

It should be noted that the above formulas for characteristic impedances were obtained assuming no loss in the radiator. In determining γ and ρ_a, one should take into account that part of the energy of the current wave induced by an incident wave goes into the creation of a secondary field (backward radiation) and into resistive losses in the wire. In this case, the wave propagation constant becomes a complex quantity: $\gamma = \beta - i\alpha$, where $\beta = \omega/v$ is the wave number of the guiding structure and α is the attenuation ratio. If the period-average power transferred by the current wave to a point z is given by

$$P(z) = 0.5|I|^2 \rho\exp(-2\alpha z) \tag{6.32}$$

the difference in power between the points z and $z + \Delta z$ is equal to the power lost on a section of length Δz

$$\Delta P = P(z) - P(z + \Delta z) \tag{6.33}$$

Dividing both sides of (6.33) by Δz and letting Δz tend to zero, we find the lost power per unit length as

$$P' = \lim_{\Delta z\to 0}\frac{P(z) - P(z + \Delta z)}{\Delta z} = -\frac{\partial P(z)}{\partial z} = \alpha|I|^2\rho\exp(-2\alpha z) \tag{6.34}$$

On the other hand, we have $P' = |I|^2 R'\exp(-2\alpha z)$, where $R' = R'_\Omega + R'_\Sigma$ is the linear loss resistance of the wire including the ohmic loss resistance R'_Ω and back-scattering loss resistance R'_Σ per unit length. Thus, we have

$$\alpha = -\frac{R'}{2\rho} \tag{6.35}$$

The linear ohmic loss resistance, taking into account the skin effect, is equal to the ratio of the surface resistance of the wire, R'_S, to the perimeter of its cross section, p:

$$R'_\Omega = \frac{R'_S}{p} = \frac{1}{p}\sqrt{\frac{\omega\mu_0}{2\sigma}} \tag{6.36}$$

where σ is the conductivity of the wire material.

In calculating R'_Σ, we proceed from that the radiated power is equal to the power lost by the current wave due to backscattering in the receiving wire:

$$P_\Sigma = \frac{1}{2}\int_S \frac{|\mathbf{E}|^2}{120\pi}\,ds = \frac{1}{2}\int_0^L |I(z)|^2\,R'_\Sigma\,dz \tag{6.37}$$

Assuming that R'_Σ is uniformly distributed along the wire, we have

$$R'_\Sigma = \frac{\int_S |\mathbf{E}|^2\,ds}{\int_0^L |I(z)|^2\,dz} = \frac{k^2 120\pi \int_0^\pi \left|\int_0^L I(z)\exp(i\xi z)\,dz\right|^2 \sin^3\theta\,d\theta}{8\pi \int_0^L |I(z)|^2\,dz} \tag{6.38}$$

From (6.38), it can be seen that R'_Σ depends on the current distribution, which, in turn, depends on R'_Σ. Therefore, in calculating R'_Σ, one has to use an iterative technique, taking for the initial approximation $\alpha = (R'_\Sigma + R'_\Omega)/2\rho$, where $R'_\Sigma = R_\Sigma/L$ is the linear backscattering loss resistance. For a radiator with a traveling current wave, we have [14]

$$R_\Sigma = 60\left[\ln(2kL) - \text{Ci}(2kL) + \frac{\sin(2kL)}{2kL} - 0.423\right] \tag{6.39}$$

The iterative process is rapidly converging, so that two or three iterations are sufficient to calculate R'_Σ.

The propagation constant of a traveling current wave is determined proceeding from that the energy density w is equal to the period-average power P transferred by the current wave divided by the wave propagation velocity $w = P/v$. In view of $w^m + w^e = 2w^e = 2w^m$ and $\gamma = \omega/v$, for a lossless guiding structure we have

$$\gamma = \omega\frac{2w^m}{P} = \omega\frac{|I|^2\,L'}{|I|^2\,\rho} = \omega\frac{L'}{\rho} = \omega\sqrt{L'C'} = \beta_0 \tag{6.40}$$

where $\beta_0 = k$ is the wave number of the structure. The losses can be taken into account by putting L' and C' as complex quantities:

$$\tilde{L} = L'\left(1 - i\frac{R'}{\omega L'}\right), \quad \tilde{C} = C'\left(1 - i\frac{G'}{\omega C'}\right) \tag{6.41}$$

where G' is the linear conductivity of the medium surrounding the guiding structure. If the medium is lossless, we have $G' = 0$, and, in view of the ohmic and backscattering losses in the wire, we obtain

$$\gamma = \omega\sqrt{L'C'\left(1 - i\frac{R'_\Omega + R'_\Sigma}{\omega L'}\right)} = \beta_0\sqrt{1 - i\frac{R'_\Omega + R'_\Sigma}{\rho\beta_0}} = k\sqrt{1 - i\frac{2\alpha}{k}}. \tag{6.42}$$

$$\rho_a = \sqrt{\frac{L'C'\left(1 - i\dfrac{R'_\Omega + R'_\Sigma}{\omega L'}\right)}{C'}} = \rho\sqrt{1 - i\frac{R'_\Omega + R'_\Sigma}{\rho\beta_0}} = \rho\sqrt{1 - i\frac{2\alpha}{k}} \tag{6.43}$$

Thus, even in the absence of ohmic losses, part of the field energy is lost due to backscattering. As a result, the characteristic impedance of the receiving wire and the current wave propagation constant become complex and frequency-dependent.

6.1.4   The Transfer Function of a Straight Receiving Wire

To determine the transfer function of a receiving wire, it is necessary to know the voltage across the antenna load, $U_a = I(0)Z_L$, where $I(0)$ is the current at the beginning of the wire. To determine this current, it suffices to put $z = 0$ in (6.18); then the voltage across the load Z_L is determined as

$$U_a = \frac{E_0 Z_L \sin\theta (1 + R_1)}{2\rho_a [\exp(i\gamma L) - R_1 R_2 \exp(-i\gamma L)]}$$
$$\times \int_0^L \{\exp(i\gamma L)\exp[i(\xi - \gamma)z'] + R_2 \exp(-i\gamma L)\exp[i(\xi + \gamma)z']\}\, dz' \tag{6.44}$$

For a straight wire located in the polarization plane of the incident wave and oriented along the z-axis, we obtain the following expression for the transfer function:

$$H_a = \frac{Z_L \sin\theta (1 + R_1)}{2\rho_a [\exp(i\gamma L) - R_1 R_2 \exp(-i\gamma L)]}$$
$$\times \int_0^L [\exp(i\gamma L)\exp(-i\gamma z') + R_2 \exp(-i\gamma L)\exp(-i\gamma z')]\exp(-i\xi z')\, dz' \tag{6.45}$$

The integral in (6.45) can be easily evaluated, and, in view of that for a traveling current wave

$$R_1 = \frac{(\rho_a - Z_L)}{(\rho_a + Z_L)}, \quad R_2 = \frac{(\rho_a - Z_2)}{(\rho_a + Z_2)} \tag{6.46}$$

we finally obtain

$$H_a = \frac{Z_L \sin\theta \cdot \left[(\rho_a\xi + Z_2\gamma)[\exp(i\xi L) - \cos\gamma L] - i(\rho_a\gamma + Z_2\xi)\sin\gamma L\right]}{i(\xi^2 - \gamma^2) \cdot (\rho_a\cos\gamma L + iZ_2\sin\gamma L) \cdot (Z_L + Z_a)} \tag{6.47}$$

where $Z_a\rho_a[Z_2\cos\gamma L + i\rho_a\sin\gamma L]/[\rho_a\cos\gamma L + iZ_2\sin\gamma L]$ has the meaning of the intrinsic impedance of the receiving antenna. Equation (6.47) indicates that the transfer function of a straight receiving wire depends on the direction of wave arrival, wire length L, impedances Z_L and Z_2, propagation constant γ, and characteristic impedance ρ_a.

The transfer function of an end-matched receiving wire can be represented as

$$H_a = \frac{-iZ_L}{Z_L + \rho_a} \frac{\sin[0.5L(k\cos\theta - \gamma)]}{k\cos\theta - \gamma}\sin\theta \cdot \exp[i0.5L(k\cos\theta - \gamma)] \tag{6.48}$$

Figure 6.9 shows the shape of the pattern of an end-matched straight receiving wire ($Z_L = Z_2 = \rho_a$, $a = 0.05L$) versus the wire electric length.

The frequency dependence of the pattern shape has the result that the transfer function depends on the direction of wave arrival. This dependence for $L = 0.5$ m, $Z_L = \rho_a$, and $a = 0.05L$ is plotted in Figure 6.10. It can be seen that for an oblique incidence of the wave, the frequency dependence of the transfer function is less

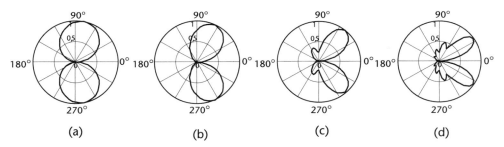

(a)   (b)   (c)   (d)

Figure 6.9 Normalized patterns of receiving wires of different length: $L = 0.25\lambda$ (a), 0.5λ (b), λ (c), and 2λ (d).

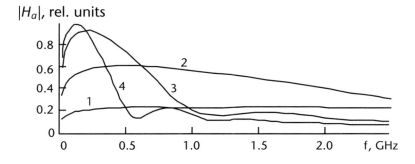

Figure 6.10 The modulus of the transfer function of a receiving wire of length 0.5 m versus the direction of wave arrival ($Z_L = Z_2 = \rho_a$): $\theta = 10°$ (curve 1), $30°$ (curve 2), $50°$ (curve 3), and $90°$ (curve 4).

pronounced, and at small angles of wave arrival, measured from the wire axis, H_a remains constant in a frequency range with a coverage ratio of 6 to 1 or more.

The transfer function of an open-ended wire ($Z_2 \to \infty$) is obtained from (6.45), and it can be represented as

$$H_a = \frac{Z_L \sin\theta}{(Z_L + Z_a)\sin\gamma L} \frac{\gamma[\cos(kL\cos\theta) - \cos\gamma L] + i[\gamma\sin(kL\cos\theta) - k\cos\theta\sin\gamma L]}{\gamma^2 - k^2\cos^2\theta} \quad (6.49)$$

In this case, the antenna impedance $Z_a = -i\rho_a\cot\gamma L$ and the pattern shape, described by the function

$$f(\theta) = \frac{\gamma\sin\theta[\cos(kL\cos\theta) - \cos\gamma L] + i[\gamma\sin(kL\cos\theta) - k\cos\theta\sin\gamma L]}{\gamma^2 - k^2\cos^2\theta} \quad (6.50)$$

substantially depend on frequency.

Figure 6.11 shows the patterns of an open-ended wire of different length for $Z_L = \rho_a$ and $a = 0.02L$.

Figure 6.12 presents the normalized modulus of the transfer function of a wire of length 0.5 m versus the direction of wave arrival.

The modulus of the normalized transfer function of a short monopole ($L = 0.05$ m, $a = 0.1L$) is plotted versus the load impedance at $\theta = 90°$ in Figure 6.13. The plots are similar to those obtained for a short dipole of length $2L$.

The above results show that if the wire length is comparable to or greater than the wavelength of the incident wave, the form of the transfer function is weakly

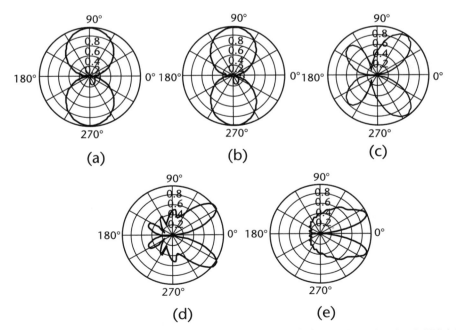

Figure 6.11   Pattern shape versus wire length for an open-ended receiving wire: $L = 0.25\lambda$ (a), 0.5λ (b), λ (c), 2λ (d), and 4λ (e).

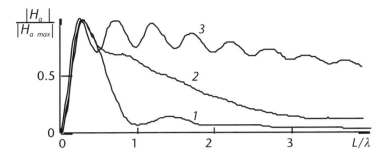

Figure 6.12 The normalized modulus of the transfer function of a wire versus the direction of wave arrival ($L = 0.5$ m, $Z_L = \rho_a$, and $a = 0.02L$): $\theta = 90°$ (curve 1), $45°$ (curve 2), and $22°$ (curve 3).

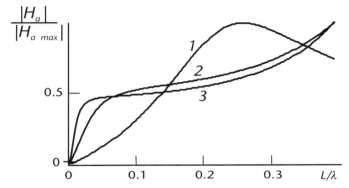

Figure 6.13 The transfer function of a short wire ($L = 0.05$ m) versus load impedance: $Z_L = 0.5\rho_a$ (curve 1), $5\rho_a$ (curve 2), and $20\rho_a$ (curve 3).

affected by the load impedance, but noticeably affected by the direction of arrival of the wave. If the wire length is much smaller than the wavelength, the transfer function weakly depends on the wave arrival direction, but substantially depends on the load impedance.

6.1.5 The Transfer Function of a Curvilinear Receiving Wire

The transfer function of a receiving loop depends on the way in which it is connected to the receive path. The possible ways to connect the loop are shown in Figure 6.6. The currents at the loop terminals are determined by the equations

$$I(0) = \frac{bE_o \exp(-ikb\cos\varphi\sin\theta)(1 + R_1)}{2\rho_l[\exp(i\gamma L) - R_1 R_2 \exp(-i\gamma L)]} \int_0^{2\pi} \exp[ikb\sin\theta\cos(\varphi - \varphi')]$$
$$\times \{\exp[i\gamma(L - b\varphi')] + R_2 \exp[-i\gamma(L - b\varphi')]\}\cos(\varphi - \varphi')d\varphi' \tag{6.51}$$

$$I(2\pi) = \frac{bE_o \exp(-ikb\cos\varphi\sin\theta)(1 + R_2)}{2\rho_l[\exp(i\gamma L) - R_1 R_2 \exp(-i\gamma L)]} \int_0^{2\pi} \exp[ikb\sin\theta\cos(\varphi - \varphi')]$$
$$\times [\exp(i\gamma b\varphi') + R_1 \exp(-i\gamma b\varphi')]\cos(\varphi - \varphi')d\varphi' \tag{6.52}$$

The evaluation of the integrals in (6.52) involves some difficulties. Therefore, we restrict our consideration to the characteristics of small loops for which the condition $kb \ll 1$ is fulfilled. In this case, the currents at the loop terminals are determined as

$$I(0) = \frac{bE_o \exp(-ikb\cos\varphi\sin\theta)\left[i\gamma b\left(\rho_l + iZ_2 \tan\frac{\gamma L}{2}\right)S_1(\varphi,\theta) + \left(Z_2 + i\rho_l \tan\frac{\gamma L}{2}\right)S_2(\varphi,\theta)\right]}{\left(\rho_l^2 + Z_1 Z_2\right)\left[1 - i\frac{\rho_l(Z_1 + Z_2)}{\rho_l^2 + Z_1 Z_2}\cot\gamma L\right]}$$

(6.53)

$$I(2\pi) = \frac{bE_o \exp(-ikb\cos\varphi\sin\theta)\left[i\gamma b\left(\rho_l + iZ_2 \tan\frac{\gamma L}{2}\right)S_1(\varphi,\theta) - \left(Z_2 + i\rho_l \tan\frac{\gamma L}{2}\right)S_2(\varphi,\theta)\right]}{\left(\rho_l^2 + Z_1 Z_2\right)\left[1 - i\frac{\rho_L(Z_1 + Z_2)}{\rho_l^2 + Z_1 Z_2}\cot\gamma L\right]}$$

(6.54)

where $S_1(\varphi,\theta) \approx -i(\sin\theta + 2\gamma b\cos\varphi)/2\gamma b$ and $S_2(\varphi,\theta) \approx \sin\varphi(1 + 0.5ikb\cos\varphi\sin\theta)$ are the functions describing the pattern of the receiving loop.

By the transfer function of a receiving loop we imply the quantity $H_L = I_L Z_L/E_0$, where I_L is the current in the antenna load and Z_L is the load impedance. Depending on the way of connection of the loop in the receive path, we have the following formulas for the transfer function:

- for an in-phase connection [Figure 6.6(b)] with $Z_1 = Z_2 = Z$, the load impedance is $Z_e = 0.5Z$, and then

$$H_{Le} = \frac{bZ_e \exp(-ikb\cos\varphi\sin\theta)}{Z_e - i\rho_l \cot\frac{\gamma L}{2}}\sin\varphi\left(1 + i\frac{kb}{2}\cos\varphi\sin\theta\right)$$

(6.55)

- for an antiphase connection [Figure 6.6(c)] with $Z_1 = Z_2 = Z$, the load impedance is $Z_0 = 2Z$, and then

$$H_{Lo} = \frac{bZ_o \tan\frac{\gamma L}{2}\exp(-ikb\cos\varphi\sin\theta)}{Z_o + i\rho_l \tan\frac{\gamma L}{2}}(\sin\theta + i2\gamma b\cos\varphi)$$

(6.56)

- for an unbalanced connection [Figure 6.6(a)] with an arbitrary Z_2, we assume that the load impedance is $Z_L = Z_1$. In this case, we have

$$H_L = \frac{bZ_1 \exp(-ikb\cos\varphi\sin\theta)}{\left(\rho_l^2 + Z_1 Z_2\right)\left[1 - i\frac{\rho_l(Z_1 + Z_2)}{\rho_l^2 + Z_1 Z_2}\cot\gamma L\right]}\left[0.5\left(\rho_l + iZ_2 \tan\frac{\gamma L}{2}\right)(\sin\theta + i2\gamma b\cos\varphi)\right.$$

$$\left. + \left(Z_2 - i\rho_l \tan\frac{\gamma L}{2}\right)\left(1 + i\frac{kb}{2}\cos\varphi\sin\theta\right)\sin\varphi\right]$$

(6.57)

Thus, when connected in phase to the receive path, a small receiving loop ($kb \ll 1$) is similar in characteristics to a monopole of effective length b and intrinsic impedance $Z_{Le} = -i\rho_l \cot(\gamma L/2)$, loaded by an impedance $Z_{Le} = 0.5Z$.

When the connection is antiphase, the effective length of a small loop is $b\tan(\gamma \pi b) \approx k\pi b^2$, the intrinsic impedance $Z_{L0} = i\rho_l\tan(\gamma L/2)$, and the load impedance $Z_{L0} = 2z$; that is, the characteristics of the receiving loop are the same as those of an elementary current loop.

In the case of an unbalanced connection, when the signal is taken off the impedance Z_1, the directivity and impedance of the loop substantially depend on Z_2. In the loop plane ($\theta = 90°$), the pattern can be described by the function

$$f(\varphi) \approx 1 + \frac{2}{A}\sin\varphi + i2\gamma b\cos\varphi\left(1 + \frac{1}{A}\sin\varphi\right) \qquad (6.58)$$

and in the orthogonal plane ($\varphi = 90°$), by the function $f(\theta) \approx 1 + (A\sin\theta)/2$, where $A = \left(\rho_l + iZ_2\tan(\gamma L/2)\right) / \left(Z_2 + i\rho_l\tan(\gamma L/2)\right)$.

The above equations show that the pattern function can vary over wide limits on varying A. For $A \ll 1$, we have $f(\varphi) \approx \sin\varphi$ and; that is, the pattern of the loop coincides with that of a monopole. For $A \gg 1$, we have $f(\varphi) \approx 1$ and $f(\theta) \approx \sin\theta$, that is, the pattern of a current loop. When $A \approx 2$, the pattern of a small loop has a shape similar to a cardioid. The pattern shapes in the plane of a receiving loop, calculated for an unbalanced connection using (6.57) for different values of ka are given in Figure 6.14 (A and B).

Figure 6.15 presents the pattern of a ring loop for $Z_2 = \rho_l$. The dependence of the pattern shape on the impedance Z_2 for $kb = 0.125$ is illustrated in Figure 6.16.

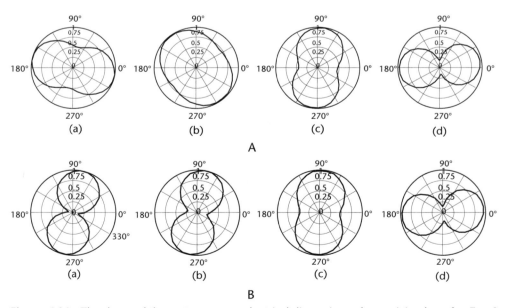

Figure 6.14 The shape of the pattern versus electrical dimensions of a receiving loop for $Z_2 = 0$ (A) and $Z_2 \to \infty$ (B); $Z_1 = \rho_l$, $a = 0.1b$; $kb = 0.063$ (a), 0.125 (b), 0.25 (c), and 0.5 (d).

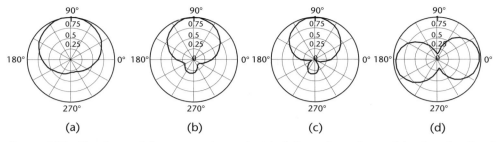

Figure 6.15   The shape of the pattern versus electrical dimensions of a receiving loop for $Z_2 = Z_1$ = ρ_l and $a = 0.1b$: $kb = 0.063$ (a), 0.125 (b), 0.25 (c), and 0.5 (d).

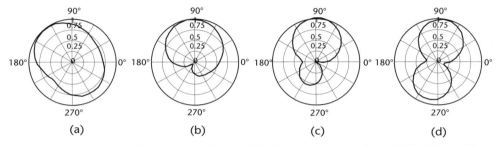

Figure 6.16   The shape of the pattern of a receiving loop versus impedance Z_2 for $Z_1 = \rho_l$, $kb = 0.125$, and $a = 0.1b$: $Z_2 = 0.1\rho_l$ (a), $0.5\rho_l$ (b), $2\rho_l$ (c), and $10\rho_l$ (d).

The calculation results suggest that an unbalanced connection of a loop antenna to the receive path should provide high noise immunity of the path in a wide frequency range due to an increase in the directivity of the antenna.

The transfer functions of a receiving loop connected in phase and in anti-phase to the receive path have different frequency dependences. Figure 6.17(a) presents the normalized modulus of the transfer function of a receiving loop of diameter 36 mm with **E** operating in an anti-phase mode as a function of the load impedance. Similar plots for the loop operating in an in-phase mode are given in Figure 6.17(b).

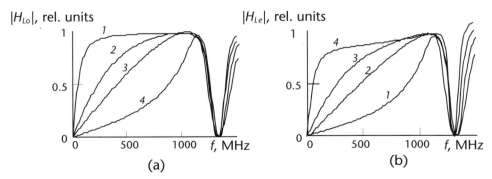

Figure 6.17   The modulus of the transfer function of a receiving loop of diameter 36 mm versus load impedance for antiphase (a) and in-phase operation (b): $Z_L = 0.1\rho_l$ (curve 1), $0.5\rho_l$ (curve 2), $1.0\rho_l$ (curve 3), and $10\rho_l$ (curve 4).

The calculation results suggest that with the load impedance close to the characteristic impedance of the receiving loop, the loop directivity should increase, however with the transfer function of the loop rather strongly depending on frequency. A weak frequency dependence of the transfer function should be provided only with $Z \ll \rho_l$ in an antiphase mode or with $Z \gg \rho_l$ in an in-phase mode. However, the directivity of the receiving loop antenna will decrease in both cases.

6.2 Distortion of Ultrawideband Electromagnetic Pulses by a Receiving Antenna

In studying the effect of the characteristics of a receiving antenna on the waveform of a recorded electromagnetic pulse, it should be borne in mind that an electromagnetic pulse propagating in space as a free wave must satisfy the equilibrium condition and the function $E(t)$ describing the pulse waveform should have the first and second derivatives not becoming zero simultaneously.

If a perfectly transmitting antenna is excited by a monopolar current pulse $I_a(t)$ whose waveform is a gaussoid,

$$I_a(t) = I_o \exp\left[-\frac{(t - t_0)^2}{\tau_p^2}\right] \tag{6.59}$$

where t_0 is a fixed time and τ_p is the pulse duration, the waveform of the electromagnetic pulse is determined by the time derivative of the pulsed current as

$$E_1(t) : \frac{dI_a(t)}{dt} = \frac{2(t - t_0)^2}{\tau_p^2} I_o \exp\left[-\frac{(t - t_0)^2}{\tau_p^2}\right] \tag{6.60}$$

A bipolar pulse satisfying the equilibrium condition can be described by other formulas, such as

$$E_1(t) = \left[\exp\left(-\frac{(t - t_0 + 0.5\tau_p)^2}{\tau_p^2}\right) - \exp\left(-\frac{(t - t_0 - 0.5\tau_p)^2}{\tau_p^2}\right)\right] \tag{6.61}$$

or

$$E_1(t) = -\sin\left(\frac{(t - t_0)}{2\tau_p}\right) \exp\left[\frac{-(t - t_0)^2}{\tau_p^2}\right] \tag{6.62}$$

For a bipolar pulse applied to the input of a transmitting antenna, the waveform of the radiated electromagnetic pulse can be described by the formulas

$$E_2(t): \frac{d}{dt}\left[\frac{(t-t_0)}{\tau_p^2}\exp\left[\frac{-(t-t_0)^2}{\tau_p^2}\right]\right] \text{ or } E_2(t):\left[C-\cos\left(\frac{(t-t_0)}{\tau_p^2}\right)\right]\exp\left[\frac{-(t-t_0)^2}{\tau_p^2}\right] \quad (6.63)$$

where C is a constant having a certain value at which the equilibrium condition is fulfilled.

The spectrum of a bipolar pulse occupies a wider frequency band. Therefore, in studying the factors affecting the distortion of received electromagnetic pulses, we will use a bipolar pulse for the reference electromagnetic pulse.

The distortion of a signal is convenient to estimate by the RMSD σ of the waveform of the voltage induced by the received electromagnetic pulse across the antenna load, $U_a(t)$, from the waveform of the electromagnetic pulse $E(t)$. If the time interval occupied by the pulse is broken into N samples, we have

$$\sigma = \sqrt{\frac{\sum_n\left[E(t_n) - MU_a(t-\Delta t)\right]^2}{\sum_n E^2(t_n)}} \quad (6.64)$$

where n is an integer number corresponding to the order number of a sample, $M = [\max E(t) - \min E(t)]/[\max U_a(t) - \min U_a(t)]$ is a scale factor, and $\Delta t = t - t_n$ is the time delay of the received signal. The quantities M and Δt are introduced because a change in signal strength and a time shift do not distort the pulse waveform. We assume that a signal is received with small distortion if $\sigma \le 0.2$.

6.2.1   Receiving of Ultrawideband Electromagnetic Pulses by a Dipole

The transfer function of a single receiving wire (monopole) depending on the wire dimensions, on the load impedance, and on the direction of arrival of an electromagnetic wave was investigated in detail in Section 6.1. Analysis of the obtained results shows that the transfer function of a long wire with $L \ge c\tau_p$, where τ_p is the pulse duration, weakly depends on the load impedance, but substantially depends on the direction of wave arrival. For a short wire, such that $L \ll c\tau_p$, the form of the transfer function weakly depends on the wave arrival direction, but strongly depends on the load impedance.

The waveform of the voltage pulse across the load of a receiving wire is found using an inverse Fourier transform. If the wire is polarization-matched to the field of the incident wave, we have

$$U_a(t) = \frac{1}{2\pi}\int_{-\infty}^{\infty} E_0(\omega)|H_a(\omega)|\exp\left[i(\omega t + \Phi_a)\right]d\omega \quad (6.65)$$

Figure 6.18 shows the waveform of the voltage pulse across the load of a long receiving wire as a function of the direction of arrival of an electromagnetic pulse, whose waveform is depicted by a dashed line. When the load impedance is varied, the RMSD σ for each direction is less than 0.1.

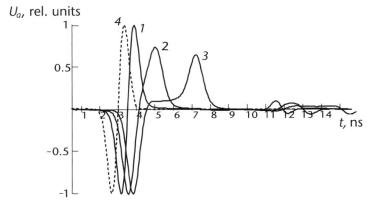

Figure 6.18 The waveform of the voltage pulse across the load of a long wire ($L = 2c\tau_p$) versus wave arrival direction: $\theta = 20°$ (curve 1), $50°$ (curve 2), $80°$ (curve 3), and the reference electromagnetic pulse waveform (curve 4).

The dependence of the pulse waveform on the load impedance of a short wire is plotted in Figure 6.16. To distinguish between the curves corresponding to the waveforms of received pulses, they are depicted with a time shift. Within the pattern width, σ for each value of Z_L is no more than 0.1.

For UWB pulses received by a dipole, the pulse waveform not only depends on the dipole length, wave arrival direction, and load impedance but also is affected by some additional factors. This is because the current distributions in the dipole arms may differ significantly from each other. As shown in Section 6.2, this may have the result that the load of the antenna will carry an anti-phase current I_o and an in-phase current I_e. These currents have different dependences on frequency and wave arrival direction, which results in an additional distortion of the received pulse waveform.

Typically, to transfer a signal from a dipole to a receiver, a feeder is used, and its influence should be taken into account in investigating the directivity of the dipole,

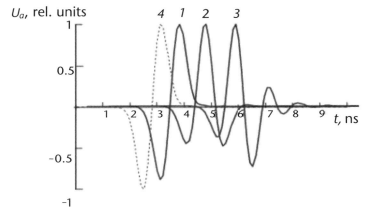

Figure 6.19 The waveform of the voltage pulse across the load of a short wire ($L = 0.1c\tau_p$) versus load impedance: $Z_L = 100\rho_a$ (curve 1), $1.0\rho_a$ (curve 2), and $0.1\rho_a$ (curve 3), and the waveform of the reference electromagnetic pulse (curve 4).

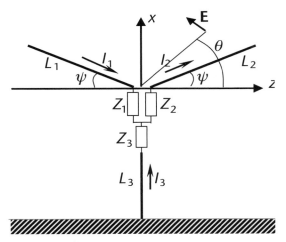

Figure 6.20   Equivalent circuit of a dipole including a feeder.

as the currents induced in the feeder by the vertical component of the field flow in the dipole arms and affect the current distribution. To account for the influence of the feeder, a three-wire model of the receiving antenna (Figure 6.20) should be used, which is more realistic than the two-wire model considered above.

It can be assumed that in the general case, the dipole arms are not collinear and set at an angle ψ to the dipole axis. In this case, the right-arm current is given by $I_1 \sim E_0\sin(\theta - \psi)$ and the left-arm current by $I_2 \sim E_0\sin(\theta + \psi)$. As a consequence, the currents $I_1(0)$ and $I_2(0)$ of a balanced dipole will be equal to each other only in the case of a normally incident wave ($\theta = \pi/2$). If the feeder is located normal to the dipole axis, a normally incident wave does not induce a current in the feeder braid; that is, we have $I_3 \to 0$. If $\theta \neq \pi/2$, a current is induced in the feeder braid, so that $|I_1(0) - I_2(0)| \neq 0$. As a result, an in-phase current I_e occurs in the load whose amplitude can be considerably greater than that of the anti-phase current I_o, leading to a distortion of the waveform of the received electromagnetic pulse. Figure 6.21 exemplifies the ratio I_e/I_o as a function of the direction of arrival of electromagnetic pulses calculated not taking into account the influence of the feeder ($I_3 = 0$) for V-shaped dipoles of different length with $\psi = 60°$.

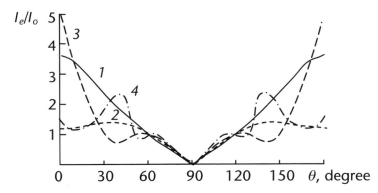

Figure 6.21   The ratio I_e/I_o versus the direction of electromagnetic pulse arrival for V-shaped dipoles with $\psi = 60°$: $L = 0.25\lambda$ (curve 1), 0.5λ (curve 2), 1.0λ (curve 3), and 1.5λ (curve 4).

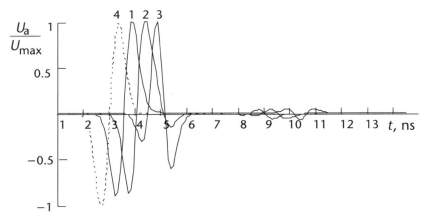

Figure 6.22 The waveform of a received pulse versus I_e/I_o: $I_e/I_o = 0$ (curve 1), 0.3 (curve 2), 3.0 (curve 3), and the waveform of the reference electromagnetic pulse (curve 4).

The plots of Figure 6.21 suggest that the effect of the in-phase current can be neglected only for θ close to 90°, but even within the pattern width, the in-phase and antiphase current amplitudes becomes comparable. If the influence of the feeder is taken into account, I_e may be substantially greater than I_o.

Figure 6.22 presents the waveform of a received pulse as a function of the ratio I_e/I_o. The waveform of the reference electromagnetic pulse is shown by a dashed line.

The RMSD of the received pulse waveform from the reference electromagnetic pulse waveform is 0.17 for $I_e = 0$ and over 0.3 for $I_e = 0.3I_o$. For $I_e = 3I_o$ and greater, the distortion of the received pulse waveform becomes substantial ($\sigma > 1$).

Thus, when using a dipole antenna whose signal is transferred to a receiver via an unbalanced feeder, it is necessary to connect an ultrawideband balun between the antenna and the feeder to attenuate the in-phase current by no less than 20 dB.

Analysis of the results shows that the receiving of ultrawideband electromagnetic pulses with low distortion by a dipole is possible

- with a long dipole, $L > 1.5c\tau_p$, having noncollinear arms the angle between which is no greater than 60°, and
- with a short dipole, $L < 0.2c\tau_p$, operated in a mismatched mode with the load impedance $R_L \gg \rho_a$.

In both cases, within the pattern width, the pulse waveform remains almost unchanged if a good balancing is provided between the dipole and the feeder.

6.2.2 Receiving of Ultrawideband Electromagnetic Pulses by a Loop Antenna

In Section 6.1.5, expressions have been derived for the transfer functions of a loop antenna with a variously connected load. The transfer functions of a receiving loop with an in-phase (6.55) and an antiphase connection of the load (6.56) have different frequency dependences, so that the waveform of the received pulse differently depends on the load impedance. The directivity and impedance of a small loop

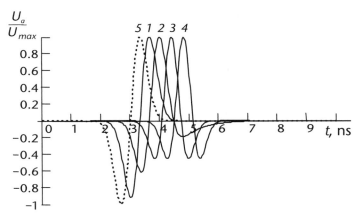

Figure 6.23 The waveforms of pulses received by a loop operated in an anti-phase mode versus load impedance: $Z_0 = 0.01\rho_l$ (curve 1), $0.1\rho_l$ (curve 2), $1.0\rho_l$ (curve 3), and $10\rho_l$ (curve 4), and the waveform of the reference electromagnetic pulse (curve 5).

operating in an in-phase mode are the same as those of a short monopole. Therefore, to reduce the pulse distortion, it is necessary to increase the load impedance corresponding to the in-phase current. For a loop operating in an antiphase mode, it is necessary, on the contrary, to reduce the load impedance corresponding to the antiphase current to reduce the distortion of the received pulse. Figures 6.23 and 6.24 give the waveforms of received pulses versus load impedance for a loop of diameter 36 mm with $b = 0.1a$ operated in an anti-phase and in an in-phase mode, respectively.

The calculation results show that when the load impedance is close to the characteristic impedance of the loop wire, the waveform of the recorded voltage pulse is markedly different from that of the reference electromagnetic pulse. Small

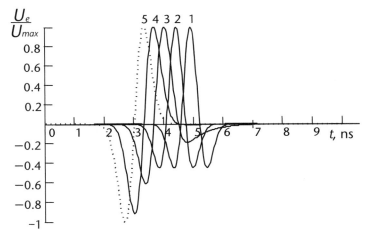

Figure 6.24 The waveforms of pulses received by a loop operated in an in-phase mode versus load impedance: $Z_0 = 0.01\rho_l$ (curve 1), $1.0\rho_l$ (curve 2), $10\rho_l$ (curve 3), and $100\rho_l$ (curve 4), and the waveform of the reference electromagnetic pulse (curve 5).

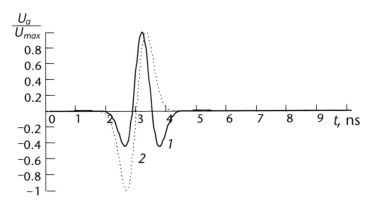

Figure 6.25 The waveforms of the pulse received by a loop in an unbalanced connection to the receive path (curve 1) and of the reference electromagnetic pulse (curve 2).

distortion of the pulse is provided only with $Z_0 \ll \rho_l$ in an antiphase mode or with $Z_e \gg \rho_l$ in an in-phase mode. An unbalanced connection of the loop to the receive path [Figure 6.6(a)] makes possible to increase the directivity of the loop antenna if $Z_2 \approx \rho_l$. In this case, the pattern shape does not depend on the impedance Z_1. The waveform of the received pulse is shown in Figure 6.25.

Figure 6.25 shows that the waveform of the voltage pulse across the load is close to the derivative of the reference electromagnetic pulse waveform. This suggests that the distortion can be reduced by using an integrating RC chain for the load of a loop antenna.

6.2.3 Proportion Between the Received Signal Power and the Dissipated Power

When a receiving antenna is used as an element of a receiving antenna array, the distortion the received signal waveform can be strongly affected by the power reradiated by the antenna. For the distortion of the received signal be small, the reradiated field strength should be significantly lower than the field strength of the incident wave [14]. In accordance with the equivalent circuit of a receiving antenna (Figure 6.1), the power absorbed by the antenna load is determined as

$$P_L = |I_a|^2 R_L = \left| \frac{E_0 l_e \chi}{Z_a + Z_L} \right|^2 R_L \tag{6.66}$$

and the power reradiated by the antenna is determined as

$$P_\Sigma = |I_a|^2 R_a = \left| \frac{E_0 l_e \chi}{Z_a + Z_L} \right|^2 R_a \tag{6.67}$$

where E_0 is the amplitude of the incident wave field, χ is the polarization field transfer efficiency, l_e is the effective length of the antenna, $R_L = \mathrm{Re}(Z_L)$, and $R_a = \mathrm{Re}(Z_a)$.

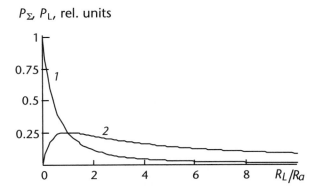

P_Σ, P_L, rel. units

Figure 6.26 Comparison between the power reradiated by a receiving antenna (curve 1) and the power absorbed by the antenna load (curve 2).

Figure 6.26 gives plots of P_L and P_Σ versus the ratio R_L/R_a for a half-wave dipole. It can be seen that a short-circuited dipole extracts the maximum power of the incident wave field, but all this power is reradiated. In the matching mode ($R_L = R_a$), only a quarter of this power is absorbed by the load and the same is reradiated. If $R_L > R_a$, the power absorbed by the load is lower but it is significantly greater than the reradiated power.

Chen and Liepa [15] have analyzed the solution of an integral equation indicating that the field reradiated by a receiving dipole can be minimized by choosing an optimum impedance of the load. This is due to that the current distribution along a receiving wire, according to (6.18), substantially depends on the impedance $Z_L = R_L + iX_L$. For the case of a normally incident wave ($\theta = 90°$), (6.18) can be rewritten as

$$I(z) = I_0\left(I_1(z) + i\frac{Z_L}{\rho_a}I_2(z) \right) = I_0\left(I_1(z) - \frac{X_L}{\rho_a}I_2(z) + i\frac{R_L}{\rho_a}I_2(z) \right) \tag{6.68}$$

where $I_0 = E_0/(Z_L + Z_a)\gamma\sin\gamma L$, $I_1(z) = \cos\gamma z - \cos\gamma L$, $I_2(z) = \sin\gamma z - \sin\gamma L + \sin\gamma(L - z)$

Equation (6.68) indicates that with an inductive load, such that $X_L > R_L$ and $R_L \ll \rho_a$, the functions $I_1(z)$ and $I_2(z)$ have different signs. Hence, for a certain value of X_L/ρ_a, the momentum of the current induced in the wire by the field of the incident wave can be made close to zero, so that the antenna becomes "invisible" to the incident field.

Figure 6.27 presents the calculated impedance of the dipole load at which the field reradiated by the dipole is a minimum. The dashed line depicts the results obtained by Chen and Liepa [15]. The frequency dependence of the reradiation factor $K_\Sigma = P_\Sigma/P_L$ for loads $Z_L = 70 + i1300$ Ω and $Z_L = R_L$ ($R_L \gg \rho_a$) is shown in Figure 6.28. It can be seen that $P_\Sigma < P_L$ for $L < 0.33\lambda$. If $L > 0.5\lambda$, then the reradiated power is comparable to or greater than the power absorbed by the load for any Z_L. It should be noted that the waveform of the received pulse is strongly distorted if the load impedance has a complex value. If the antenna load is purely real and $R_L \gg \rho_a$, then the distortion of the pulse waveform is small (RMSD is no more than 0.2), but the condition $P_\Sigma < P_L$ is fulfilled only for $L < 0.2\lambda$.

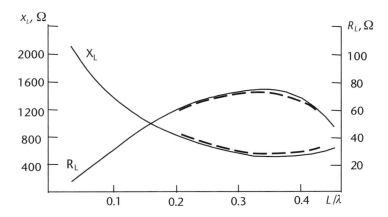

Figure 6.27 The load impedance at which the reradiated field is a minimum versus the length of the receiving dipole ($\rho_a = 250 \ \Omega$).

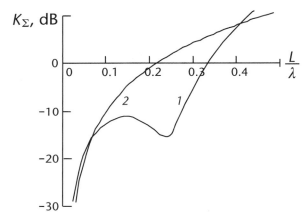

Figure 6.28 Frequency dependence of the reradiation factor for different loads: $Z_L = 70 + i1300$ Ω (curve 1) and 2500 Ω (curve 2).

6.3   Methods for Reducing Distortion of a Received Signal

Analysis of the above results allows us to formulate some design principles for receiving antennas that ensure minimum distortion of received electromagnetic pulses. In general, the frequency dependence of the transfer function of a linear receiving antenna results from the frequency dependence of its impedance and directivity. If the dimensions of a receiving dipole do not exceed the spatial length of the electromagnetic pulse, its directivity weakly depends on frequency, but its impedance has strong frequency dependence. When the dipole length is significantly greater than the electromagnetic pulse length, the pattern shape becomes strongly dependent on frequency, whereas the dipole impedance varies slightly with frequency. Therefore, the methods for reducing distortion of received pulses for short and for long dipoles are different.

6.3.1   Long Dipoles with Noncollinear Arms

In Section 6.1.3, it was shown that the frequency dependence of the transfer function of a long receiving wire decreases at a certain direction of wave arrival θ_0 close to the wire axis. This suggests that if two long wires are arranged at an angle 2φ to each other, their boresights will be oriented toward the bisector of this angle. Such a system is a dipole with noncollinear arms, which is commonly called a V-shaped antenna or a V-antenna (Figure 6.29). Unlike a dipole with collinear arms, in such an antenna, under the action of an external field, besides the currents I_τ induced in the arms by the tangential field component E_t, currents I_n may be excited by the field component normal to the conductors [16, 17]. This is due to that at small angles φ, the arms of a V-antenna are rather close to each other and can be considered a nonuniform transmission line with a characteristic impedance $\rho_L(x)$ depending on coordinates. If the V-antenna arms and the external field vector E lie in the same plane, a potential difference arise between the ends of the arms and excites a wave traveling through the line toward the load.

Thus, the tangential field component excites current waves at every point of the antenna arms (distributed excitation), whereas the normal component excites a wave at a single point corresponding to the antenna aperture (lumped excitation). In view of this, the equivalent circuit of a receiving V-antenna can be represented as two EMF sources with a common load Z_L (Figure 6.30).

One of the EMF sources, U_n, makes the V-antenna a nonuniform transforming transmission line with the wire spacing varying from $2h$ at the antenna aperture to $2b$ at the load connection point. For a line of this type, the voltage ratio is given by [18]

$$K_U = \sqrt{\frac{\rho_L(0)}{\rho_L(L)}}$$

(6.69)

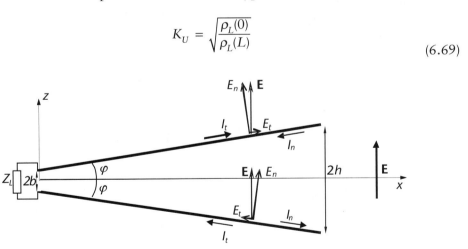

Figure 6.29   Sketch illustrating the calculation of the transfer function of a V-antenna.

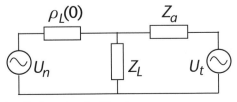

Figure 6.30   Equivalent circuit diagram of a V-antenna.

where $\rho_L(0) = 120\ln[(2b - a)/a]$ is the characteristic impedance of the line at the load connection point, $\rho_L(L) = 120\ln[(2b - a)/a]$ is the characteristic impedance at the antenna aperture, and a is the wire radius. The field component normal to the antenna arms, E_n, produces the potential difference $U_n(L) \approx 2hE_n$ between the ends of the arms. If $2h$ is no greater than half the wavelength, a TEM traveling wave propagates in the line, whose voltage and current are related as $U_n = I_n\rho_L(x)$, where x is the coordinate along the line. The distribution of the current I_n after multiple reflections of the wave from the line ends can be represented as

$$I(x) = \frac{U_n(L)}{\rho_L(L)}\left\{\exp[-ik(L - x)] + R_n\exp[-ik(2L + L - x)] + R_nR_2\exp[-ik(2L - L + x)]\right.$$

$$\left. + R_n^2R_2\exp[-ik(4L + L - x)] + R_n^2R_2^2\exp[-ik(4L - L + x)] + ...\right\} \tag{6.70}$$

where R_n is the reflection coefficient of the V-antenna load and $R_2 \approx -1$ is the reflection coefficient of the ends of the V-antenna arms. The current carried by the antenna load (at $x = 0$) is determined as

$$I_n(0) \approx \frac{2hK_UE_n\exp(-ikL)}{\rho_L(0) + Z_n}\left[1 - R_n\exp(-i2kL) + R_n^2\exp(-i4kL) + ...\right] \tag{6.71}$$

where Z_n is the impedance of the antenna load to the wave excited by the field component normal to the antenna arms and $R_n = (\rho_L - Z_n)/(\rho_L + Z_n)$ is the coefficient of reflection of this wave from the load impedance. Thus, the E_n-field transfer function can be represented as

$$H_a^n \approx \frac{2hK_U\exp(-ikL)}{\rho_L(0) + Z_n}\left[1 - R_n\exp(-i2kL) + R_n^2\exp(-i4kL) + ...\right] \tag{6.72}$$

Figure 6.31(a) presents the waveforms of pulses of different duration across the load of a V-antenna with the arm length $L = 0.9$ m, arm radius $a = 0.01L$, and the

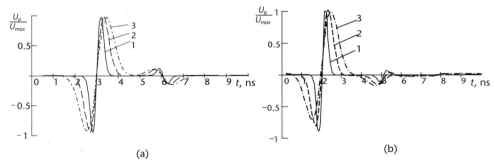

(a)   (b)

Figure 6.31 Waveforms of pulses of different duration produced by the field normal (a) and tangential component (b) across the load of a V-antenna with $L = 0.9$ m and $2h = 0.1$ m: $\tau_p = 1$ ns (curve 1), 2 ns (curve 2), and 3 ns (curve 3).

distance between the arm ends $2h = 0.1$ m. The antenna is loaded by an impedance $Z_L = \rho_L(0)$.

The other EMF source is associated with the EMF U_t generated at the V-antenna terminals by the field component tangential to the arms, E_t. The impedance of this EMF source is equal to the impedance of a dipole with noncollinear arms, $Z_a = -i\rho_a\cot\gamma L$. To calculate the E_t-field transfer function, (6.49) is used with the change $\theta \to \theta + \theta_0$, where $\theta_0 = \pi/2 - \varphi$, for one arm and the change $\theta \to \theta - \theta_0$ for the other. In this case, we have

$$H_{1a}^t = \frac{Z_L \sin(\theta + \theta_0)}{(Z_L + Z_a)\sin\gamma L}$$
$$\times \frac{\gamma\{\cos[kL\cos(\theta + \theta_0)] - \cos\gamma L\} + i\{\gamma\sin[kL\cos(\theta + \theta_0)] - k\cos(\theta + \theta_0)\sin\gamma L\}}{\gamma^2 - k^2\cos^2(\theta + \theta_0)}$$

(6.73)

$$H_{2a}^t = \frac{Z_L \sin(\theta - \theta_0)}{(Z_L + Z_a)\sin\gamma L}$$
$$\times \frac{\gamma\{\cos[kL\cos(\theta - \theta_0)] - \cos\gamma L\} + i\{\gamma\sin[kL\cos(\theta - \theta_0)] - k\cos(\theta - \theta_0)\sin\gamma L\}}{\gamma^2 - k^2\cos^2(\theta - \theta_0)}$$

(6.74)

For an antiphase signal, we have $H_a^t = (H_{1a}^t - H_{2a}^t)/2$.

The waveforms of the voltage pulses generated across the load of a V-antenna by the tangential field component for $Z_L = \rho_a$ are shown in Figure 6.31(b).

The above results show that for a V-antenna of length $L > c\tau_p$, the RMSD of the received pulse waveform is no more than 0.2, but in both cases, after a time interval $\Delta t = 2L/c$, a spurious response occurs whose amplitude can reach -15dB of the amplitude of the received pulse. This effect is not essential if a V-antenna is used in a UWB communication system. However, it becomes a considerable drawback when an antenna of this type is a component of a UWB radar system if the object has larger dimensions than the antenna.

Also, the results show that the distortion of the waveform of the voltage pulse U_n produced by the normal field component is no more than several percents ($\sigma < 0.1$) and weakly depends on the load impedance. The pulse-to-spurious response amplitude ratio is almost independent of the pulse duration, but it depends on the load impedance. The waveform of the voltage pulse U_t produced by the tangential field component is distorted more substantially, but the RMSD of the received pulse waveform is no more than 0.2 if the condition $L > c\tau_p$ is fulfilled. The amplitude of the spurious response depends on the load impedance and increases with τ_p, and its waveform may change.

The total current in the load of a V-antenna is defined as $I(0) = I_n(0) + I_t(0)$. According to the equivalent circuit, the voltage across the impedance Z_L, which can be considered a real quantity ($Z_L = R_L$), as the load of a receiving antenna is conventionally the impedance of the feeder. Hence, for the total current we have

$$I(0) = \frac{U_n(L)K_U(R_L + Z_a)}{\rho_L(0)(R_L + Z_a) + R_L Z_a} + \frac{U_t(R_L + \rho_L(0))}{Z_a(R_L + \rho_L(0)) + R_L \rho_L(0)} \tag{6.75}$$

Taking into account that the load voltage is defined as $U_L = I(0)R_L$ and putting $U_n = E_0 H_a^n$ and $U_t = E_0 H_a^t$, where E_0 is the amplitude of the incident wave field, we may represent the transfer function of a V-antenna as

$$H_V = \frac{H_a^n(R_L + Z_a)R_L}{\rho_L(0)(R_L + Z_a) + R_L Z_a} + \frac{H_a^t(R_L + \rho_L(0))R_L}{Z_a(R_L + \rho_L(0)) + R_L \rho_L(0)} \tag{6.76}$$

Examination of (6.76) suggests that the amplitude of the spurious response can be reduced by elimination of the spurious responses produced by the currents I_n and I_τ. To do this, it is necessary to optimize the proportion between R_L and $\rho_L(0)$, bearing in mind that the condition $\rho_L(0) < R_L < 2\rho_a$ must be fulfilled. The waveforms of the pulses received by a V-antenna with a nearly optimum value of R_L are shown in Figure 6.32.

In this case, the amplitude of the spurious response can be reduced by 6–12 dB and make no more than −30 dB of $U_{\max}$ if $L > 2c\tau_p$. The RMSD of the pulse waveform taking into account the spurious response is not above 0.2 for $L = c\tau_p$, and less than 0.1 for $L > 2c\tau_p$. It is impossible to fully eliminate spurious responses in antennas of this type, as the responses produced by the currents I_n and I_t differently vary with time.

The transfer function H_a^n is proportional to the voltage transformation factor K_U, which is less than unity for V-antennas with arms of constant cross section. To increase the effective length of a V-antenna due to an increase in K_U, the antenna arms are made to have a varied cross section, so that ρ_L would remain constant or vary insignificantly. In practice, V-shaped harp antennas, whose arms are configured as a set of divergent cylindrical conductors [Figure 6.33(a)], are used for more than half a century [19]. The arms of antennas intended for operation in the UWB

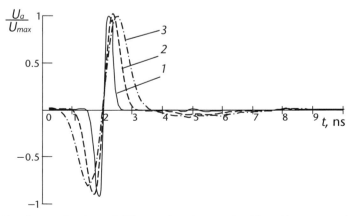

Figure 6.32 Waveforms of pulses of different duration received by a V-antenna with an optimum load resistance $\rho_L(0) < R_L < 2\rho_a$, $L = 0.9$ m, and $2h = 0.1$ m: $\tau_p = 1$ ns (curve 1), 2 ns (curve 2), and 3 ns (curve 3).

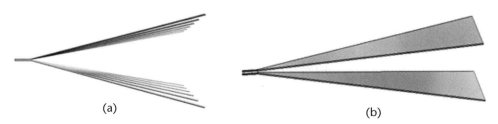

Figure 6.33 Versions of V-antennas with arms of varied cross section: a V-shaped harp antenna (a) and a TEM horn (b).

range are made as triangular plates [Figure 6.33(b)]. Antennas of this type are called TEM horns [20, 21]. It was shown [22] that the characteristic impedance ρ_L of a TEM horn with a small angular aperture can be calculated by the formulas derived to calculate the characteristic impedance of a strip line.

A V-antenna should be connected to the receiver via a balanced line or a balancing device, which can restrict the waveband in the microwave range, that is, additionally distort the waveform of the received UWB pulse. Therefore, to receive UWB electromagnetic pulses, an unbalanced TEM horn [23] is often used, which is connected to the receiver via a coaxial cable. This type of TEM horn can be considered as a segment of an unbalanced strip line with a strip width w and a strip–screen distance h. In this case, it should be taken into account that the screen width should be at least twice that of the strip, and the strip width and the strip–screen distance should be less than half the shortest wavelength in the spectrum of the received signal [24]. If these conditions are not fulfilled, higher-order waves may arise and additionally distort the pulse waveform.

The characteristics of TEM horns of different geometric dimensions (Figure 6.34) [25] were investigated by Andreev et al. [26]. The antenna dimensions were calculated using the well-known formula for the characteristic impedance of an unbalanced strip line supporting a TEM wave [27]

$$\rho_L = \frac{119.904\pi}{\left(\dfrac{w}{h}\right) + 2.42 - 0.44\left(\dfrac{h}{w}\right) + \left(1 - \left(\dfrac{h}{w}\right)\right)^6} \tag{6.77}$$

where w and h are, respectively, the width and height of the strip above the ground plate. The characteristic impedance of the line was put equal to the characteristic

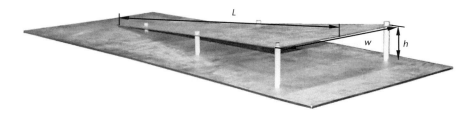

Figure 6.34 The appearance of a TEM horn. (With permission from Pleiades Publishing, Ltd.)

Table 6.1   Geometric Dimensions of TEM Horns.

	L, mm	w, mm	h, mm
TEM 1	503	220	45
TEM 2	905	400	80
TEM 3	903	100	20
TEM 4	1505	100	20

impedance of the output feeder (50 Ω). To simplify the design, the width of the antenna strip was approximated by a linear function.

The dimensions of the test antennas are given in Table 6.1. The antennas differed by a factor of four in aperture height (h) and width (w) and by a factor of three in length (L). The investigation results have shown that the effective length of a TEM horn can be estimated as $l_e \approx h/2$ and it remains almost unchanged as long as $h < 0.5\lambda_{\min}$. The frequency dependence of the horn effective length is presented in Figure 6.35.

The RMSD of the waveform of the received pulse on the waveform of the reference electromagnetic pulse (without regard for spurious response) is not over 0.1 for $L > c\tau_p$ and can be 0.2 and greater for $L < 0.5c\tau_p$.

Thus, a TEM horn can receive a UWB electromagnetic pulse with minimum distortion of its waveform, and therefore it can be used as a reference antenna in investigating the characteristics of newly developed antennas.

6.3.2 Unmatched Short Dipoles

For a short monopole ($kL \ll 1$), we can assume that $\gamma = k$ and rewrite (6.49) as

$$H_a = \frac{1}{2} \cdot \frac{L}{\left(1 + \dfrac{Z_a}{Z_L}\right)} \sin\theta \tag{6.78}$$

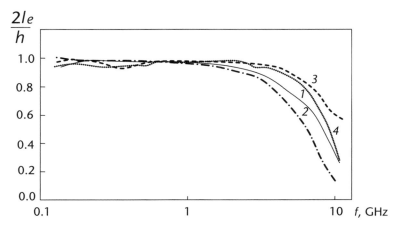

Figure 6.35   Normalized effective length of a TEM horn versus frequency for TEM 1 (curve 1), TEM 2 (curve 2), TEM 3 (curve 3), and TEM 4 (curve 4).

where Z_a is the intrinsic impedance of the monopole and Z_L is the impedance of the load.

For a short dipole of length $2L$, we have

$$H_d = \frac{L}{\left(1 + \dfrac{Z_d}{Z_L}\right)} \sin\theta \tag{6.79}$$

where Z_d is the intrinsic impedance of the dipole and $Z_L = 2Z_1$ is the impedance of the dipole load. Equations (6.78) and (6.79) indicate that the frequency dependence of the transfer function of a short straight antenna can be weakened substantially if

$$|Z_L| \gg |Z_a| \quad \text{or} \quad \frac{Z_a}{Z_L} \approx const \tag{6.80}$$

As the load of a receiving antenna is most often a feeder, whose impedance is a real quantity, and for a short monopole or dipole we have $\mathrm{Re}(Z_a) \ll \mathrm{Im}(Z_a)$, it is necessary, first of all, to reduce $\mathrm{Im}(Z_a)$. This can be done by reducing ρ_a, that is, by increasing transverse dimensions of the antenna. However, to make $\mathrm{Re}(Z_a) \ge \mathrm{Im}(Z_a)$, that is, to reduce the effect of the reactance, is possible only in about an octave frequency band, which is too narrow to receive ultrawideband electromagnetic pulses, whose spectrum may occupy a frequency range of more that two octaves.

Figure 6.36 presents the waveform of a bipolar electromagnetic pulse ($U_a/U_{\max}$) received by a short dipole ($L < c\tau_p$, where τ_p is the duration of the bipolar pulse) of characteristic impedance 160 Ω as a function of the load impedance. It can be seen that the waveform of the voltage pulse across the load is almost the same as that of the reference electromagnetic pulse if $Z_L > 20000$ Ω. However, this condition can hardly be satisfied over a wide frequency range, especially in the decimeter and centimeter wavelength ranges. The matter is that to connect an antenna to a receiver, balanced or unbalanced feeders are used whose characteristic impedance is tens of

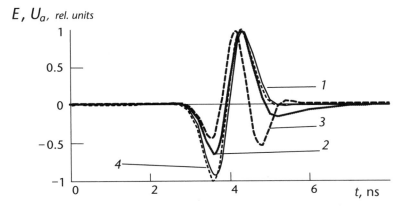

Figure 6.36 The waveform of a pulse received by a short dipole ($L < 0.1 c\tau_p$) versus load impedance: $Z_l = 20000$ (curve 1), 2000 (curve 2), and 200 Ω (curve 3), and the waveform of the reference electromagnetic pulse (curve 4).

ohms for an unbalanced and hundreds of ohms for a balanced feeder. It seems that this problem could be resolved by connecting an impedance transformer between the antenna and the feeder. However, the high-transformation-ratio transformers have a too narrow bandwidth at frequencies over 0.6 GHz, because of the effect of stray parameters (winding capacitance and scattered inductance), and increased losses. Really, a transformer bandwidth of more than two octaves can be achieved if the impedance transformation ratio is no more than four. Thus, even with a balanced feeder having a characteristic impedance of 200–300 Ω, it is difficult to realize a dipole load of impedance more than 1200 Ω.

Another method to reduce distortion of received pulses was proposed by the authors of [28] and [26]. The idea of the method is that the dipole arms are made of low-conductivity resistive material. This significantly increases the current wave attenuation ratio, and therefore no standing wave is generated in the dipole. Figure 6.37 shows the waveform of a pulse received by a short dipole as a function of the loss resistance per unit length, R'_Ω, for $Z_L = 600\ \Omega$. This method can be implemented at almost all radio frequencies, but it has a major drawback: the strength of the signal received by a resistive dipole is several times lower compared with that received by a metallic dipole. Thus, for instance, for $R'_\Omega = 50000\ \Omega/m$, the pulse waveform is distorted slightly, but the signal strength decreases by 20 dB. Balzovsky et al. [30] proposed a practical design of a resistive dipole. The results of its experimental investigations are in good agreement with calculations.

Small loop antennas with a perimeter not over $0.2c\tau_p$ exhibit increased directivity, but the conditions for receiving undistorted UWB pulses are at variance with those for achieving maximum directivity. However, an unbalanced connection of a loop with a balancing capacitor (Figure 6.38) can significantly reduce the distortion of received pulses while maintaining a reasonably high protective action factor (strength ratio of a signal received from the direction of the antenna boresight to that received from the opposite direction).

Figure 6.39 presents the waveforms of pulses received by a loop with a balancing capacitor of different capacitance. In the case of no capacitor, the waveform of the received pulse is close to the derivative of the reference pulse. The capacitance

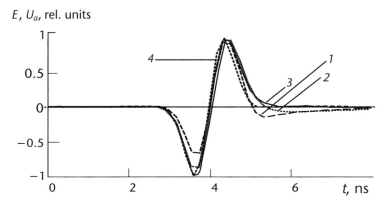

Figure 6.37 The waveform of a pulse received by a short dipole ($L < 0.2c\tau_p$) versus the loss resistance per unit length: $R'_\Omega = 5$ (curve 1), 15 (curve 2), and 50 $k\Omega/m$ (curve 3), and the waveform of the reference electromagnetic pulse (curve 4).

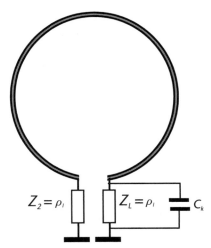

Figure 6.38 Schematic of a receiving loop with a balancing capacitor.

C_k connected in parallel with the load and the loop impedance form an integrating chain that corrects the waveform of the incoming pulse. Increasing C_k reduces the pulse distortion; however, the signal amplitude decreases. Thus, as C_k was increased from 20 to 80 pF, the signal amplitude decreased by 12 dB, whereas the protective action factor decreased insignificantly (from 14 to 12 dB). Additional resistive losses in the loop deteriorate its directivity and do not reduce pulse distortion.

Thus, short antennas can well be used for receiving ultrawideband pulses with low distortion. However, a lower distortion may cost a marked decrease in the signal amplitude.

6.3.3 Active Antennas

Low distortion of received UWB pulses is provided by TEM horns. However, their dimensions are generally greater than the spatial length of a recorded electromagnetic pulse. This limits the use of TEM horns in antenna arrays and makes

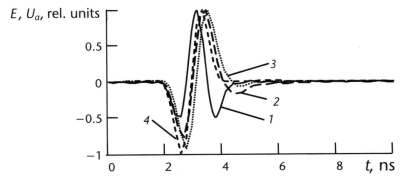

Figure 6.39 The waveform of a pulse received by a loop ($L = 0.18c\tau_p$, $\rho_l = 140\ \Omega$) versus balancing capacitance: $C_k = 0$ (curve 1), 20 (curve 2), and 80 pF (curve 3), and the waveform of the reference electromagnetic pulse (curve 4).

impossible their use as elements of receiving vector antennas. More preferred are short dipole antennas that are shorter than the minimum wavelength in the spectrum of a recorded pulse. These antennas have angular and polarization patterns weakly depending on frequency. However, the strong frequency dependence of the impedance of a metallic dipole results in a distortion of output signals. The use of dipoles whose arms are made of materials having high resistivity per unit length extends the antenna bandwidth and allows receiving of a short UWB pulse almost without distortion of its waveform. However, the amplitude of the output signal is significantly decreased [30].

A study was performed [31] to investigate whether the waveband of a receiving antenna can be extended with an insignificant decrease in received signal amplitude by including controlled energy sources (active elements, AEs), for which voltage amplifiers are generally used. Antennas of this type [32–34] are called active antennas (AAs).

The simplest AA is a short dipole of length $2L$ connected to a load via an AE with voltage gain K_u, input impedance Z_{in}, and output impedance Z_{out}. In the equivalent circuit (Figure 6.40), the dipole can be represented by a source of EMF ε with intrinsic impedance Z_a loaded by Z_{in}. For a short dipole ($L < 0.1\lambda$), we can put $\varepsilon \cong E_t L$; then the voltage across the AA load is given by

$$U_a = \frac{E_t L K_u Z_{in}}{Z_a + Z_{in}} \tag{6.81}$$

where E_t is the field component tangential to the dipole.

Typically, the antenna load is a feeder, whose input impedance can be considered a real quantity, and for the AE, an active nonreciprocal four-terminal or three-terminal network is used. As an AE is a nonreciprocal and unidirectional, its output impedance weakly depends on Z_a and can be matched, over a wide frequency range, to the intrinsic impedance of the feeder that connects the antenna and the receiver. In what follows, it is assumed that the match of the antenna impedance to Z_L is provided in the AA transmission band, by which we mean the frequency range for which the following condition is fulfilled:

$$\frac{|U_a|}{|E_t|} = |H_{AA}| = const, \; \arg(H_{AA}) = t_0\omega \tag{6.82}$$

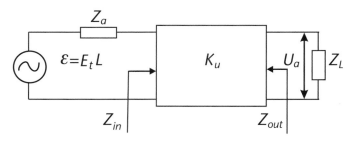

Figure 6.40 The equivalent circuit diagram of an active antenna.

where t_0 is a constant, $H_{AA} = LK_u/(1 + Z_a/Z_{in})$ is the transfer function of the AA for a normally incident wave whose vector **E** is collinear to the dipole.

Condition (6.82) can be satisfied if

1. $K_u = const$, $|Z_{in}| >> |Z_a|$. This cannot be attained in the microwave range, as an AE with an input impedance modulus of several tens of kiloohms capable of operating at frequencies over 0.5 GHz is unfeasible;

2. $K_u = const$, $|Z_a|/|Z_{in}| \cong const$, $\arg(Z_{in}) - \arg(Z_a) \to 0$. This can be attained in a wider frequency range, as the ratio $|Z_a|/|Z_{in}|$ can be even greater than unity; it is only required that Z_a and Z_{in} had identical frequency dependences. As Z_a of a short dipole is capacitive in nature, Z_{in} of the AE must also be capacitive and the condition $\text{Re}Z_{in} << \text{Im}Z_{in}$ should be satisfied. In this case, the dipole is unmatched to the AE, but the mismatch loss can be compensated by the gain. Using an unmatched mode also reduces the distortion of the field spatial structure arising due to reradiation of the field by the current induced in the antenna. The momentum of the current induced in the dipole is lower compared with a matched mode by a factor greater than $(|Z_a|^2 + |Z_{in}|^2)/2\text{Re}(Z_a)$ [35]. The reradiated field strength is lower by the same factor;

3. $K_u/(1 + Z_a/Z_{in}) = const$. This ensures constancy of H_{AA} when the dipole is not short for a desired frequency range and its length at the upper frequency reaches half the wavelength. The effective dipole length becomes frequency dependent and the transmission ratio of the voltage divider formed by Z_a and Z_{in} increases. In this case, the excess in dipole effective length can be compensated by a proper choice of the frequency dependence of K_u. With increasing frequency, K_u should decrease in proportion to the increase in dipole effective length and in the transmission ratio of the voltage divider formed by Z_a and Z_{in}.

The above approach was used to design an AA for receiving UWB pulses of duration 0.5–3 ns. The external view of the AA is shown in Figure 6.41. The antenna was fabricated using printing technology on a plate (1) cut off one-sided fiberglass

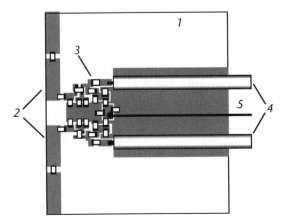

Figure 6.41   External view of the active antenna: 1—dielectric foil plate, 2—dipole arms, 3—AE, 4—screened symmetric line, and 5—power wire. (With permission from Springer.)

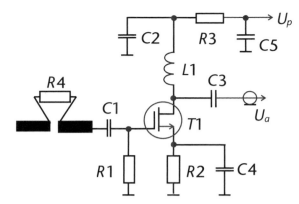

Figure 6.42   Circuit diagram of one channel of the active element. (With permission from Springer.)

foil of thickness 1 mm. The plate dimensions are 4.5×4 cm. The arms of the dipole of length $2L = 4.5$ cm and width 0.3 cm (2) are connected to an AE (3). The AE comprises two identical channels whose terminals are loaded by coaxial cables forming a screened balanced line (4) through which the AE output signal, U_a, is transmitted to the input of the receiver. The power supply voltage U_p is applied to the AE via an individual wire (5). The principal circuit diagram of one AE channel is given in Figure 6.42. The ATF-38143 field-effect transistor $T1$ is connected as in a common-source circuit, the resistor $R2$ sets up a drain current of 20 mA at a voltage of 3 V. The capacitors C1 and C4 restrict the bandwidth to a frequency below 200 MHz to suppress high-power broadcast signals. To extend the bandwidth, 200-Ω resistors $R4$ are mounted in the middle of each dipole arm.

The frequency dependence of K_u and the deviation of the phase-frequency response from a linear function (ΔPFR) for one AE channel are presented in Figure 6.43. The measurements were performed using an Agilent Technologies 8719ET transmission/reflection measuring set. The K_u decreased with frequency, and the ΔPFR had no abrupt jumps in the range 0.05–5 GHz and was not above $\pi/8$ in the

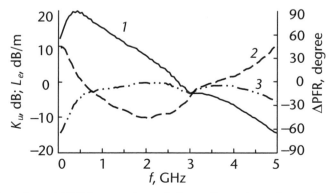

Figure 6.43   The gain and ΔPFR for one channel of an active antenna (curve 1 and curve 2, respectively) and the normalized transfer function of the antenna (curve 3). (With permission from Springer.)

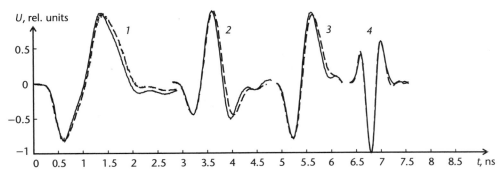

Figure 6.44   The waveforms of pulses recorded using the AA (solid curves) and the TEM horn (dashed curves). Curves 1 and 3 were obtained with the transmitting antenna excited by a monopolar voltage pulse of duration 1 and 0.5 ns, respectively; curves 2 and 4 were obtained with the transmitting antenna excited by a bipolar voltage pulse of duration 1 and 0.5 ns, respectively. (With permission from Springer.)

range 0.5–4.6 GHz. The H_{AA} normalized to a maximum is shown in Figure 6.43 (curve 3). In the frequency band 0.6–4.6 GHz, the variation in H_{AA} was no more than 3 dB.

In measuring the waveforms of received electromagnetic pulses, combined antennas KA [36] were used as transmitting antennas. In measuring UWB pulse waveforms, a TEM horn of length 90 cm and aperture height 8 cm was used as a reference.

The pulse field measurements are presented in Figure 6.44. The RMSD of the waveforms of pulses recorded using the AA and the TEM horn was not over 0.1.

Thus, the inclusion of an active four-terminal network with proper parameters in a short monopole can significantly extend the antenna bandwidth. However, any active four-terminal network, comprising electronic or semiconductor devices, introduces additional linear and nonlinear noise, which can significantly impair the sensitivity of the radio receiver.

In examining the noise characteristics of an AA, we restrict our consideration to the linear part of the radio channel that includes the transmission path, the receiving AA, the feeder circuit (F), and the circuit input of the receiver (R). The sketch diagram of the radio channel that explains the accumulation of noise is shown in Figure 6.45.

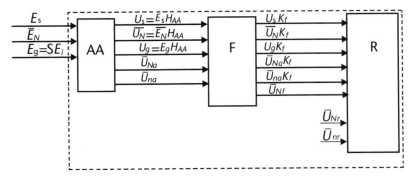

Figure 6.45   Sketch diagram of a radio channel explaining the accumulation of noise.

A signal of field strength E_S and an ambient noise of time-average field strength $\bar{E}_N$ arrive at the input of an AA with transfer function H_{AA}. In the case of a near-omnidirectional antenna, the SNR

$$q = \frac{|E_S|^2}{|\bar{E}_N|^2} \tag{6.83}$$

determines the noise-limited sensitivity of the receiving radio channel. At the AA output, we have an effective voltage of the signal, $U_S = E_S H_{AA}$, and a time–average effective voltage of the ambient noise, $\bar{U}_N = \bar{E}_N H_{AA}$, enhanced by the AE. Besides the input signal and the ambient noise, an ultrawideband antenna is affected by signals of other radio stations transmitted in the frequency band of the AA (station interference). The overall station interference acting on the AA can be represented as a generalized signal of field strength [37]

$$E_g = \sum_{i=0}^{N} E_i \cos(\omega_i t + \varphi_i) \tag{6.84}$$

which is subsequently referred to as a group signal. The group signal voltage at the AA output is given by $U_g = E_g H_{AA}$.

In the AE incorporated in the AA, additive fluctuation noises (thermal noise, current distribution noise, shot noise, etc.) arise. Denote the time-average noise voltage at the AA output by $\bar{U}_{Na}$. An actual AE is not a perfect linear four-terminal network. Therefore, when its input is exposed to station interference (whose strength may be greater than the ambient noise by 80–100 dB), nonlinearity products arise in the AE whose strength may be comparable to or even be greater than the fluctuation (linear) noise. When the number of station interferences is large ($N > 5$), the number of second-order and third-order nonlinearity products is over ($N^2 + N^3$) and their frequency spectrum becomes almost continuous. Therefore, the nonlinearity products can be considered to be noises of nonlinear origin (nonlinear noise). Denote their amplitude at the AA output by $\bar{U}_{na}$. Then the SNR at the AA output is determined as

$$q_A = \frac{|U_s|^2}{|\bar{U}_N|^2 + \bar{U}_{Na}^2 + \bar{U}_{na}^2} = \frac{q}{1 + \dfrac{\bar{U}_{Na}^2 + \bar{U}_{na}^2}{|U_N|^2}} \tag{6.85}$$

where $q = |U_S|^2/|\bar{U}_N|^2$ is the SNR at the AE input.

In examining the sensitivity of a radio channel containing an AA, the noise of the feeder and receiver should also be taken into account, as the noise immunity of the receive path is determined not by the absolute value of the noise occurring in the AA, but by its relative contribution to the noise accumulated in the channel. The influence of the feeder shows up in that its transmission coefficient K_f is less than unity, and so all signals arriving at its input and the noise are attenuated due to losses by a factor K_f. In addition, a thermal noise $\bar{U}_{Nf}$ arises in the feeder due to resistive losses. The influence of the receiver input circuits, comprising the first amplification stage, shows up in that linear and nonlinear noises occur in them. These noises referred to the receiver input are denoted as $\bar{U}_{Nr}$ and $\bar{U}_{nr}$, respectively.

The SNR q_R at the receiver input, determining the actual sensitivity of the radio receiver, as is determined

$$q_R = \frac{q}{1 + \dfrac{\bar{U}_{Na}^2 K_f + \bar{U}_{na}^2 K_f + \bar{U}_{Nf}^2 + \bar{U}_{Nr}^2 + \bar{U}_{nr}^2}{|U_N|^2 K_f}} \tag{6.86}$$

It is almost impossible to calculate exactly the level of nonlinear noise. Therefore, to estimate its effect, we represent a group signal as a set of equal-amplitude harmonics uniformly distributed over the frequency band of UWB signal. We assume that the nonlinearity of the AE transmission ratio is so small that when expanding it in a Maclaurin series, it suffices to consider the first three terms (to take into account the second-order and third-order nonlinearity). Under these assumptions, the level of nonlinear noise at the AA output is determined as

$$\bar{U}_{nna}^2 = g^2 Y_{(2)}^2(b,\omega) \frac{|U_g|^4}{|K_u|^2} + v^2 Y_{(3)}^2(b,\omega) \frac{|U_g|^6}{4|K_u|^4} \tag{6.87}$$

where $g = (1/K_U)(\partial K_U/\partial U_{in})$ and $v = (1/K_U)(\partial^2 K_U/\partial U_{in}^2)$ are the linearity parameters of the AE; $U_{(2)}(b,\omega)$ and $Y_{(3)}(b,\omega)$ are functions describing the frequency distribution of the second-order and third-order nonlinearity products, respectively; $b = f_{max}/f_{min}$, and U_{in} is the voltage across the AE input.

In analyzing the effect of the AA linear noise on the immunity of the receiving channel, we disregard the nonlinear noise and write q'_R as

$$q'_R = q \frac{\bar{E}_n^2 H_{AA}^2}{\bar{E}_n^2 H_{AA}^2 + \bar{U}_{na}^2 + \bar{U}_{nr}^2} \tag{6.88}$$

Equation (6.88) shows that q'_R is always less than q, and for a given ambient noise level $\bar{E}_n^2$, the effect of the AA noise is the less pronounced, the higher the receiver noise level and the greater the transfer function of the antenna,

$$H_{AA} = \frac{LK_u}{2(1 + Z_n/Z_{in})} \tag{6.89}$$

which increases with dipole length, AE input impedance, and voltage transfer ratio. If the length of the dipole is given, the choice of an AE must be dictated by that it should have high-input impedance and a large gain. This implies that preference should be given to field-effect transistors, which, additionally, have a low noise figure.

To estimate the efficiency and noise properties of an AA, it can be compared with a passive antenna by using as a criterion the quotient of the division of the signal/noise ratio of the radio path including the AA by the SNR of the radio path including the passive antenna:

$$q = \frac{H_{AA}}{H_a} \frac{\bar{E}_N^2 H_a^2 + \bar{U}_{Nr}^2}{\bar{E}_N^2 H_{AA}^2 + \bar{U}_{Na}^2 + \bar{U}_{Nr}^2} \qquad (6.90)$$

where $H_a = L\rho_f/2(Z_a + \rho_f)$ is the transfer function of the passive dipole of length $2L$ loaded by the characteristic impedance of the feeder, ρ_f.

It was shown [38] that when a dipole AA of length 0.1λ is operated at frequencies above 100 MHz, the gain in the SNR is 15–25 dB. If the passive dipole is matched, the gain decreases to 2–5 dB, but in this case, the bandwidth of the antenna–feeder path may decrease by a factor of tens.

The nonlinear noise level can be calculated only if the level and frequency distribution of the station interference are known exactly. However, their effect on the immunity of the radio receiver can be estimated taking into consideration the features of appearance of nonlinearity products. If the transfer function of an AA is approximated by a polynomial of the third degree

$$H_{AA}(E_g) \approx H_{AA} + gY_{(2)}(b,\omega)\left(\frac{E_g H_{AA}}{K_u}\right)^2 + vY_{(3)}(b,\omega)\left(\frac{E_g H_{AA}}{K_u}\right)^3 \qquad (6.91)$$

the second-order nonlinearity has the result that each pair of station interferences with frequencies ω_i and ω_k generates harmonics at frequencies $2\omega_i$ and $2\omega_k$ and combinative components at frequencies $\omega_i \pm \omega_k$. The third-order nonlinearity results in that when an AA receives three signals with frequencies ω_i, ω_k, and ω_l, combinative components with frequencies $\omega_i \pm \omega_k \pm \omega_l$, $2\omega_i \pm \omega_k$, $\omega_i \pm 2\omega_k$, $2\omega_k \pm \omega_l$, $\omega_k \pm 2\omega_l$, $2\omega_i \pm \omega_l$, and $\omega_i \pm 2\omega_l$ arise besides harmonics with frequencies $3\omega_i$, $3\omega_k$, and $3\omega_l$. Thus, even when only three signals are received, 46 nonlinearity products arise in the AE. If the number of station interferences is large, the spectrum of combinative components and harmonics becomes almost continuous, and, therefore, they can be considered noises. The decrease in the group signal level by 1 dB reduces the level of the second-order nonlinear noise by 2 dB and the level of the third-order nonlinear noise by 3 dB. The frequency distributions of the second-order and the third-order nonlinear noise are significantly different. If the frequency range occupied by station interference is less than an octave, all second-order nonlinearity products are outside this range. The envelope of the spectrum of third-order nonlinearity products is almost uniform in the frequency range occupied by station interference, irrespective of the width of this range. Especially dangerous are the combinative components of frequency $2\omega_i - \omega_k$, because they may arise when there is only one signal. If $\omega_i \approx \omega_k$, the frequency of the nonlinearity product coincides with the frequency of the signal and its phase differs from the phase of the signal by π. This results in a decrease in the AE gain. The reference data for transistors usually contain a parameter P_{1db}, which determines the input signal level at which the gain decreases by 1 dB.

Based on the foregoing, we can offer the following methods for reducing the level of nonlinear noise:

- to reduce the second-order nonlinear noise, it is necessary to use a balanced amplifier circuit (if $b \geq 2$), which reduces the nonlinearity products of even orders by 20–40 dB;
- to reduce the third-order nonlinear noise arising in an AE, it is necessary to reduce the amplitude of signals at the AE input (by decreasing the dipole length or $|Z_{in}|$) and to use transistors with a large value of P_{1db}, which must be 10–15 dB above the maximum level of station interference at the AE input.

Thus, the requirements for an AE to provide a maximum q'_R are contradictory to the requirement to provide a maximum signal/nonlinear noise ratio. To increase q'_R, it is desirable to increase the length of the monopole, L, and the AE input impedance Z_{in} and gain K_u, whereas to increase q_R, these quantities should be decreased, as an increase in signal amplitude at the receiver input increases the level of its nonlinear noise. In developing an AA, to attain an optimal SNR, it is necessary to take into account the radio environment and the receiver parameters. The use of an ultra-wideband AA in a radio range heavily occupied with station interferences results in that the dependence of the noise immunity of the radio path on the AA parameters becomes extreme. The deviation of the AA parameters in either direction from the optimal values can reduce the sensitivity of the radio receive path several times.

An important characteristic of an active antenna is the dynamic range, which is defined as the ratio of the maximum signal amplitude (A_m) at which it is distorted to the sensitivity of the receive path determined by its intrinsic noise. Typically, for harmonic signals, by A_m is meant the signal level at which the signal amplitude decreases by 1 dB due to saturation of the AE. In recording received pulses, it is proposed to estimate the A_m by the pulse waveform distortion $\sigma = 0.1$ in reference to its waveform at a low field strength.

The dynamic range was measured for a dipole antenna with the antenna AE consisting of amplifiers based on low-noise FET ATF-38143 field-effect transistor (FET) that were connected in a common source circuit [39]. At a supply voltage of 3 V, the drain current was 25 mA. The field strength at the location of the AA was varied in the range 30–250 V/m. The maximum field strength E at which the pulse amplitude decreased by 1 dB was 100 V/m, whereas when estimated by the waveform distortion criterion ($\sigma = 0.1$), it was 130 V/m. Thus, the dynamic range of the AA was at least 90 dB. To measure the electric strength of receiving antennas, a high-voltage generator of bipolar pulses of duration 1 ns and a combined transmitting antenna KA were used. It was observed that when the test antennas were exposed to pulses of high field strength (up to 6 kV/m) at a repetition rate of 1 kHz, the waveforms of the pulses received by the antennas were significantly distorted due to saturation of the AE. However, when the pulse exposure was removed, the service capability of the antennas was restored with all their parameters retained.

6.4   Vector Antennas for Recording the Space–Time Structure of Ultrawideband Electromagnetic Pulses

To study the polarization structure (PS) of UWB electromagnetic pulses, a measuring tool is required to measure the three orthogonal components of the pulse field vector **E**. In this connection, it is of interest to study the possibility of measuring, with sufficient accuracy, the PS of a pulsed UWB electromagnetic field using a so-called vector receiving antenna (VRA). This type of antenna allows one to measure simultaneously and independently the coordinate components of electromagnetic field vectors [40]. A PS characteristic is the hodograph of the electric field vector **E**, which is the projection of the space curve traced by the end of the vector **E** onto the plane of the wave front. If the hodograph of an electromagnetic pulse degenerates into a straight line, the pulse polarization is naturally called linear. In general, the hodograph of a UWB electromagnetic pulse is a flat curve that is not an ellipse or circle. In what follows, this polarization is called nonlinear. It should be noted that for a nonlinearly polarized electromagnetic field, a VRA offers a possibility to determine the direction of arrival of an input signal [41], as the normal to the hodograph coincides with the wave arrival direction. The use of short UWB electromagnetic pulses in radar makes possible not only detection, but also recognition of sensed objects, as a substantial portion of the information about an object is contained in the PS of the electromagnetic pulse scattered by the object.

6.4.1 Design Concepts of Vector Receiving Antennas

A design of an UWB VRA was proposed by Koshelev et al. [42]. The antenna consists of two balanced metallic dipoles, one of which is oriented along the x-axis and the other along the y-axis of the Cartesian coordinate system associated with the antenna. The VRA third element, oriented along the z-axis, is an unbalanced metallic dipole (monopole). For one of its arms, a counterpoise is used, which formed by in-phase-connected balanced dipoles. However, the waveforms of the recorded pulses are distorted because of the strong frequency dependence of the impedances of the short metallic dipoles. It was shown [30] that a balanced dipole, whose arms are made of a resistive material with a linear resistance of 20 kΩ/m, is capable of receiving a short UWB pulse with almost no distortion of its waveform. The use of resistive dipoles as VRA elements reduces distortion of the recorded pulses and increases the accuracy of measuring E_x and E_y. However, the accuracy of measuring E_z decreases. This is because the balanced resistive dipoles cease to work as a counterpoise, and the braid of the feeder connecting the VRA with the receiver becomes the second arm of the unbalanced dipole. As a result, the shape of the radiation pattern of the unbalanced dipole is significantly distorted and the region where signals are recorded with acceptable accuracy is confined to the angular sector $0 < \theta < 60°$, where the angle θ is measured from the axis of the monopole. To reduce the influence of the feeder on the characteristics of the unbalanced dipole, it was proposed to use a locking sleeve, which would simultaneously serve as the dipole arm [43, 44].

In a vector antenna, the dipoles are located in close proximity to each other. Therefore, when calculating the VRA characteristics, one should consider the interaction between them, as the secondary field reradiated by a dipole distorts the structure of the recorded field near the neighboring dipoles. This may result in distortion of the received pulse waveform and in deterioration of the polarization isolation between the VRA channels. For instance, the current $I(x)$ induced by a plane incident wave in the arms of a balanced dipole of length $2L$ arranged along the x-axis generates a secondary (reradiated) field $\mathbf{E}^\Sigma$ whose component tangential to the dipole arranged along the y-axis is given by

$$E_y^\Sigma = -i\frac{Z_0}{k}\left(\frac{\partial^2 A_x}{\partial x\,\partial y} + \frac{\partial^2 A_y}{\partial y^2} + k^2 A_y\right) \tag{6.92}$$

where Z_0 and k are the free space impedance and wave number, respectively, and $\mathbf{A}$ is the vector potential produced by the current $I(x)$. Equation (6.92) indicates that if the dipoles are balanced and strictly orthogonal to each other, and their phase centers are co-located, we have $\mathbf{E}_y^\Sigma = 0$; that is there is no interaction between the dipoles. However, these conditions are difficult to fulfill in practice. Therefore, in the subsequent calculations, we use a model (Figure 6.46) in which two crossed balanced dipoles of length $2L$ are excited by a plane wave whose vector $\mathbf{E}$ lies in the xOy plane. Dipole 1 is arranged along the straight line η passing at an angle α to the x-axis and dipole 2, whose center is at the origin of coordinates, is arranged along the y-axis. The distance between the terminals of each dipole is 2δ. The center of dipole 1 is shifted by m relative to the center of dipole 2.

In this case, the vector potential is parallel to the straight line η and has two components: $A_x = A_\eta\cos\alpha$ and $A_y = A_\eta\sin\alpha$. Therefore, the secondary field component tangential to dipole 2 is nonzero:

$$E_y^\Sigma = -i\frac{Z_0}{k}\left(\frac{\partial^2 A_\eta}{\partial x\,\partial y}\cos\alpha + \frac{\partial^2 A_\eta}{\partial y^2}\sin\alpha + k^2 A_\eta\sin\alpha\right) \tag{6.93}$$

To calculate the current distribution along the dipole arm, the method of superposed traveling waves considered above was used which was modified to calculate the dipoles with the arms made of a resistive material.

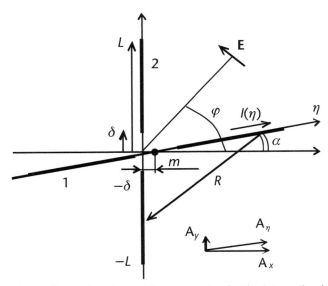

Figure 6.46 Sketch illustrating the problem geometry for the interaction between two crossed dipoles. (With permission from Pleiades Publishing, Ltd.)

When a dipole is connected to a collinear wire via a balancing devise, the signal voltage at the feeder input has generally an anti-phase (U_o) and an in-phase component (U_e): $U_a = U_o + U_e$, where $U_o = 0.5[I_\eta(+\eta_1) + I_\eta(-\eta_1)]K_o\rho_f$, $U_e = 0.5[I_\eta(+\eta_1) - I_\eta(-\eta_1)]K_e\rho_f$; K_o and K_e are the antiphase and the in-phase current transfer ratio of the balancing device, respectively; ρ_f is the characteristic impedance of the feeder; $\eta_1 = \delta + m$, and $\eta_2 = L + m$. The anti-phase and in-phase transfer functions have different frequency dependences. Therefore, the interaction between the dipoles may affect not only the strength, but also the waveforms of the recorded pulses.

If dipole 1 is in the field of a plane wave incident at an angle j whose vector $\mathbf{E}$ is located in the xOy plane, the field component tangential to the dipole is given by $E_\eta(\eta') = E_0\sin(\varphi - \alpha)\exp[ik\eta'\cos(\varphi - \alpha)]$. In this case, for $\eta > 0$ we have

$$
\begin{aligned}
I_\eta(+\eta) = &\frac{E_0\sin(\varphi - \alpha)}{\rho[\exp(i\gamma l) + G_1\exp-(i\gamma l)]}\Big\{\sin\big[\gamma(\eta_2 - \eta)\big] \\
&\times\left[\frac{\exp[i(\xi + \gamma)\eta] - \exp[i(\xi + \gamma)\eta_1]}{\xi + \gamma} + G_1\frac{\exp[i(\xi - \gamma)\eta] - \exp[i(\xi - \gamma)\eta_1]}{\xi - \gamma}\right] \\
&+ \frac{1}{2i}\big\{\exp[i\gamma(\eta - \eta_1)] + G_1\exp[-i\gamma(\eta - \eta_1)]\big\}\Big[\exp(i\gamma l)\frac{\exp[i(\xi - \gamma)\eta_2] - \exp[i(\xi - \gamma)\eta]}{\xi - \gamma} \\
&- \exp(-i\gamma l)\frac{\exp[i(\xi + \gamma)\eta_2] - \exp[i(\xi + \gamma)\eta]}{\xi + \gamma}\Big]\Big\}
\end{aligned}
$$

$$(6.94)$$

where $\xi = k\cos(\varphi - \alpha)$. The current distribution in the second arm of the unbalanced dipole with $\eta < 0$ is found in a similar way:

$$
\begin{aligned}
I_\eta(-\eta) = &\frac{E_0\sin(\varphi - \alpha)}{\rho[\exp(i\gamma l) + G_1\exp(-i\gamma l)]}\Big\{\sin\big[\gamma(\eta_2 - \eta)\big] \\
&\times\left[\frac{\exp[-i(\xi - \gamma)\eta] - \exp[-i(\xi - \gamma)\eta_1]}{\xi - \gamma} + G_1\frac{\exp[-i(\xi + \gamma)\eta] - \exp[-i(\xi + \gamma)\eta_1]}{\xi + \gamma}\right] \\
&+ \frac{1}{2i}\big\{\exp[i\gamma(\eta - \eta_1)] + G_1\exp[-i\gamma(\eta - \eta_1)]\big\}\Big[\exp(i\gamma l)\frac{\exp[-i(\xi + \gamma)\eta_2] - \exp[-i(\xi + \gamma)\eta]}{\xi + \gamma} \\
&- \exp(-i\gamma l)\frac{\exp[-i(\xi - \gamma)\eta_2] - \exp[-i(\xi - \gamma)\eta]}{\xi - \gamma}\Big]\Big\}
\end{aligned}
$$

$$(6.95)$$

The tangential component of the electric field near dipole 2 is the sum of the tangential components of the incident plane wave field and of the field reradiated by dipole 1: $E_y(y) = E_0\cos\varphi\exp(iky\sin\varphi) + E_y^\Sigma(y)$.

The calculation of $E_y^\Sigma(y)$ is performed by formula (6.93) in view of $\eta_1 = \delta + m$ and $\eta_2 = L + m$, where

$$
A_\eta = \int_{\delta+m}^{L+m} I_\eta(+\eta)\frac{\exp(ikR_+)}{4\pi R_+}\,d\eta + \int_{-L+m}^{-\delta+m} I_\eta(-\eta)\frac{\exp(ikR_-)}{4\pi R_-}\,d\eta \qquad (6.96)
$$

$$R_+ = \sqrt{\eta^2 + y^2 - 2\eta y \sin\alpha}, \quad R_- = \sqrt{\eta^2 + y^2 + 2\eta y \sin\alpha} \tag{6.97}$$

The currents $I_y(+\delta)$ and $I_y(-\delta)$ at the terminals of dipole 2 are calculated by the formula

$$I_y(\pm\delta) = \frac{(1 + q_1)i}{\rho[\exp(i\gamma L) + q_1 \exp(-i\gamma L)]} \int_{\pm\delta}^{\pm L} E_y(y') \sin\gamma(L - y' \pm \delta)\, dy' \tag{6.98}$$

where the upper sign corresponds to the current at $y = +\delta$ and the lower one to the current at $y = -\delta$.

The equations obtained allow us to take into account the effect of the reradiated field at an arbitrary angle φ. The dipole interaction has a maximum effect when the wave is incident at $\varphi = 90°$. Then the current in the arms of dipole 2 is induced only by the secondary field $E_y^\Sigma(y)$ produced by the current flowing in dipole 1, which, in this case, is a maximum. The current induced in dipole 2 as a result of the interaction, $I_y^\Sigma(\pm\delta)$, is also a maximum. Therefore, we take the ratio of the voltage at the input of the feeder of dipole 2 to the voltage at the input of the feeder of dipole 1 as a measure of the interaction between the dipoles, M_{21}. If the baluns of the dipoles are identical, we have

$$M_{21} = \frac{\left[I_y^\Sigma(+\delta) + I_y^\Sigma(-\delta)\right] + \dfrac{K_{ye}}{K_{yo}}\left[I_y^\Sigma(+\delta) - I_y^\Sigma(-\delta)\right]}{\left[I_\eta(+\eta_1) + I_\eta(-\eta_1)\right] + \dfrac{K_{\eta e}}{K_{\eta o}}\left[I_\eta(+\eta_1) - I_\eta(-\eta_1)\right]} \tag{6.99}$$

For $\alpha = 0$ and $m \neq 0$, we have $E_y^\Sigma = -i(Z_0/k)(\partial^2 A_\eta/\partial x \partial y)$; that is, the anti-phase current in dipole 2 is equal to zero, and the in-phase current can be greater than the anti-phase current in dipole 1. For $\alpha \neq 0$ and $m = 0$, we have $E_y^\Sigma = -i\dfrac{Z_0}{k}\left(\dfrac{\partial^2 A_\eta}{\partial y^2}\sin\alpha + k^2 A_\eta \sin\alpha\right)$, and there is no in-phase current in dipole 2.

The dependence of M_{21} on the angle α calculated for $m = 0$ is plotted in Figure 6.47(a). The dependence of M_{21} on the shift between the dipole centers, m, in percentage of the arm length, calculated for $\alpha = 0$ is plotted in Figure 6.47(b). The calculations were performed for metallic dipoles (solid curves) and for resistive dipoles with the linear resistance $R' = 2 \cdot 10^4\ \Omega/\text{m}$ (dashed curves). The dipole length was $2L = 20$ cm, the distance between the terminals of each dipole $2\delta = 10$ mm, and the load impedance $Z_L = 600\ \Omega$.

The results obtained suggest that at the admissible level of interaction, $M_{21} = -30$ dB, the deviation from the perpendicular position should not be over $2°$ for metallic dipoles and over $4°$ for resistive dipoles. The shift between the arm centers should be no more than 3% of the arm length for both dipole types. This precision of fabrication of VRAs is quite practicable.

To calculate the characteristics of the unbalanced dipole of a VRA shown schematically in Figure 6.48(a), the equivalent circuit shown schematically in Figure

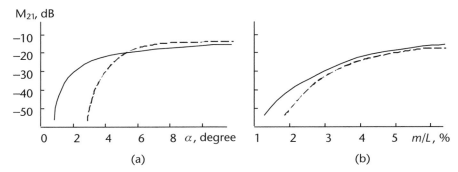

(a)   (b)

Figure 6.47 The measure of the interaction between metallic dipoles (solid curves) and between resistive dipoles with $R' = 2 \cdot 10^4$ Ω/m (dashed curves) versus angle α for $m = 0$ (a) and versus shift m for $\alpha = 0$ (b). (With permission from Pleiades Publishing, Ltd.)

6.48(b) was used. A resistive arm of length L_1 with characteristic impedance ρ_1 and transfer constant γ_1 (1) is connected, via a load of impedance Z_L, with a metallic arm of length L_2 with characteristic impedance ρ_2 and transfer constant γ_2 (2). The end load of the second arm is a locking sleeve of impedance Z_s connected in series with a long wire of characteristic impedance ρ_k whose diameter is equal to that of the feeder braid.

The current distribution in the resistive arm ($z > 0$) is calculated by formula (6.94) after the following change of variables: $\eta \to z$, $\eta_1 \to 0$, $\eta_2 \to L_1$, $\varphi - \alpha \to \theta$, $\gamma = \gamma_1$, and $Z_1 = Z_L + Z_{2in}$, where Z_{2in} is the input impedance of the second (metallic) arm:

$$Z_{2in} = \rho_2 \frac{Z_s + \rho_k + i\rho_2 \tan\gamma_2 L_2}{\rho_2 + i(Z_s + \rho_k)\tan\gamma_2 L_2} \tag{6.100}$$

To calculate the current distribution in the second arm, formula (6.95) is used after the following change of variables: $\eta \to z$, $\eta_1 \to 0$, $\eta_2 \to L_2$, $\varphi - \alpha \to \theta$, $\gamma = \gamma_2$, and $Z_1 \to Z_L + Z_{1in}$, where Z_{1in} is the input impedance of the first (resistive) arm: $Z_{1in} = -i\rho\cot\gamma L_2$. The coefficient of wave reflection from the end of this arm is not equal to -1 and should be calculated by the formula

$$\Gamma_2 = \frac{\rho_2 - (\rho_k + Z_s)}{\rho_2 + \rho_k + Z_s} \tag{6.101}$$

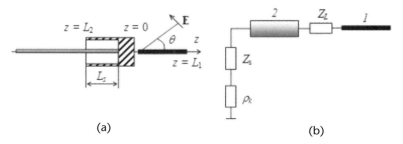

(a)   (b)

Figure 6.48 Schematic diagram of an unbalanced dipole (a) and its equivalent circuit (b). (With permission from Pleiades Publishing, Ltd.)

The impedance Z_s of the locking sleeve, which is a short-circuited segment of a coaxial line whose inner conductor of diameter d is the feeder braid and the outer conductor of diameter D is the wall of a metallic cylinder, is determined by the formula

$$Z_s = i60 \ln \frac{D}{d} \tan kL_s \qquad (6.102)$$

To reduce the influence of the feeder and make the characteristics of the unbalanced dipole close to those of a balanced dipole, it is necessary to have $\Gamma_2 \rightarrow -1$ and $Z_S \gg \rho_k$. Therefore, the locking sleeve depth L_s and diameter D are chosen from the condition that Z_s should be a maximum at the midband frequency of received pulses. The calculations have shown that the length L_2 of the metallic arm should make about 1/4 of the wavelength at the signal midband frequency.

Figure 6.49 presents the waveform of the voltage pulse across the load ($Z_L = 600\ \Omega$) of an unbalanced dipole with the resistive arm of length $L_1 = 10$ cm and linear resistance $R' = 2 \cdot 10^4\ \Omega/m$ and the metallic arm ($D = 10$ cm, $L_2 = 8$ cm) with a locking sleeve of length $L_s = 4$cm for a pulse field incident at an angle $\theta = 75°$ (curve 1). For $\theta > 75°$, the distortion of recorded pulses is lower, and for $0 < \theta < 60°$, the RMSD of the waveform of a recorded pulse from that of the incident field pulse is not over 0.15. Herein, the waveform of the voltage pulse across the load of an unbalanced dipole with the same parameters but with no locking sleeve is given for a pulse field incident at an angle $\theta = 75°$ (curve 2). The incident field, a pulse of duration 3 ns with three time lobes, is represented by curve 3.

The calculations have shown that adding a locking sleeve to the antenna structure and using it as a second arm should bring the pattern function of the unbalanced dipole in the half-space $z > 0$ closer to that of a balanced dipole and reduce the distortion of the received pulse.

The obtained results were used to design a VRA with resistive dipoles. The external view of the VRA is shown schematically in Figure 6.50(a). The arms of the horizontal dipoles X and Y of length $2L = 28$ cm and of the vertical monopole Z of length 10 cm are made of a graphite-based material of linear resistance $2 \cdot$

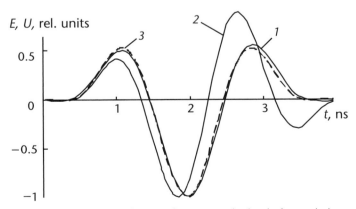

Figure 6.49 The waveforms of the voltage pulses across the load of an unbalanced resistive dipole with (curve 1) and without a locking sleeve (curve 2) for $\theta = 75°$, and the waveform of the incident field (curve 3). (With permission from Pleiades Publishing, Ltd.)

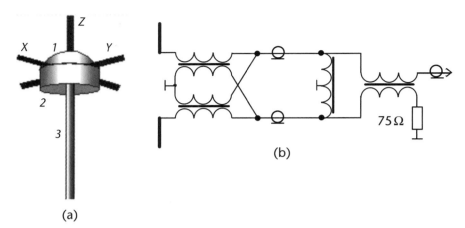

Figure 6.50 Schematic design of a vector receiving antenna (a) and the equivalent circuit of the balancing device (b). (With permission from Pleiades Publishing, Ltd.)

10^4 Ω/m. A hemispherical metal cup (1) combined with a hollow cylinder (2) of diameter 10 cm and length 4 cm form an additional arm for the monopole. The locking sleeve is a segment of a short-circuited coaxial line formed by the braid of the feeder (3) and the hollow cylinder. Cup *1* serves to screen the input circuits. All VRA channels contain identical balancing devices made according to the circuit shown schematically in Figure 6.50(b).

Figure 6.51 presents the measured patterns (solid curves) of the vertical (a) and of the horizontal channel (b). In measuring the patterns, the angular dependence of the voltage was determined at fixed points of time during the pulse at a given distance from the radiation source. Besides, the theoretical patterns (dashed curves) of short dipoles are given. The pattern of the channel Y coincides, to the turn by 90°, with the pattern of the channel X.

In processing the measurements of the polarization structure of received electromagnetic pulses, the deviation of the pattern of the vertical channel Z from the theoretical one is eliminated by introducing a correction factor $p_z(\theta)$ and the deviations of the patterns of the channels X and Y from the theoretical ones by introducing correction factors $p_x(\varphi)$ and $p_y(\varphi)$. In what follows, by the signal in the ith VRA channel $(i = x,y,z)$ is meant $U_{ai}(t) = p_i U'_{ai}(t)$, where $U'_{ai}(t)$ is the measured voltage.

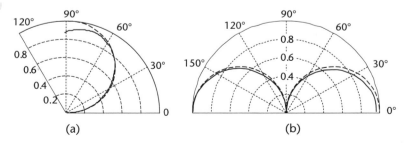

Figure 6.51 Measured patterns (solid curves) of the vertical (a) and of the horizontal channels (b) subjected to the action of a bipolar pulse of duration 2 ns and the corresponding theoretical patterns (dashed curves). (With permission from Pleiades Publishing, Ltd.)

The waveforms of the pulses recorded by the channel X at different incidence angles φ and by the channel Z at different incidence angles θ are given in Figure 6.52.

The experimental investigations have shown that a VRA with resistive dipoles allows one to measure the PS of pulsed electromagnetic radiation in the upper half-space $0 < \theta < 90°$ with an RMSD of error in recorded pulse waveforms not over 0.15.

It is possible to increase the amplitude of signals received by a VRA, enhance the VRA sensitivity, extend its bandwidth and reduce its dimensions by making the antenna active. The design of an active VRA and the results of its testing are described in detail elsewhere [45]. The active VRA, like a VRA with resistive arms, consists of two crossed balanced dipoles and an unbalanced dipole arranged perpendicular to both. The dipoles are enclosed in a common case mounted in the central part of the antenna. The AEs are located inside the case. The general view and the arrangement of the elements inside the case are shown in Figure 6.53.

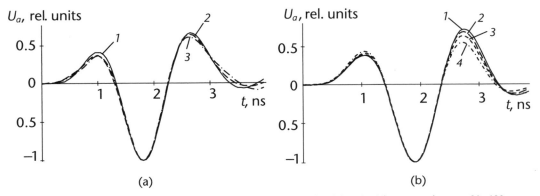

Figure 6.52 Pulse waveforms recorded by the channel X (a) at incidence angles $\varphi = 0°$, $60°$, $75°$ (curves 1–3, respectively) and by the channel Z (b) at incidence angles $\theta = 45°$, $60°$, $75°$, $90°$ (curves 1–4, respectively). (With permission from Pleiades Publishing, Ltd.)

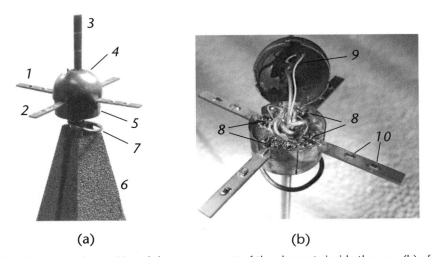

Figure 6.53 The general view (a) and the arrangement of the elements inside the case (b) of an active vector receiving antenna.

The antenna consists of two horizontal balanced dipoles (*1* and *2*) of length 7.6 cm (Figure 6.53) and a vertical unbalanced dipole, whose upper arm (3) is a strip of length 3 cm and the lower arm is a hemispherical metal case (4) of diameter 2.5 cm combined with a hollow cylinder (5) of height 0.7 cm. The feeder lines and the power wire of the AE are located inside a metal tube that serves as the antenna support. The tube is covered with bulk pieces of absorbent (6). Cylinder *5* and tube 6 are connected to an inductive element (7). The AEs (8) of the horizontal channels are located on a board inside the VRA case. The board (9) on which the AE of the vertical channel is mounted is fastened to the hemispherical cup of the antenna case. The dipole arms have cutouts in which resistors (10) are soldered.

The electric circuit of the horizontal channel AE is shown schematically in Figure 6.54(a). The AE consists of two identical field-effect transistor amplifiers. The transistors $T1$ and $T2$ are connected according to a common-source circuit. The circuit elements $C11(C12)$, $L1(L2)$, $R41(R42)$, $C41(C42)$ set up the required frequency dependences of the AE input impedance and transmission ratio. The AE output made of two coaxial cables is balanced. The AEs of the two horizontal channels are identical. The arm of the unbalanced vertical dipole is loaded by an AE, which is half the AE of the horizontal channel and has an unbalanced output. The block diagram of the VRA is given in Figure 6.54(b). To connect the balanced outputs of the horizontal channels, $X - X'$ and $Y - Y'$, to the recording unit, whose output is unbalanced, a two-channel UWB balancing device is used. The unbalanced output of the vertical channel (Z) is directly connected to the recording unit.

The electronic components are placed on a printed board made of one-sided fiberglass foil of thickness 1 mm [Figure 6.55(a)]. The dipole arms (1) are mounted on the same board. Electrical contact between the dipole arms and the VRA case is provided at points (2) close to the AE inputs. The feeder lines (3) made of 0.8-mm coaxial cable and the power wire mounted inside the tubular support of the antenna. To reduce losses in thin cables, the 0.8-mm coaxial cables are changed over by a 2-mm cable in crossovers placed on a separate board at a distance of 25 cm from the antenna. The topology of the printed board and the layout of the AE

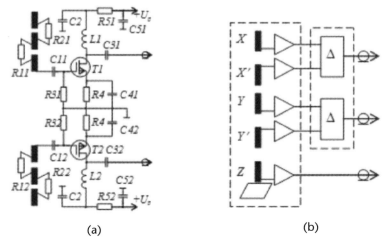

(a)   (b)

Figure 6.54 Schematic circuit diagram of the active element of the horizontal channel (a) and the block diagram (b) of an active vector receiving antenna.

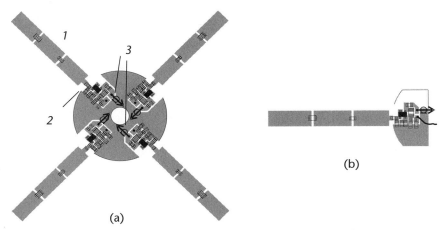

Figure 6.55 Topology and arrangement of the elements of the horizontal (a) and vertical (b) channels.

vertical channel are shown in Figure 6.55(b). The VRA supply voltage is 3 V; the total consumed current is less than 90 mA.

To reduce the effect of the currents induced on the outer surface of the antenna metal support and influencing the vertical dipole, a locking sleeve (5) made as a hollow cylinder [Figure 6.53(a)] and a spiral inductive element (Figure 6.56) are used. The dimensions of this element were determined experimentally by trial and error. Investigations have shown that the antenna with the spiral element made of copper wire 2 mm in diameter [Figure 6.56(a)] allowed a more accurate recording of the pulse waveform than the antenna with the spiral element made of a copper foil strip [Figure 6.56(b)].

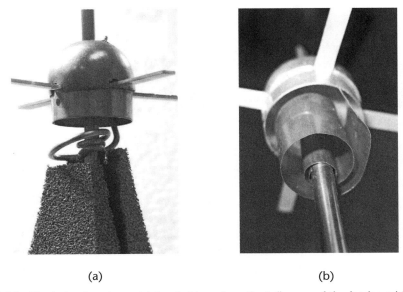

Figure 6.56 The inductive element intended to reduce the influence of the feeder cable.

A combined antenna KA excited by bipolar voltage pulses of duration 0.5 ns was used as a UWB radiation source [46]. The waveforms of pulses recorded by one of the horizontal channels and by the vertical channel are given in Figure 6.57 (curve 1 and curve 2, respectively). A TEM horn of effective length l_e = 4 cm was used as a reference antenna. The waveform of the signal at the output of the TEM horn is given in Figure 6.57 (curve 3). The effective length of the VRA was 1.2 cm for the horizontal channels and 1 cm for the vertical channel.

The RMSD of error of recording the waveform of a pulse of duration 0.5 ns within the pattern half-power width was not over 0.2, and it was even smaller for pulses of longer duration (1–3 ns).

To estimate the dynamic range of the active VRA, the maximum voltage at the antenna output ($U_{\alpha max}$) was determined at which the voltage pulse waveform was distorted. The sensitivity of the receiving antenna is dictated by the intrinsic noise level equal to 50 μV. The voltage $U_{\alpha max}$ was estimated by the waveform distortion factor σ = 0.1 in reference to the voltage pulse waveform corresponding to a low field strength. The field strength at the location of the active VRA was varied in the range 50–250 V/m. The maximum field strength at which σ = 0.1 was 145 V/m for each horizontal channel of the VRA and 60 V/m for the vertical channel. Thus, the dynamic range of the active VRA was estimated to be no less than 88 dB for the horizontal channels and 80 dB for the vertical channel.

6.4.2 Investigation of the Polarization Structure of a Pulsed Electromagnetic Field

Using a VRA, the PS of the radiation produced by a UWB combined antenna KA was investigated [47]. A KA version optimized for excitation by a monopolar voltage pulse of duration 1 ns or a bipolar voltage pulse of duration 2 ns was used. The VRA was oriented so that the dipoles X and Y were arranged in the plane of the incident wave front.

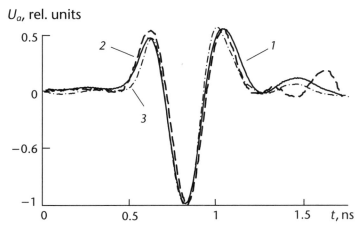

Figure 6.57 The waveforms of pulses recorded by one of the horizontal channels (curve 1), by the vertical channel (curve 2), and by the TEM horn (curve 3).

The radiation of the KA was linearly polarized in the boresight direction. The cross-polarized component that occurred in the KA radiation field on deflection of the polarization direction from the boresight direction caused the vector **E** to rotate. Figure 6.58 presents the hodographs of the vector **E** in the xOy plane for different angles of its deflection from the KA boresight. The point "s" in the plots corresponds to the beginning of the pulse; the consecutive points are separated in time by 0.5 ns. Within the width of the KA half-power pattern ($\leq 90°$), the radiation polarization is nonlinear with the projection of the vector **E** rotating relative to the boresight clockwise for positive angles and counterclockwise for negative angles. Beyond the width of the half-power pattern, the radiation PS becomes complicated.

To investigate experimentally the PS of scattered UWB radiation pulses, an object was used that constituted a thin (1.5 mm) metal strip of dimensions 680×20 cm. The strip was arranged so that the directions to the radiator and to the VRA were symmetrical relative to the normal to the strip plane. In this plane, the strip could be rotated about its center by an arbitrary angle in the range $-90° < \beta \leq 90°$. When viewed from the receiving point, a positive and a negative angle β corresponded, respectively, to the strip tilt right and left from the vertical position.

The object was irradiated with bipolar pulses of linearly polarized electromagnetic field of duration $\tau_p = 2$ ns. The maximum dimension of the strip was comparable to the spatial length of the radiated pulses, $c\tau_p$, where c is the velocity of light. The transmitting and receiving antennas were oriented so that the vector **E** in the incident field and the y-axis of the VRA were arranged vertically, and the xy-plane of the VRA was perpendicular to the direction toward the object. The signals of

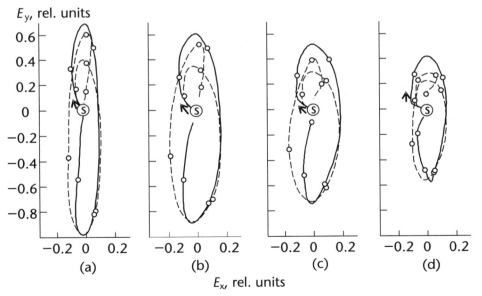

Figure 6.58 The polarization structure of the radiation of a combined antenna with the field vector deflected from the boresight by 15° (a), 30° (b), 45° (c), and 60° (d). The antenna was excited by a monopolar voltage pulse of duration 1 ns (solid curves) and by a bipolar voltage pulse of duration 2 ns (dashed curves). (With permission from Pleiades Publishing, Ltd.)

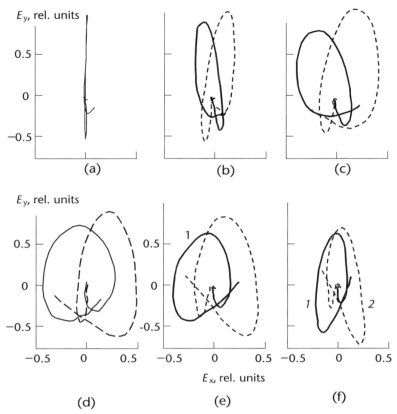

Figure 6.59 Polarization structure of the reflected pulse field for $\beta = 0$ (a), $\pm15°$ (b), $\pm30°$ (c), $\pm45°$ (d), $\pm60°$ (e), and $\pm75°$ (f): dashed curves—positive β and solid curves—negative β. (With permission from Pleiades Publishing, Ltd.)

the channels X and Y of the VRA were recorded simultaneously with a Tektronix TDS 7404 digital oscilloscope.

When the strip was in vertical position ($\beta = 0°$), the reflected pulse field remained linearly polarized [Figure 6.59(a)]. When the strip was rotated by an angle $\beta = 15°$, a cross-polarized field component arose [Figure 6.59(b)] that changed the hodograph of the vector **E** (dashed curve). The slope of the hodograph and the direction of rotation of the vector **E** became opposite, as the strip was tilted by an angle $\beta = -15°$ (solid curve). For angles $\beta = \pm30°$, $\pm45°$, $\pm60°$, and $\pm75°$ [Figure 6.59(c–f), respectively], to each position of the strip there corresponded an individual PS of the reflected pulse field. When the strip was in horizontal position ($\beta = 90°$), the reflected pulse field was linearly polarized. However, the hodograph of the vector **E** was different from that occurred when the TEM strip was in vertical position in that it was equally tilted toward both the positive and the negative values of E_y.

The experimental results demonstrate a relationship between the PSs of the reflected pulses and the orientation of an elongated object and the possibility of using a VRA for investigating the PSs of reflected pulses. It is expected that when probing extended objects of complex shape with linearly polarized radiation, the polarization of the reflected pulses will be nonlinear.

6.4.3   Determination of the Direction of Arrival of Ultrawideband Electromagnetic Pulses

If the field of an electromagnetic pulse is linearly polarized, the use of a VRA allows one to determine the orientation of the vector **E**, but the direction of arrival of the wave cannot be uniquely determined. When the polarization of the incident pulsed radiation is nonlinear, the hodograph of **E** is a spatial curve specified parametrically by three functions: $U_{ax}(t)$, $U_{ay}(t)$, and $U_{az}(t)$. The direction of the normal to the hodograph plane coincides with the direction toward the radiation source.

As the patterns of the VRA channels are different from the theoretical ones, the direction of wave arrival was determined using two algorithms [43, 45]. The first algorithm is fast, but insufficiently exact and serves to tentatively determine the direction of wave arrival and to find the correction factors p_i for the VRA channels from the measured voltages $U'_{ai}(t)$. It is based on finding the plane of the wave front by three points: the origin of the coordinate system, the maximum hodograph point of the vector **E**, and a hodograph point that is found from the condition that the straight line segments connecting the two points and the origin are normal to each other.

The second algorithm uses the values of $U_{ai}(t) = p_i U'_{ai}(t)$. It is based on minimizing the RMSD of all points of the hodograph from the sought-for plane during the pulse or its portion. The direction of the normal to the found plane coincides with the direction toward the radiation source. This is a more exact but more resource-intensive method.

To produce nonlinearly polarized radiation, a combined transmitting antenna was used [47] tilted at 45° to the main direction in the H-plane. The peak value of the cross-polarized electromagnetic field component was 0.25 of the peak value of the main polarization component. The voltages at the three outputs of the VRA, proportional to the coordinate components of **E**, were recorded simultaneously with a TDS 6604 oscilloscope. Figure 6.60 presents the hodograph of **E** reconstructed by its three projections.

The results of measuring the direction of radiation arrival by the above method are depicted in Figure 6.61. The circular arcs represent the horizontal angle φ

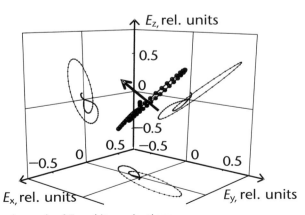

Figure 6.60   The hodograph of **E** and its projections.

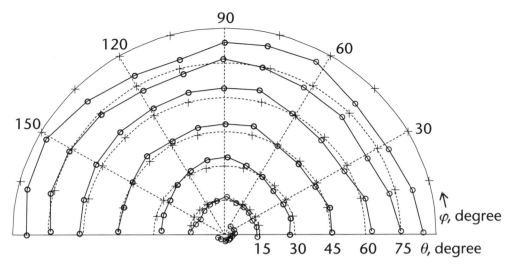

Figure 6.61   The measured directions of arrival of pulsed electromagnetic radiation. (With permission from Pleiades Publishing, Ltd.)

measured from the x-axis of the coordinate system associated with the VRA and the radial lines represent the vertical angle θ measured from the z-axis. The incidence directions set up in the experiment are marked by crosses, and the measured ones are marked by circles. The radiation was bipolar pulses of duration 2 ns corresponding to the main polarization axis with the cross-polarized-to-main component amplitude ratio equal to 0.25.

For the angles θ close to 90°, the accuracy of determination of the direction of arrival of pulsed radiation was lower because of the influence of the cables. The RMSD of error for all points was no more than 4° for θ and 1° for φ.

Conclusion

The basic requirements for the antennas intended for receiving UWB electromagnetic pulses with low distortion have been formulated. With the traveling wave superposition method, the factors affecting the distortion of the pulse waveform by linear receiving antennas have been revealed and investigated. The notion of the transfer function of a receiving antenna has been introduced and calculation formulas have been derived for the transfer functions of dipole and loop receiving antennas.

It has been shown that electromagnetic radiation pulses can be received with low distortion, such that the RMSD of the waveform of the voltage pulse across the antenna load from the waveform of the received electromagnetic pulse is less than 0.2, by using

- short dipoles whose arm length is no more than half the spatial length of the received pulse that are loaded by a load whose impedance is several times greater than the dipole impedance modulus;

- short dipoles whose arms are made of a resistive material of linear resistance no less than 10^3 Ω/m;
- dipoles with noncollinear arms (V-shaped antennas) whose length is greater than the spatial length of the received pulse and the angle between the arms is less than 20°;
- TEM horns whose length is greater than half the spatial length of the received pulse and the aperture is no more than a quarter of the pulse spatial length, and
- active antennas, which are short dipoles or monopoles with integrated controlled power sources (active elements) for which voltage amplifiers can be used.

The design principles and the methodology of calculation of the characteristics of vector receiving antennas that allow recording simultaneously and independently the three projections of the vector **E** of the received electromagnetic pulse have been discussed. It has been shown that using a vector antenna, not only the polarization structure of the radiated UWB pulse or of the pulse scattered by an extended object can be investigated, but the direction of its arrival can also be determined.

Problems

6.1 What is meant by the transfer function of a receiving antenna?

6.2 How does the shape of the pattern of a receiving antenna depend on the way of connecting the load?

6.3 Why the frequency dependence of the transfer function of a short dipole becomes less pronounced as the load impedance is increased?

6.4 Which is the difference between a TEM horn and a V-antenna?

6.5 How the received-to-reradiated power ratio depends on the load impedance?

6.6 How many times the attenuation factor of a current wave in a receiving wire of length 0.25λ and diameter 0.001λ will change if the wire diameter is increased 30 times?

6.7 Use formula (6.18) to show that a value of the load impedance exists at which the reradiated power is a minimum.

References

[1] Il'ushenko, V. N., et al., Picosecond Pulse Technology, Moscow: Energoatomizdat, 1993 (in Russian).

[2] Markov, G. T., and D. M. Sazonov, Antennas, Moscow: Energia, 1975 (in Russian).

[3] Helmholtz, H., "Über Einige Gesetze der Vertheilung Elektrischer Ströme in Körperlichen Leitern mit Anwendung auf die Thierisch-Elektrischen Versuche," Ann. der Physik und Chemie, Bd. 89, No. 6, 1853, S. 211–233.

[4] Thévenin, L., "Extension de la loi d'Ohm aux Circuits Electromoteurs Complexes," Annales Télégraphiques (3eme série), Vol. 10, No. 5, 1883, pp. 222–224.

[5] Lavrov, G. A., and A. S. Knyazev, Near-Surface and Subsurface Antennas, Moscow: Sov. Radio, 1965 (in Russian).

[6] Drabkin, A. L., and I. L. Zuzenko, Antenna-Feeder Devices, Moscow: Sov. Radio, 1961 (in Russian).

[7]   Nadenenko, S. I., *Antennas*, Moscow: Svyazizdat, 1959 (in Russian).

[8]   Schelkunoff, S. A., and H. T. Friis, *Antenna Theory and Practice*, New York: Wiley, 1952.

[9]   Stratton, J. A., and L. J. Chu, "Steady-State Solutions of Electromagnetic Field Problems," *J. Appl. Phys.*, Vol. 12, No. 3, 1941, pp. 230–248.

[10] Kessenikh, V. N., "On the Characteristic Impedance of a Long Single-Wire Line," *Dokl. AN SSSR*, Vol. 27, No. 4, 1940, pp. 558–562.

[11] King, R. W. P., and G. S. Smith, *Antennas in Matter: Fundamentals, Theory, and Applications*, Cambridge, MA: MIT Press, 1981.

[12] Kessenikh, V. N., *Propagation of Radio Waves*, Moscow: Gostekhteorizdat, 1953 (in Russian).

[13] Kalantarov, P. L., and L. A. Tseitlin, *Calculation of Inductances*, Leningrad: Energoatomizdat, 1986 (in Russian).

[14] Ramo, S., and J. R. Whinnery, *Fields and Waves in Modern Radio*, New York: Wiley, 1944.

[15] Chen, K.-M., and V. V. Liepa, "Minimisation of the Back-Scattering of a Cylinder by Control Loading," *IEEE Trans. Antennas Propagat.*, Vol. 12, No. 5, 1964, pp. 576–582.

[16] Podosenov, S. A., A. A. Sokolov, and S. V. Al'betkov, "Excitation of a V-antenna by a Pulse Electromagnetic Field," *IEEE Trans. Electromagn. Compat.*, Vol. 38, No. 1, 1996, pp. 31–42.

[17] Podosenov, S. A., A. A. Potapov, and A. A. Sokolov, *Pulse Electromagnetics of Ultrawideband Radio Systems and Coupled Structure Fields*, Moscow: Radiotekhnika, 2003 (in Russian).

[18] Litvinenko, O. N., and V. I. Soshnikov, *Theory of Inhomogeneous Lines and Their Applications in Radio Engineering*, Moscow: Sov. Radio, 1964 (in Russian).

[19] Rothammel, K., *Antennenbuch*, Berlin: Deutscher Militarverlag, 1966.

[20] Farr, E. G., and C. E. Baum, "A Simple Model of Small-Angle TEM Horns," In *Sensor and Simulation Notes 340*, C. E. Baum (ed.), Kirtland, NM: Air Force Research Laboratory, Directed Energy Directorate, 1992.

[21] Lee, R. T., and G. S. Smith, "A Design Study for the Basic TEM Horn Antenna," *IEEE Antennas Propagat. Mag.*, Vol. 46, No. 1, 2004, pp. 86–92.

[22] Lee, R. T., and G. S. Smith, "On the Characteristic Impedance of the TEM Horn Antenna," *IEEE Trans. Antennas Propagat.*, Vol. 52, No. 1, 2004, pp. 315–318.

[23] Farr, E. G., et al., "Multifunction Impulse Radiating Antennas: Theory and Experiment." In *Ultra-Wideband, Short-Pulse Electromagnetics 4*, pp. 131–144, E. Heyman, J. Shiloh, and B. Mandelbaum (eds.), New York: Plenum Press, 1999.

[24] Izyumova, T. I., and V. T. Sviridov, *Waveguides, Coaxials, and Strip Lines*, Moscow: Energia, 1975 (in Russian).

[25] Andreev, Yu. A., et al., "A High-Performance Source of High-Power Nanosecond Ultrawideband Radiation Pulses," *Instrum. Exp. Tech.*, Vol. 54, No. 6, 2011, pp. 794–802.

[26] Andreev, Yu. A., V. I. Koshelev, and V. V. Plisko, "Characteristics of TEM Antennas in Receiving and Radiating Modes," *Proc. 5th All-Russian Science-and-Technology Conf. Radar and Radio Communications*, Moscow, Nov. 21–25, 2011, pp. 77–82.

[27] Gunston, M. A. R., *Microwave Transmission-Line Impedance Data*, London: Van Nostrand, 1972.

[28] Wu, T. T., and R. W. P. King, "The Cylindrical Antenna with Nonreflecting Resistive Loading," *IEEE Trans. Antennas Propagat.*, Vol. 13, No. 5, 1965, pp. 369–373.

[29] Esselle, K. P., and S. S. Stuchly, "Pulse-Receiving Characteristics of Resistively Loaded Dipole Antennas," *IEEE Trans. Antennas Propagat.*, Vol. 38, No. 10, 1990, pp. 1677–1683.

[30] Balzovsky, E. V., Yu. I. Buyanov, and V. I. Koshelev, "An Ultrawideband Dipole Antenna with Resistive Arms," *J. Commun. Technol. Electron.*, Vol. 49, No. 4, 2004, pp. 426–431.

[31] Balzovskii, E. V., Yu. I. Buyanov, and V. I. Koshelev, "An Active Antenna for Measuring Pulsed Electric Field," *Russ. Phys. J.*, Vol. 50, No. 5, 2007, pp. 503–508.

[32] Meinke, H., "Aktive antennen," *Nachrichten Technische Zeitschrift.*, Bd. 19, No. 12, 1966, S. 697–704.

[33] Meinke, H., and G. Flachenecker, "Active Antennas with Transistors," *Broadcast Commun.*, Vol. 3, No. 3, 1967, pp. 18–24.

[34] Meinke, H., "Zur Definition Einer Aktiven Antenne," *Nachrichten Technische Zeitschrift.*, Bd. 26, No. 4, 1973, S. 179–180.

[35] Andreev, Yu. A., et al., "A Receiving Antenna for Investigating the Space-Time Structure of Ultrawideband Electromagnetic Pulses," *Electromagnitnye Volny i Elektronnye Sistemy*, Vol. 6, No. 2–3, 2001, pp. 69–75.

[36] Andreev, Yu. A., Yu. I. Buyanov, and V. I. Koshelev, "A Combined Antenna with Extended Bandwidth," *J. Commun. Technol. Electron.*, Vol. 50, No. 5, 2005, pp. 535–543.

[37] Chelyshev, V. D., *Receiving Radio Centers*, Moscow: Svyaz, 1975 (in Russian).

[38] Yoon, I. J., et al., "Active Integrated Antenna for Mobile TV Signal Reception," *Microwave Opt. Technol. Lett.*, Vol. 49, No. 12, 2007, pp. 2998–3001.

[39] Balzovskii, E. V., Yu. I. Buyanov, and V. I. Koshelev, "Dual Polarization Receiving Antenna Array for Recording of Ultra-Wideband Pulses," *J. Commun. Technol. Electron.*, Vol. 55, No. 2, 2010, pp. 172–180.

[40] Knyaz, A. I., and V. A. Katorgin, "Vector Receiving Antennas," *Zarubezhnaya Radioelektronika*, No. 8, 1984, pp. 36–42.

[41] Bulakhov, M. G., Yu. I. Buyanov, and V. P. Yakubov, "Polarization of the Interference Field during Reflection of Electromagnetic Waves from an Intermedia Boundary," *Russ. Phys. J.*, Vol. 39, No. 10, 1966, pp. 956–960.

[42] Koshelev, V. I., E. V. Balzovsky, and Yu. I. Buyanov. "Investigation of Polarization Structure of Ultrawideband Radiation Pulses," *IEEE Pulsed Power Plasma Science Conf.*, Las Vegas, June 17–22, 2001, Vol. 2, pp. 1657–1660.

[43] Balzovskii, E. V., Yu. I. Buyanov, and V. I. Koshelev, "A Vector Receiving Antenna for Measuring the Polarization Structure of Ultrawideband Electromagnetic Pulses," *J. Commun. Technol. Electron.*, Vol. 50, No. 8, 2005, pp. 863–872.

[44] Koshelev, V. I., et al., "Radar Signal Polarization Structure Investigation for Object Recognition," pp. 707–714. In *Ultra-Wideband, Short-Pulse Electromagnetics 7*. F. Sabat et al. (eds.), New York: Springer, 2007.

[45] Balzovsky, E. V., Yu. I. Buyanov, and V. I. Koshelev, "Characterization of Active Vector Receiving Antenna," *Proc. 16th Inter. Symposium on High Current Electronics*, Tomsk, Sep. 19–24, 2010, pp. 451–454.

[46] Efremov, A. M., et al. "High-Power Sources of Ultra-Wideband Radiation with Subnanosecond Pulse Lengths," *Instrum. Exp. Tech.*, Vol. 54, No. 1, 2011, pp. 70–76.

[47] Gubanov, V. P., et al., "Sources of High-Power Ultrawideband Radiation Pulses with a Single Antenna and a Multielement Array," *Instrum. Exp. Tech.*, Vol. 48, No. 3, 2005, pp. 312–320.

CHAPTER 7

Transmitting Antennas

Introduction

The waveform of a radiated UWB pulse, whose spectrum can occupy a frequency range of more than two octaves, can be distorted due to the following factors:

1. As shown in chapter 2 for an electromagnetic pulse radiated by a short electric dipole, the electric field strength in the far-field region, $E(t)$, is proportional to the time derivative of the exciting current $I(t)$. As any radiator can be represented by a set of elementary electric dipoles, the radiated electromagnetic pulse may be different in waveform from the current pulse exciting the radiator.

2. The waveform of an electromagnetic pulse must satisfy the equilibrium condition $\int_{-\infty}^{\infty} E(t)\,dt = 0$. From this condition, it also follows that the shortest radio pulse able to exist in space as an electromagnetic pulse, is a bipolar pulse, that is, a radio pulse whose duration is one oscillation period.

3. The pass band of any antenna is finite, and in the general case, an antenna may be considered as a filter. If the antenna pass band is narrower than the frequency band of the pulse spectrum, the pulse waveform of the current exciting the radiator is also inevitably distorted.

The first of the above factors is associated with the physical processes that occur during the radiation of an electromagnetic field, and it cannot be eliminated. As to the second and third factors, the distortion of the radiated pulse waveform caused by them can be substantially reduced by exciting an antenna with current pulses satisfying the equilibrium condition. In addition, the current pulse waveform should correspond to a spectrum where most of the pulse energy is concentrated in a rather narrow frequency range. The radiation efficiency of a radiator (the ratio of the radiated energy to the energy of the pulse applied to the radiator input) is also determined not only by the properties of the radiator, but also by the exciting pulse waveform. The matter is that an antenna can be matched to the feeder in a restricted frequency range. If the match band is narrower than the frequency band of the current pulse spectrum, a portion of the pulse energy will be reflected from the antenna input.

Thus, to radiate a pulse with maximum efficiency and minimum waveform distortion, the antenna should have a pass band comparable to the frequency range occupied by the current pulse spectrum (e.g., a band in which no less than 90% of the electromagnetic pulse energy at the antenna input is concentrated). By the pass band of a radiator, we mean the frequency range in which the main characteristics of the antenna that are responsible for its operability remain unchanged or vary within tolerance limits.

If it is required to ensure only maximum radiation efficiency, the main characteristics are:

- the shape of the antenna pattern and
- the input impedance of the antenna or the degree of its match to the feeder.

If, besides, it is required to minimize the distortion of the radiated pulse waveform, the main parameters should be complemented with

- the polarization characteristic and
- the presence of a phase center and its position.

If a radiator is intended for use as an element of a scanning antenna array, rather stringent requirements are imposed on the radiator dimensions and pattern shape.

The requirements for a radiator of UWB pulses can be formulated only if the following parameters are known:

- the duration and waveform of the exciting pulse,
- the required radiation efficiency, and
- the admissible degree of distortion of the pulse waveform.

Moreover, it is necessary to consider the correlation between the pulse waveform, the radiation efficiency, and the degree of distortion of the radiated pulse. In addition to broadbandedness, a radiator of UWB pulses should also possess; in some cases, high electric strength, as the up-to-date pulse formers make feasible voltage pulses having amplitudes of some hundreds of kilovolts.

In developing the design principles for radiators of UWB pulses, it is necessary, first of all, to reveal and investigate the factors affecting the bandwidth of an antenna.

7.1   The Transfer Function of a Transmitting Antenna

Any radio engineering system intended for generation of UWB radiation pulses comprises, at least, a generator of voltage or current pulses, a feeder, and an antenna radiating a pulse in a desired direction. The radiation of electromagnetic waves involves the conversion of the energy of alternating currents and charges into the energy of electromagnetic waves freely propagating in the surrounding medium. This process is inertial, and therefore, the time dependence of the electromagnetic field components does not coincide with the time dependence of its exciting currents and charges. The pulse waveform can be additionally distorted due to the energy losses in the antenna-feeder path.

7.1.1   The Transfer Function of a Radiation Source

To reveal the factors responsible for the distortion of the waveforms of electromagnetic radiation pulses, we consider a simplified structural diagram of a UWB radiation source (Figure 7.1).

The pulse generator produces at the output (terminals 1–1′) a voltage pulse $U(t)$, having a spectral function $U(\omega)$, that, via a feeder with characteristic impedance ρ_f, arrives at the antenna input (terminals 2–2′). The antenna converts the voltage (or current) pulse into an electromagnetic field pulse, which propagates as a free wave with a spectral function $\mathbf{E}(\mathbf{r},\omega)$. Subsequently, it is supposed that the generator is matched to the feeder and that the feeder does not introduce distortions in the voltage pulse waveform, that is, its transmission ratio is equal to unity and does not depend on frequency. In this case, it can be assumed that the antenna is excited by a source of EMF $U(t)$ with internal impedance ρ_f. As the antenna is the load of the feeder, it can be represented as a two-terminal network whose impedance is equal to the antenna input impedance $Z_a(\omega) = R_a(\omega) + iX_a(\omega)$. In this case, we have $U_a(\omega) = U(\omega)Z_a/(Z_a(\omega) + \rho_f)$ and the current at the antenna input $I_0(\omega) = U_a(\omega)/Z_a(\omega)$. The equivalent circuit of a transmitting antenna is given in Figure 7.2.

The electric field strength produced by a radiator in the far-field region is determined by the expression [1]

$$\mathbf{E}(\mathbf{r},\omega) = -i\omega\mu_0 I_0(\omega)\frac{\exp\left(\dfrac{-i\omega r}{c}\right)}{4\pi r}\int_{V_a} \mathbf{j}(\omega,\mathbf{r}')\exp\left(i\omega\frac{r'}{c}\cos\alpha\right)d^3\mathbf{r}' \tag{7.1}$$

where $\mathbf{j}(\omega,\mathbf{r}')$ is the current distribution function normalized to $I_0(\omega)$, $\mathbf{r}$ is the radius vector of the observation point, $\mathbf{r}'$ is the radius vector of a point on the radiator, and

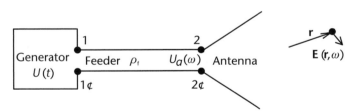

Figure 7.1   Structural diagram of a UWB radiation source.

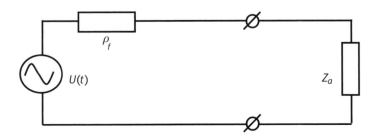

Figure 7.2   Equivalent circuit of a transmitting antenna.

α is the angle between $\mathbf{r}$ and $\mathbf{r}'$. The factor $i\omega$ accounts for the fact that for each spectral component, the radiated field strength is proportional to $d[\exp(i\omega t)]/dt$.

Then, the transfer function of the radiation source is determined by the equation

$$H_t(\mathbf{r},\omega) = \frac{\mathbf{E}(\mathbf{r},\omega)}{U(\omega)} = -\frac{i\omega\mu_0}{Z_a(\omega) + \rho_f} \frac{\exp\left(\dfrac{-i\omega r}{c}\right)}{4\pi r} \int_{V_a} \mathbf{j}(\omega,\mathbf{r}')\exp\left(i\omega\frac{r'}{c}\cos\alpha\right)d^3\mathbf{r}' \quad (7.2)$$

The multiplier $\exp(-i\omega r/c)/4\pi r = H_0(r,\omega)$ can be treated as the transfer function of free space, which does not depend on the characteristics of the radiation source and does not introduce distortions to the pulse waveform. In this case, we have

$$H_t(\mathbf{r},\omega) = H_0(\mathbf{r},\omega)H_a(\alpha,\omega) \quad (7.3)$$

where $H_a(\alpha,\omega)$, having the meaning of the space-frequency vector transfer function of the antenna, can be represented as

$$H_a(\alpha,\omega) = \frac{-i\omega\mu_0}{Z_a(\omega) + \rho_f} \int_{V_a} \mathbf{j}(\omega,\mathbf{r}')\exp\left(i\omega\frac{r'}{c}\cos\alpha\right)d^3\mathbf{r}' \quad (7.4)$$

If the transfer function of an antenna, $H_a(\alpha,\omega)$ is known, the waveform of an electromagnetic pulse $\mathbf{E}(t)$ radiated in a direction $\mathbf{r}$ is determined as the inverse Fourier transform of the product of $H_a(\alpha,\omega)$ and the spectral function $U(\omega)$ of the exciting pulse $U(t)$. For a pulse to be radiated without distortion of its waveform, the following condition for undistorted pulse transmission should be satisfied:

$$\left|H_a(\alpha,\omega)\right| = const, \;\; \arg\left(H_a(\alpha,\omega)\right) = -t_0\omega, \;\; \left(\omega_{\min} < \omega < \omega_{\max}\right) \quad (7.5)$$

where t_0 is a constant having the dimension of time; $\omega_{\min}$ and $\omega_{\max}$ are, respectively, the lower and the upper bound of the frequency band occupied by the signal spectrum. The last condition is called "linear phase."

In the general case, the transfer function of an antenna can be expressed in terms of the antenna parameters as

$$H_a(\alpha,\omega) = -ikA(\omega)\left|f(\alpha,\omega)\right|\exp\left[i\psi(\alpha,\omega)\right]\mathbf{p}(\alpha,\omega) \quad (7.6)$$

where $k = \omega/c$ is the wave number of free space, $A(\omega) = Z_0/(\rho_f + Z_a(\omega))$ is an amplitude factor independent of spatial coordinates, Z_0 is the characteristic impedance of free space, $|f(\alpha,\omega)|$ is the amplitude pattern function, $\psi(\alpha,\omega)$ is the phase pattern function, and $\mathbf{p}(\alpha,\omega)$ is the polarization characteristic of the antenna.

The frequency dependence of $A(\omega)$ is determined in the main by the frequency dependence of the antenna input impedance $Z_a(\omega)$ that is responsible for the distortion of the signal spectral function in the case of a mismatch between the antenna and the feeder. For radiators of finite dimensions, the pattern shape depends on frequency, and for the majority of antennas, the width of the pattern main lobe decreases and the boresight direction can vary with increasing frequency. As a result, the spectral function of a radiated signal becomes spatially dependent; that is, signals of different waveform are radiated in different directions. Additional space-frequency distortions of the radiated signal spectrum can arise due to the frequency dependence of the polarization characteristic of the antenna. The frequency dependence of the phase pattern function is mainly caused by the frequency dependence of the antenna input impedance, and its spatial dependence is related to the antenna dimensions and design. In the general case, we have $\psi(\alpha,\omega) \neq -t_0\omega$, and this also violates the condition for undistorted pulse transmission.

Now, several types of antennas and tens of their design versions are used to produce UWB signals. Almost for each antenna type, a special calculation method is available or is being developed. This substantially complicates a quantitative analysis of the time or frequency dependences of the characteristics of antennas of various types. However, to reveal the main factors responsible for the distortion of the waveform of a radiated pulse, it suffices to perform a qualitative analysis of the antenna characteristics. For doing this, it is reasonable to classify all existing antennas into three groups:

- linear radiators (antennas whose traverse dimensions are considerably less than their longitudinal dimensions, and the radiation field is created by a linear current that can vary only along the radiator);
- aperture radiators (antennas whose radiation field is created by currents distributed over some surface), and
- combined radiators (antennas consisting of two and more nonidentical radiators).

According to the superposition principle, the field of a combined radiator is the sum of the fields of its constituent linear or aperture radiators. The field of an aperture radiator can be determined as the sum of the fields of linear radiators if the radiating surface is represented as a set of linear radiators. Thus, to reveal the factors affecting the frequency dependence of the transfer function of an antenna, it suffices to consider the frequency dependences of the input impedance and pattern functions of the linear radiator that are determined by the current distribution along the radiator.

7.1.2   The Current Distribution in a Linear Radiator

Let a straight wire of length L with characteristic impedance ρ_a be arranged along the z-axis. A source of an EMF U with internal impedance Z_1 is included in the wire at the origin of coordinates, and the wire end is loaded with an impedance Z_2 (Figure 7.3). The EMF generates a current wave of amplitude $I_0 = U/(\rho_a + Z_1)$ propagating along the wire with velocity v in the direction of positive z. Having reached the end of the wire, the wave is reflected from the load Z_2 with a reflection coefficient R_2, its amplitude being multiplied by $R_2\exp[-i\omega(L - z)/v]$ and propagates in the opposite direction. Then, having reflected from the impedance Z_1 with a reflection coefficient R_1 and a phase incursion $\beta L = \omega(L - z)/v$, it propagates again in the direction of positive z, etc. Thus, the current at a point z is equal to the sum of the currents of all multiply reflected waves having arrived at this point.

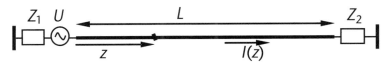

Figure 7.3   Sketch of a linear radiator.

For a harmonic regime $U(\omega) = U_0 \exp(i\omega t)$, we have

$$I(z) = I_0 \Big[\exp(-i\gamma z) + R_2 \exp(-i2\gamma L)\exp(i\gamma z) + R_1 R_2 \exp(-i2\gamma L)\exp(-i\gamma z)$$
$$+ R_1 R_2^2 \exp(-i4\gamma L)\exp(i\gamma z) + R_1^2 R_2^2 \exp(-i4\gamma L)\exp(-i\gamma z) + \ldots \Big] \tag{7.7}$$

where $\gamma = \omega/v$ is the propagation constant of a current wave in the wire. The square-bracketed terms in (7.7) can be grouped in two infinite geometrical progressions whose summation yields

$$I(z) = \frac{U}{Z_1 + \rho_a} \frac{\exp(-i\gamma z) + R_2 \exp(-i2\gamma L)\exp(i\gamma z)}{1 - R_1 R_2 \exp(-i2\gamma L)} \tag{7.8}$$

Supposing that the field of the current wave along the wire is similar to that of a TEM wave, we have for the reflection coefficients

$$R_1 = \frac{(\rho_a - Z_1)}{(\rho_a + Z_1)}, \quad R_2 = \frac{(\rho_a - Z_2)}{(\rho_a + Z_2)} \tag{7.9}$$

and then equation (7.8) can be represented as

$$I(z) = \frac{U}{Z_1 + Z_a} \left(\cos(\gamma z) - i\frac{Z_a}{\rho_a}\sin(\gamma z) \right) \tag{7.10}$$

where Z_a is the input impedance of the radiator given by

$$Z_a = \rho_a \frac{Z_2 + i\rho_a \tan(\gamma L)}{\rho_a + iZ_2 \tan(\gamma L)} \tag{7.11}$$

The obtained equations describe the current distribution in an unbalanced linear radiator (monopole) excited at one end. Equation (7.10) allows one to investigate the frequency dependence of the radiator impedance and radiation characteristics in various operation modes, which are determined by the conditions at the ends of the radiator, its characteristic impedance ρ_a, and the current wave propagation constant γ. Let the generator be matched to the feeder and, hence, $Z_1 = \rho_f$. It should also be taken into account that ρ_a and γ are complex quantities due to radiation losses. The method of their calculation is discussed in Section 6.1.3.

If $Z_2 = \rho_a$, the current distribution in a linear radiator is given by

$$I(z) = \frac{U}{\rho_f + Z_a}\exp(-i\gamma z) \tag{7.12}$$

where $Z_a \cong \rho_a$; that is, a traveling current wave is formed in the radiator. A radiator of this type is termed a traveling-wave antenna (TWA).

If the end load is absent ($Z_2 \to \infty$), we have $Z_a = -i\rho_a \cot(\gamma L)$ and (7.10) becomes

$$I(z) = \frac{U}{Z_1 + Z_a}\frac{\sin[\gamma(L - z)]}{\sin(\gamma L)} \tag{7.13}$$

indicating that a standing current wave is formed along the wire. A radiator of this type is termed a standing-wave antenna (SWA).

A balanced version of SWA (dipole antenna) is shown in Figure 7.4. The antenna is designed by adding a linear radiator of length L oriented in the direction of negative z. The current distribution along this direction is described by the formula

$$\begin{aligned}
I(-z) &= \frac{0.5U}{\rho_a + 0.5Z_1}\frac{\exp[i\gamma(L + z)] - \exp[-i\gamma(L + z)]}{\exp[i\gamma L] + R_1\exp[-i\gamma L]} \\
&= \frac{0.5U}{0.5Z_1 - i\rho_a\cot(\gamma L)}\frac{\sin[\gamma(L + z)]}{\sin(\gamma L)}
\end{aligned} \tag{7.14}$$

In this case, we obtain the well-known expression for a balanced dipole with a sinusoidal current distribution:

$$I(-z) = \frac{U}{Z_1 - i\rho_d\cot(\gamma L)}\frac{\sin[\gamma(L - |z|)]}{\sin(\gamma L)} \tag{7.15}$$

where $\rho_d = 2\rho_a$ is the characteristic impedance of the dipole. To calculate γ and ρ_d, an iterative process (Section 6.1.3) is also used, but to attain its fast convergence, the radiation resistance in the initial approximation is calculated by the formula for a dipole with a sinusoidal current distribution without regard for radiation loss ($\gamma = k$):

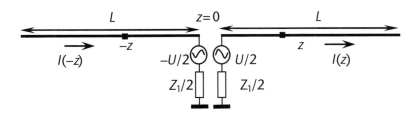

Figure 7.4   Sketch of a dipole radiator.

$$R_\Sigma = 120 \int\limits_0^\pi \frac{[\cos(kL\cos\theta) - \cos(kL)]^2}{\sin\theta} \, d\theta \tag{7.16}$$

The above method can readily be extended to the case of a curvilinear radiator, as the progression sums in (7.7) do not depend on the radiator shape. Hence, it suffices to substitute the coordinate ζ along the curvilinear wire for z in the current distribution and, when calculating the radiated field, to take into account the wire curvature in finding the phase incursion between a point on the radiator and the observation point. For a ring radiator (loop) of radius b with unbalanced excitation (Figure 7.5), the coordinate of a point on the radiator is defined as $\zeta = b\delta$ and the radiator length as $L = 2\pi b$.

For this case, the current distribution is described by the expression

$$I_l(\zeta) = \frac{U}{Z_1 + Z_l}\left(\cos(\gamma\zeta) - i\frac{Z_l}{\rho_l}\sin(\gamma\zeta)\right) \tag{7.17}$$

where $Z_l = \rho_l[Z_2 + i\rho_l\tan(2\gamma\pi b)]/[\rho_l + Z_2 i\tan(2\gamma\pi b)]$ is the input impedance of the loop, $\rho_l = 60(\ln(8b/a) - 2)$ is the characteristic impedance of the loop wire, and a is the wire radius.

The obtained expressions describe the current distribution in a linear radiator and allow one to investigate the radiator characteristics in various operation modes, which are determined by the conditions at the ends of the radiator, its characteristic impedance ρ_a, and the current wave propagation constant γ.

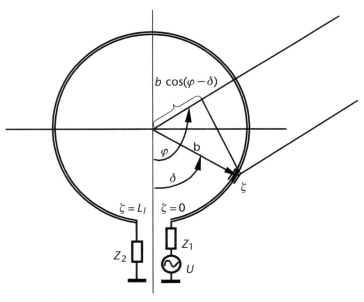

Figure 7.5   Sketch of a ring radiator.

7.1.3   The Transfer Function of a Linear Radiator

The field radiated by a linear radiator arranged along the z-axis has a frequency-independent linear polarization characteristic, and using formula (7.10), we can be described it as

$$
\begin{aligned}
E_\theta &= -\frac{i\omega\mu_0 \exp(-ikr)\sin\theta}{4\pi r}\int_0^L I(z)\exp(ikz\cos\theta)\,dz \\
&= -\frac{ikUZ_0\sin\theta\exp(-ikr)}{4\pi r(\rho_f + Z_a)}\int_0^L\left(\cos(\gamma z) - i\frac{Z_a}{\rho_a}\sin(\gamma z)\right)\exp(ikz\cos\theta)\,dz
\end{aligned}
\tag{7.18}
$$

The transfer function (7.6) in this case reads

$$
\mathbf{H}_a(\theta,\omega) = A(\omega)|f(\theta,\omega)|\exp[i\psi(\theta,\omega)]\mathbf{p}(\theta,\omega)
\tag{7.19}
$$

where $A(\omega) = -ikZ_0/[\rho_f + Z_a(\omega)]$, $Z_a(\omega) = \rho_a[Z_2 + i\rho_a\tan(\gamma L)]/[\rho_a + iZ_2\tan(\gamma L)]$ is the antenna impedance, $\mathbf{p}(\theta,\omega) = \boldsymbol{\theta}_0$ is the polarization characteristic ($\boldsymbol{\theta}_0$ is the basis vector of the spherical coordinate system), $|f(\theta,\omega)| = \left|\sin\theta\int_0^L(\cos(\gamma z) - i(Z_a/\rho_a)\sin(\gamma z))\exp(ikz\cos\theta)\,dz\right|$ is the amplitude pattern function, and $\psi(\theta,\omega) = \arg[f(\theta,\omega)]$ is the phase pattern function, which, for a linear radiator, satisfies the condition of undisturbed pulse transmission.

Thus, the frequency dependence of the transfer function of a linear radiator is determined by the factor $i\omega$ and by the frequency dependences of the input impedance and of the amplitude pattern shape that depend on the conditions at the ends of the radiator. For a hypothetical perfect radiator of linearly polarized radiation matched to the feeder and having a frequency-independent pattern shape, the transfer function is given by

$$
H_a(\omega) = -i\omega B_a, \text{ where } B_a = const
\tag{7.20}
$$

In Figure 7.6, the waveforms of the voltage (current) pulses at the input of a perfect antenna with an unlimited pass band are depicted by dashed lines and the waveforms of the radiated pulses are depicted by solid lines.

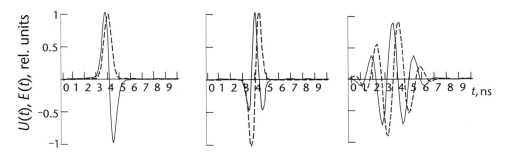

Figure 7.6   The waveforms of the voltage (current) at the input of a perfect antenna (dashed lines) and the radiated pulse waveforms (solid lines).

In an antenna, the radiated pulse undergoes additional distortions due to the frequency dependence of the antenna characteristics.

If $R_2 = 0$ ($Z_2 = \rho_a$), a traveling wave of current $I(z) = I_0 \exp(-i\gamma z)$ propagates along the radiator; that is, the radiator is a traveling-wave antenna. In this case, the input impedance, $Z_a \cong \rho_a$ is almost independent of frequency. The pattern function of a TWA is given by

$$f(\theta) = \sin\theta \frac{\sin[0.5L(k\cos\theta - \gamma)]}{0.5L(k\cos\theta - \gamma)} \tag{7.21}$$

The dependence of the shape of the normalized pattern of a TWA on its electric length is shown in Figure 7.7.

The transfer function of a TWA, H_T, is given by

$$H_T = \frac{Z_0 L \sin\theta}{2(\rho_f + \rho_a)} \frac{\sin[0.5L(k\cos\theta - \gamma)]}{[0.5L(k\cos\theta - \gamma)]} \tag{7.22}$$

The frequency dependence of the TWA transfer function is determined by the frequency dependence of the pattern shape and weakly depends on the impedance of the voltage source. Figure 7.8 gives plots of the frequency dependence of $|H_T|$ corresponding to different directions for a TWA of length 0.6 m with $\rho_f = \rho_a$.

If $Z_2 \to \infty$, then $R_2 \to -1$, and in a radiator whose length of no greater than one or two wave lengths, a standing wave is typically formed, and the radiator becomes a SWA. For this case, the transfer function H_S is given by

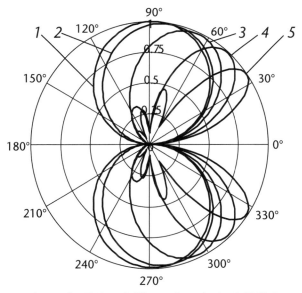

Figure 7.7 The pattern shapes for TWAs of different length: $L = 0.125\lambda$ (curve 1), 0.25λ (curve 2), 0.5λ (curve 3), 1.0λ (curve 4), and 2.0λ (curve 5).

$$H_S = \frac{Z_0}{Z_a + \rho_f} \frac{\left[\cos(kL\cos\theta) - \cos\gamma L\right]}{\sin\theta} \tag{7.23}$$

The input impedance of an SWA is given by $Z_a = -i\rho_a\cot(\gamma L)$ and rather strongly depends on frequency. As an example, Figure 7.9 presents a plot of the dependence of Z_a on the electric length of a cylindrical wire of diameter $0.1L$. For a radiator whose length is less than half the wave length, the coefficient R_f of wave reflection in a feeder with characteristic impedance $\rho_f = 140\ \Omega$ can vary within the limits $-1 < R_f < 0.6$. If $L > \lambda$, we have $|R_f| < 0.3$.

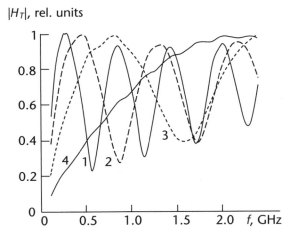

Figure 7.8 The modulus of the transfer function of a TWA for different directions: $\theta = 90°$ (curve 1), 70° (curve 2), 50° (curve 3), and 30° (curve 4).

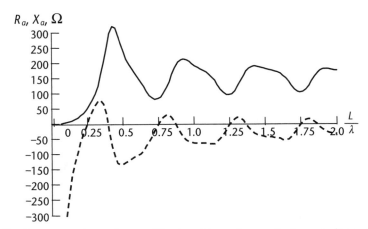

Figure 7.9 The frequency dependence of the input impedance of a linear radiator with a standing current wave.

The normalized pattern functions for SWAs of different length are given in Figure 7.10.

The transfer functions of an SWA of length 0.6 m corresponding to different directions are shown in Figure 7.11.

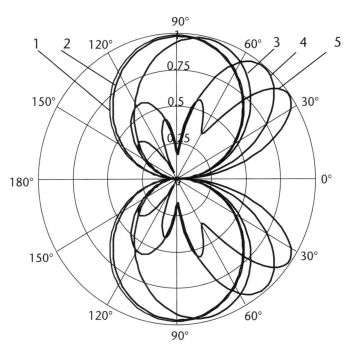

Figure 7.10 The pattern shapes for SWAs of different length: $L = 0.125\lambda$ (curve 1), 0.25λ (curve 2), 0.5λ (curve 3), 1.0λ (curve 4), and 2.0λ (curve 5).

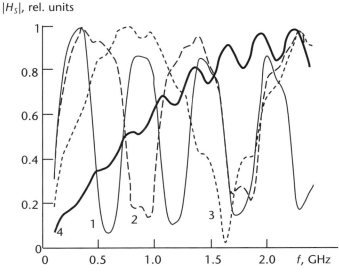

Figure 7.11 The SWA transfer function modulus corresponding to different directions: $\theta = 90°$ (curve 1), $70°$ (curve 2), $50°$ (curve 3), and $30°$ (curve 4).

The input impedance of an balanced SWA (dipole), $Z_d = 2Z_a$, has the same frequency dependence as that of a monopole. The pattern shape of a dipole is given by the equation

$$f(\theta, \omega) = \frac{[\cos(kL\cos\theta) - \cos\gamma L]}{\sin\theta} \tag{7.24}$$

Its dependence on the electric length of a dipole is shown in Figure 7.12.

These results show that when the dipole length is less than the wave length, the pattern shape weakly depends on frequency, but the input impedance varies strongly. If the dipole length is greater than the wave length, the variations in input impedance, on the contrary, become inappreciable, but the position of the pattern maximum and the pattern width depend on frequency.

The field strength in the far-field region and the transfer function of a ring radiator (Figure 7.5) located in the plane $\theta = \pi/2$, for the case of its unbalanced excitation, are determined by the expressions

$$E_\varphi = \frac{ikbZ_0U}{Z_1 + Z_l}\frac{\exp(-ikr)}{4\pi r}$$

$$\times \int_0^{2\pi}\left[\cos(\gamma b\delta) - i\frac{Z_l}{\rho_l}\sin(\gamma b\delta)\right]\exp[ikb\sin\theta\cos(\varphi - \delta)]\cos(\varphi - \delta)d\delta \tag{7.25a}$$

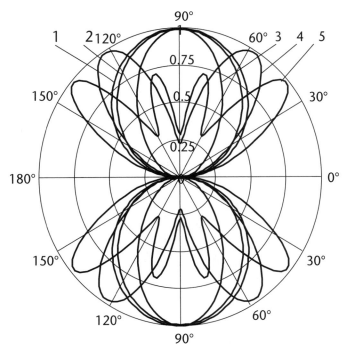

Figure 7.12 The pattern shape of a dipole versus its length: 0.125λ (curve 1), 0.25λ (curve 2), 0.5λ (curve 3), 1.0λ (curve 4), and 2.0λ (curve 5).

$$H_l = \frac{-ikbZ_0}{Z_1 + Z_l}$$

(7.25b)

$$\times \int_0^{2\pi} \left[\cos(\gamma b\delta) - i\frac{Z_l}{\rho_l}\sin(\gamma b\delta) \right] \exp[ikb\sin\theta\cos(\varphi - \delta)]\cos(\varphi - \delta)d\delta$$

where φ and δ are the coordinates of the observation point and of a point on the radiator, respectively, Z_l is the input impedance of the loop, and ρ_l is the characteristic impedance of the loop. In this case, unlike a linear radiator, two standing wave modes can be realized: a short-circuit mode for $Z_2 = 0$ and a no-load mode for $Z_2 \to \infty$. The input impedance of the loop for these modes is determined by the formulas

$$Z_{lo} = -i\rho_l\cot(\gamma 2\pi b), \quad Z_{ls} = i\rho_l\tan(\gamma 2\pi b)$$

(7.26)

The pattern shapes for different values of kb are given in Figure 7.13(a) (for $Z_2 = 0$) and in Figure 7.13(b) (for $Z_2 \to \infty$).

The above results show that for a ring radiator, the pattern shape weakly depends on frequency as long as $L_l < 0.25\lambda$. Unlike a linear radiator, for $L_l > 0.5\lambda$, there is no direction for which the position of the pattern main lobe would be retained in a wide frequency range.

Figure 7.14 presents the patterns of a traveling-wave ring radiator ($Z_2 = \rho_l$).

A distinguishing feature of a traveling-wave ring radiator is its ability to form a cardioid pattern. However, the radiator efficiency is reduced, as the radiation power is partly absorbed by the impedance Z_2.

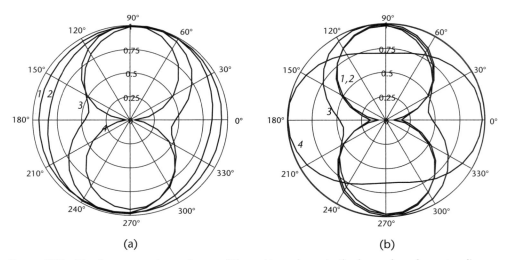

(a)   (b)

Figure 7.13 The frequency dependence of the pattern shape in the loop plane for a standing-wave loop with $Z_1 = \rho_l$ in a short-circuit mode ($Z_2 = 0$) (a) and in a no-load mode ($Z_2 \to \infty$) (b): $kb = 0.05$ (curve 1), 0.1 (curve 2), 0.25 (curve 3), and 0.5 (curve 4).

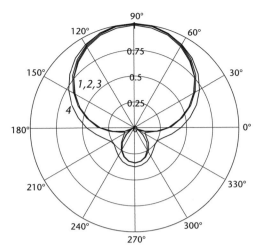

Figure 7.14   The frequency dependence of the pattern shape in the loop plane for a traveling-wave loop: kb = 0.05 (curve 1), 0.1 (curve 2), 0.25 (curve 3), and 0.5 (curve 4).

7.2   Distortion of Ultrawideband Electromagnetic Pulses during Radiation

As shown earlier, the waveform of a radiated pulse is varied during radiation even if the radiator is a perfect antenna with an unlimited pass band. Therefore, to investigate the effect of a parameter of an antenna on the pulse waveform, it is appropriate to compare the pulse radiated by the antenna with a pulse whose waveform is determined by the time derivative of the voltage pulse $U(t)$ exciting the antenna. If the admissible RMSD of the radiated pulse waveform from $dU(t)/dt$ is no more than 0.2, in most cases, it suffices to provide conditions under which the variation of the modulus of the transfer function (AFR) within the frequency range occupied by the pulse spectrum would not be over 3 dB and the deviation of the argument of the transfer function (ΔPFR) from a linear function would not be over $\pm\pi/16$.

7.2.1   The Radiated Pulse Waveform for a Monopole and a Collinear Dipole

By pulse distortion, we mean the deviation of the waveform of a pulse radiated by a given antenna from that of a pulse radiated by a perfect antenna, which can be considered as a reference pulse. In what follows, for a reference pulse, we use the pulse radiated by a hypothetical perfect antenna with an unlimited pass band, having a frequency-independent pattern shape and constant real input impedance. It is assumed that the antenna is excited by a bipolar pulse (Gauss monocycle) of duration 2 ns.

Figure 7.15 shows how the waveform of a pulse radiated by a standing-wave monopole (SWA) in the direction normal to its axis depends on the monopole length. In the calculations, it was assumed that the monopole diameter makes 0.1 of its length. The monopole length was measured in terms of the pulse spatial length $c\tau_p$, where c is the velocity of light and τ_p is the pulse duration.

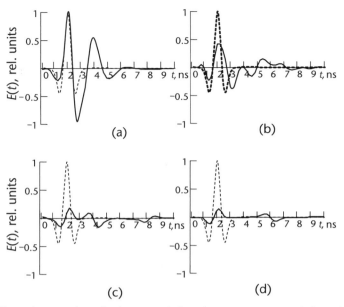

Figure 7.15 The pulse waveform in the normal direction versus monopole length: $L = 0.25c\tau_p$ (a), $0.5c\tau_p$ (b), $1.0c\tau_p$ (c), and $2.0c\tau_p$ (d); the reference pulse waveform is depicted by a dashed line.

The waveforms of pulses radiated by a monopole of length $1.5c\tau_p$ in different directions are shown in Figure 7.16. In the angular domain $15° < \theta < 30°$, the RMSD of the pulse waveform from that of the reference pulse is no more than 0.2. However, in a time $t = 2L/c$, a spurious response appears due to the reflection of the current wave from the open end of the monopole.

The calculation results show that in the direction normal to a linear radiator, the radiation field consists of two pulses spaced by the length of the radiator. The

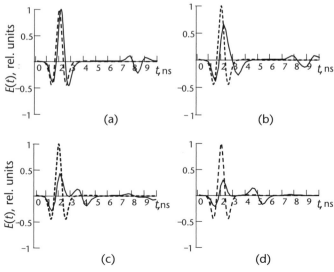

Figure 7.16 The waveforms of pulses radiated by a monopole in different directions: $\theta = 30°$ (a), $50°$ (b), $70°$ (c), and $90°$ (d).

pulse waveforms coincide with the plot of the time dependence of the charge density at the radiator ends that corresponds to the waveform of the exciting current pulse. In the boresight direction, the electromagnetic pulse waveform is close to the time derivative of the current pulse waveform. The obtained results are supported by experimental data [2].

The pulse distortion by a linear traveling-wave radiator (TWA) is similar to the distortions introduced by an SWA whose length is over $c\tau_p$. The least pulse waveform distortions are observed in the angular domain $15° < \theta < 40°$. In this case, there is no spurious response, as there is no wave reflection from the end of the radiator. A feature of a traveling-wave radiator is that the pulse waveform is distorted slightly even if the radiator length is small. However, if the radiator length is less than 0.75λ, its efficiency sharply decreases, as the power absorbed by the terminating load, P_2, becomes greater than the radiated power P_Σ.

As an example, Figure 7.17 presents the AFR, ΔPFR, and the waveform of the pulse radiated by a TWA of length $L = 1.5c\tau_p$ in the 30° direction together with

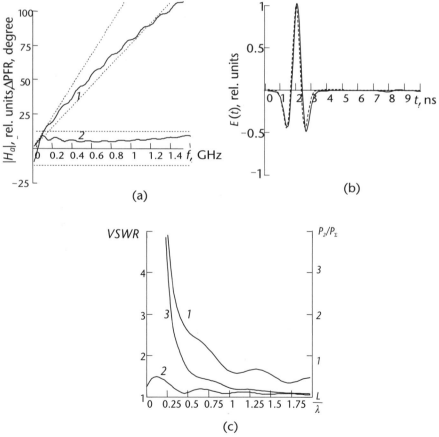

Figure 7.17 Plots of the frequency dependences of AFR (curve *1*) and ΔPFR (curve *2*) for a TWA of length $L = 1.5c\tau_p$ in the 30° direction (the limits of admissible values at which the RMSD of the pulse waveform is not over 0.2 are shown by dashed lines) (a); the radiated pulse waveform (the reference pulse waveform is depicted by a dashed line) (b), and plots of the frequency dependences of the VSWR (c) for an SWA (curve 1) and the TWA (curve 2), and of the ratio P_2/P_Σ (curve 3).

the reference pulse waveform. Herein, the voltage standing-wave ratio (VSWR) is given as a function of the radiator length for a standing-wave and a traveling-wave radiator (curves 1 and 2, respectively), and the radiator length dependence of P_2/P_Σ is also shown (curve 3).

The waveform distortion of a pulse radiated by a long dipole with the arm length $L > c\tau_p$ is the least in the directions at $\theta = 15°–40°$ and $\theta = 140°–165°$ and weakly depends on the internal impedance of the voltage pulse source. If the dipole is short ($L < 0.2c\tau_p$), the radiated pulse waveform weakly depends on direction, but strongly depends on the source resistance R_{in}. Figure 7.18 presents the waveforms of pulses radiated by a dipole with the arm length $L = 0.1c\tau_p$ in the direction at $\theta = 90°$ for different values of R_{in}.

As the input impedance of a short dipole strongly depends on frequency, these modes can be realized only by connecting the signal source immediately to the dipole. This is due to that a TEM feeder with a characteristic impedance of several thousands of ohms is unfeasible.

7.2.2   The Waveform of a Pulse Radiated by a V-shaped Radiator

The transfer function of a dipole with collinear arms depends on frequency, mainly, due to the strong frequency dependence of the dipole input impedance if $2L < c\tau_p$ and due to the frequency dependence of the pattern shape if $2L > c\tau_p$. This results in a substantial distortion of the waveform of the pulse radiated in the direction normal to the dipole. However, for long dipoles, there are θ_0 directions in which

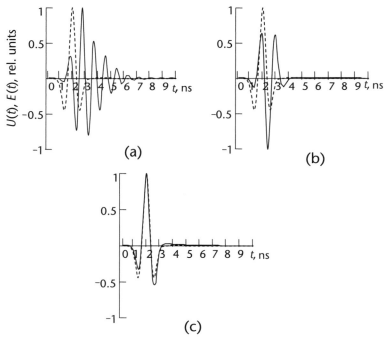

Figure 7.18   The waveform of a pulse radiated by a short dipole as a function of the resistance of the voltage pulse source: $R_{in} = 0.1\rho_a$ (a), $1.0\rho_a$ (b), and $10\rho_a$ (c); the reference radiation pulse waveform is depicted by a dashed line.

pulse distortions are inappreciable. Thus, if the arms of a dipole are arranged at an angle $2\theta_0$ to each other (Figure 7.19) to form a V-shaped radiator, then for $L > c\tau_p$, the waveform distortion for a pulse radiated along the bisector of this angle will be minimal.

The frequency dependence of the pattern shape for different angles between the arms of a V-shaped radiator of length $L = 0.9$ m with $\rho_a = \rho_f$ is shown in Figure 7.20.

It can be seen that the frequency dependence of the AFR levels off with decreasing the angle between the arms. The frequency dependence of the PFR of a V-shaped standing-wave radiator is periodic due to the complex input impedance. However, for $L > \lambda$, its deviation from a linear function is not over $\pi/8$. In Figure 7.21, the AFR and ΔPFR of a V-shaped radiator of length 0.9 m with $\theta_0 = 10°$ are depicted.

The dependence of the radiated pulse waveform on the angle between the arms of a V-shaped radiator is illustrated by plots in Figure 7.22.

The RMSD of the waveform of the pulse radiated by a V-shaped radiator from that of the reference pulse for the angle between the arms less than 50° is not over 0.2. However, as a result of the reflection of the current wave from the ends of the arms, a spurious pulse arises after a time $t = 2L/c$. Its amplitude can make one-fourth of the amplitude of the main pulse. The spurious pulse can be reduced by using an absorbent applied on the arm ends. However, this reduces the radiator efficiency, as part of the energy is lost in the absorbent.

A V-antenna with small angles between the arms can be considered, in the general case, as a nonuniform line segment with an open end. In this case, the current in each arm contains two components: the current I_a determined by (7.8) and the current in the line, I_L, resulting from the interaction between the arms. When the V-antenna arms are made as plates, a line of this type with a constant characteristic

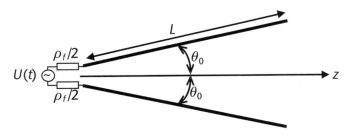

Figure 7.19 Sketch of a V-shaped radiator.

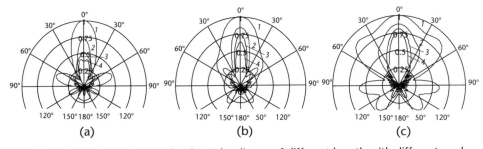

Figure 7.20 The pattern shapes of V-shaped radiators of different length with different angles between the arms: $\theta_0 = 30°$ (a), 20° (b), and 10° (c); $L = 1.5\lambda$ (curve 1), 3λ (curve 2), 4.5λ (curve 3), and 6λ (curve 4).

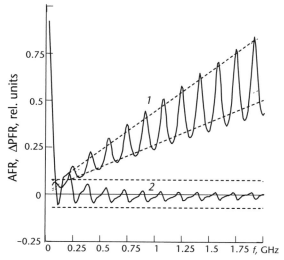

Figure 7.21 The AFR (*1*) and ΔPFR (*2*) of a V-shaped radiator of length 0.9 m; the dashed lines mark the tolerance limits in which the RMSD of the pulse waveform from the reference one is not over 0.2.

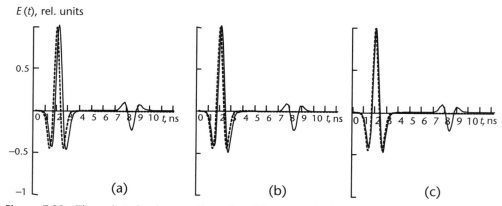

Figure 7.22 The radiated pulse waveforms for different angles between the radiator arms: $\theta_0 =$ 30° (a), 20° (b), and 10° (c); the dashed lines depict the waveform of the reference pulse.

impedance ρ_L is a regular TEM horn. If the characteristic impedance varies along the line, a TEM horn of this type is suggested [3] to term irregular. By analogy, we term a V-antenna with a constant width of the arms, d, regular if the angle α between the arms is fixed [Figure 7.23(a)] and irregular if the angle α varies along the antenna [Figure 7.23(b)].

A V-antenna of either type is an open-end line segment with smoothly varying characteristic impedance. To avoid higher modes in such a line, the distance between the ends should be no greater than $0.5c\tau_p$. The distance between the arms in the beginning of the line is chosen so that ρ_L in this cross section would be equal to the impedance of the pulse source (typically, 50 Ω). A balun can be used [4] to connect the coaxial and the strip line.

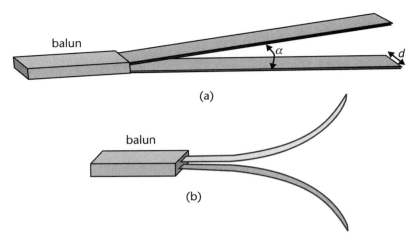

Figure 7.23   Schematic drawings of a regular (a) and an irregular V-antenna (b).

In nonuniform lines with smoothly varying characteristic impedance, undistorted pulse waveform transmission can be provided by properly choosing the law of variation of the characteristic impedance $\rho_L(z)$ along the line. It was shown [5] that distortionless lines are those in which the characteristic impedance varies by the law

$$\rho_L(z) = \rho_L(0)\left(1 + \frac{\tau}{v}\right)^{-2k} \tag{7.27}$$

where $\tau = \int_0^z d\zeta/v(\zeta)$ is the time of propagation of the wave front from the beginning of the line to a point with coordinate z, $v(\zeta)$ is the wave velocity, v is a constant characterizing the rate of variation of the characteristic impedance, and k is an integer. If $k > 0$, the line is hyperbolic ($\rho_L(z)$ decreases) and if $k < 0$, it is parabolic ($\rho_L(z)$ increases).

If the arm width of a V-antenna is considerably greater than the distance between the arms at the antenna input, the current I_L is dominant and the pulse reflected from the arm ends is not reflected from the matched input of the V-antenna. Therefore, the spurious pulse can be attenuated by 20–30 dB. For the RMSD of the waveform of a radiated pulse from that of the reference pulse to be not above 0.2, the length of a regular V-antenna should be over $3c\tau_p$, and the length of an irregular V-antenna can be no more than $c\tau_p$.

Figure 7.24 presents the waveforms of pulses of duration 0.5 ns radiated by a regular antenna with $L = 450$ mm and $d = 20$ mm and by an irregular antenna of length 150 mm.

The frequency dependence of the VSWR for these antennas is given in Figure 7.25.

These results suggest that using irregular nonuniform distortionless lines as V-antennas should shift substantially the lower bound of the match band toward the lower frequencies, that is, to reduce the antenna dimensions. A similar effect should take place for TEM horns [3].

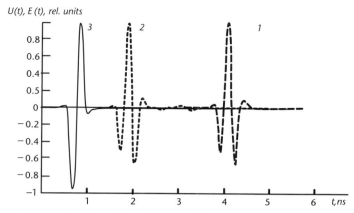

Figure 7.24   Waveforms of pulses radiated by a regular (curve 1) and an irregular V-antenna (curve 2) and of the voltage at the antenna input (curve 3).

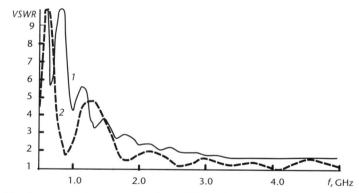

Figure 7.25   The frequency dependence of the VSWR of a regular (curve 1) and an irregular V-antenna (curve 2).

7.2.3   The Waveform of a Pulse Radiated by a Ring Radiator

A ring radiator is characterized by a stronger frequency dependence of the input impedance and pattern shape compared to a linear radiator. For a loop of small size, as long as its perimeter is not over $\lambda/4$, the pattern shape remains rather stable, but the radiation resistance is proportional ω^4, and this adds to the nonlinearity of the PFR. The effective length of a small loop is also frequency dependent, and this, in addition, strengthens the frequency dependence of the AFR. Figure 7.26 presents the AFR and ΔPFR of a loop of diameter 50 mm with unbalanced excitation by a bipolar pulse of duration 2 ns and the waveforms of the pulses radiated by the loop at $Z_1 = \rho_l$.

In a traveling wave mode, the PFR is almost linear, but the AFR shows quadratic frequency dependence. Therefore, the waveform of the radiated pulse coincides with the second time derivative of the exciting pulse waveform. This suggests that the waveform of a radiated pulse can be made close to that of the reference pulse by previously integrating the exciting pulse waveform, that is, by connecting a capacitor in parallel with the signal source. Figure 7.27 shows the corrected pulse waveform

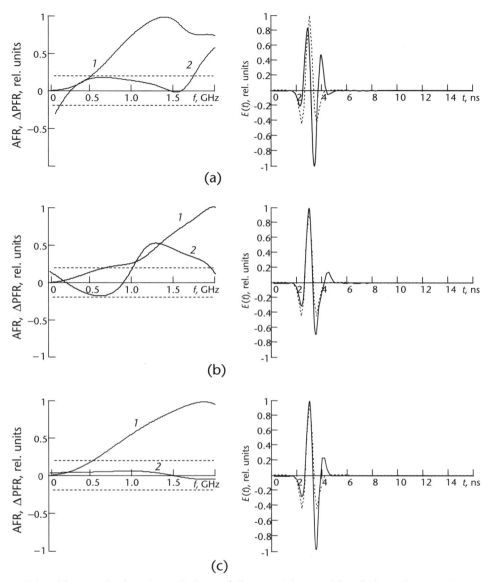

Figure 7.26 The transfer functions of a loop of diameter 50 mm with unbalanced connection and the waveforms of the pulses radiated in a short-circuit mode ($Z_2 = 0$) (*a*), in a no-load mode ($Z_2 \to \infty$) (*b*), and in a traveling-wave mode ($Z_2 = \rho_l$) (*c*); the AFR and ΔPFR are depicted by curves 1 and 2, respectively, and the reference pulse waveform is depicted by a dashed line.

obtained with a capacitor in capacitance 15 pF connected in parallel with a source of impedance $Z_1 = 240 \; \Omega$.

With this correction, the RMSD of the waveform of a radiated pulse from that of the reference pulse is not over 0.2, but the radiated pulse amplitude is 8–10 times lower.

In standing-wave modes, the variation of the source impedance from $Z_1 = \rho_l$ in any direction results in a strong distortion of the radiated pulse waveform, as illustrated by Figure 7.28.

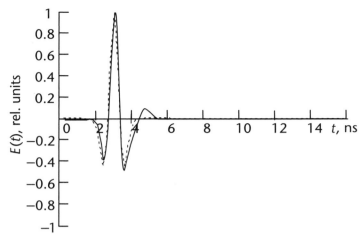

Figure 7.27 The waveform of a pulse radiated by a loop in a traveling-wave mode with a correcting capacitor.

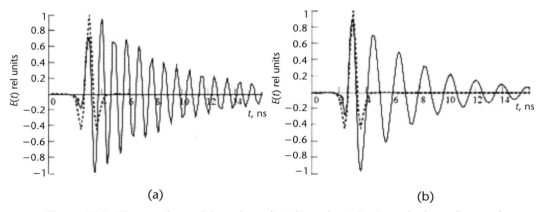

(a)   (b)

Figure 7.28 The waveform of the pulse radiated in a short-circuit mode depending on the resistance of the voltage pulse source: $Z_1 = 0.1\rho_l$ (a) and $10\rho_l$ (b).

7.3 Methods for Broadening the Pass Band of a Transmitting Antenna

As shown earlier, the radiation characteristics of a short dipole with $2L < \lambda$ weakly depend on frequency. However, the input impedance shows strong frequency dependence, and therefore, the pass band of a short dipole is mainly determined by its match band. If the length of the dipole arm is greater than the wave length, its impedance shows weak frequency dependence, but the pattern shape strongly depends on frequency. So, to broaden the pass band to two or three octaves, which is required for efficient radiation of UWB pulses, it is necessary to extend the match band toward the lower frequencies. To find a way for solving this problem, it is necessary to understand the processes occurring in the near-field region of a dipole excited by a monochromatic signal that determine the impedance characteristics of the dipole.

7.3.1   The Energy Relationships Determining the Match Band of a Radiator

The most common methods used to investigate the characteristics of any system are usually based on the energy approach. Therefore, we considered the space–time distribution of the energy density near a linear radiator [6, 7]. Using an electric dipole as an example, we have revealed some general relationships determining the energy motion and distribution, and, thus, the factors limiting the match bandwidth of short antennas.

The components E_r, E_z, and H_φ of the field of a dipole in its immediate vicinity can be described by analytic expressions derived in the approximation of a sinusoidal current distribution [8]. These expressions can be used for a qualitative description of the field behavior for the distances from the dipole axis greater than about 0.01 of the wave length, and they give good quantitative agreement for $r \geq$ 0.1 of the wave length.

Using these expressions, we obtain formulas for the average electrical ($\bar{w}^e$) and magnetic ($\bar{w}^m$) energy densities:

$$\bar{w}^e = \frac{\mu I_0^2}{64\pi^2 r^2}\left\{ d_0^2\left(\frac{1}{d_1^2} + \frac{1}{d_2^2}\right) + \frac{L(L-2z)}{d_1^2} + \frac{L(L+2z)}{d_2^2} + 2\frac{d_0^2 - L^2}{d_1 d_2}\cos k(d_1 - d_2) \right.$$
$$\left. - 4\frac{d_0^2 - zL}{d_0 d_1}\cos kL\cos k(d_1 - d_0) - 4\frac{d_0^2 + 2L}{d_0 d_2}\cos kL\cos k(d_2 - d_0) + 4\cos^2 kL \right\} \tag{7.28}$$

$$\bar{w}^m = \frac{\mu I_0^2}{32\pi^2 r^2}\left\{ 1 + \cos k(d_1 - d_2) - 2\cos kL\cos k(d_1 - d_0) \right.$$
$$\left. - 2\cos kL\cos k(d_2 - d_0) + 2\cos^2 kL \right\} \tag{7.29}$$

and for the instantaneous values of the electrical (w^e) and magnetic (w^m) energy densities:

$$w^e = \frac{\mu I_0^2}{32\pi^2 r^2}\left\{ \left[\left(2\frac{r}{d_0}\cos kL\sin kd_0 - \frac{r}{d_1}\sin kd_1 - \frac{r}{d_2}\sin kd_2\right)\cos\omega t \right.\right.$$
$$\left. - \left(2\frac{r}{d_0}\cos kL\cos kd_0 - \frac{r}{d_1}\cos kd_1 - \frac{r}{d_2}\cos kd_2\right)\sin\omega t\right]^2$$
$$+ \left[\left(\frac{z-L}{d_1}\sin kd_1 + \frac{z+L}{d_2}\sin kd_2 - 2\frac{z}{d_0}\cos kL\sin kd_0\right)\cos\omega t \right.$$
$$\left.\left. - \left(\frac{z-L}{d_1}\cos kd_1 + \frac{z+L}{d_2}\cos kd_2 - 2\frac{z}{d_0}\cos kL\cos kd_0\right)\sin\omega t\right]^2\right\} \tag{7.30}$$

$$w^m = \frac{\mu I_0^2}{8\pi^2 r^2}\left[\left(\sin k\frac{d_1 + d_2}{2}\cos k\frac{d_1 - d_2}{2} - \cos kL\sin kd_0\right)\cos\omega t \right.$$
$$\left. - \left(\cos k\frac{d_1 + d_2}{2}\cos k\frac{d_1 - d_2}{2} - \cos kL\cos kd_0\right)\sin\omega t\right]^2 \tag{7.31}$$

where μ is the absolute permeability of the medium; I_0 is the current at the wave antinode; k is the wave number; $\omega = 2\pi/T$ is the circular frequency; T is the oscillation period; d_0, d_1 and d_2 are the distances from the center and from the upper and lower ends of the dipole, respectively, to the observation point.

Owing to the symmetry of the energy density distributions about the plane $z = 0$, it suffices to restrict our consideration to $z \geq 0$. Thus, of greatest interest is to elucidate how the character of the energy motions affects the frequency bandwidth of a radiator whose dimensions are substantially less than the wave length. For radiators with the arm length from $\lambda/30$ to $\lambda/2$, the distributions of $\bar{w}^e$ and $\bar{w}^m$ along cylindrical surfaces of radius r from 0.05λ to 1.0λ coaxial with a dipole, and the distributions of w^e and w^m at different times have been calculated. As an example, Figure 7.29 presents some calculation results illustrating the space-time behavior of the energy density distributions in the near-field region of a radiator: the distribution of the period-average energy density along the z-axis at different distances and the energy density distribution at a fixed distance at different times.

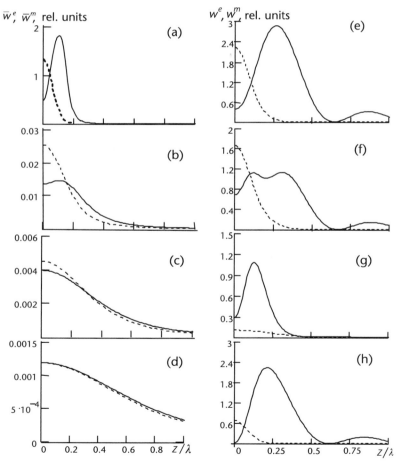

Figure 7.29 Plots of the space-time dependences of the energy density distribution in the near-field region of a dipole ($L = 0.15\lambda$, $\bar{w}^e$ -solid line, $\bar{w}^m$ -dashed line), of the average energy density distribution along the z-axis at distances $r = 0.1\lambda$ (a), 0.25λ (b), 0.5 (c), and 1.0λ (d), and of the energy density distribution at a distance 0.1λ at $t = 0$ (e), $T/8$ (f), $T/4$ (g), and $3T/8$ (h).

Analysis of the calculation results allows the following conclusions:

1. In the immediate vicinity of a radiator ($kr \ll kL$), the average electrical and magnetic energies are localized in different, partially overlapped, regions. The localization is most pronounced for a dipole of resonance length ($kL = \pi/2$) when $\bar{w}^m \gg \bar{w}^e$ at $z = 0$ and $\bar{w}^m \ll \bar{w}^e$ at $z = \pm L$, and the stored average electrical and magnetic energies are equal in any layer of the medium coaxial with the dipole. A decrease in dipole length is accompanied by violation of the localization, and for $kL \leq 1$ the stored electrical energy is considerably greater than the stored magnetic energy.

2. For a dipole of resonance length, the localization of $\bar{w}^e$ and $\bar{w}^m$ becomes less pronounced with distance from the dipole in the transverse direction, and for $kr \geq 3$, the $\bar{w}^e$ and $\bar{w}^m$ distributions almost coincide. If the dipole length is much less than the resonance length, in going away from the dipole, regions of preferential localization of $\bar{w}^e$ and $\bar{w}^m$ first appear due to the shift of the $\bar{w}^e$ maximum toward the greater $|z|$ and to the deepening of the $\bar{w}^e$ minimum near $z = 0$, and only thereafter, the smoothing of the $\bar{w}^e$ and $\bar{w}^m$ distributions along the z axis begins. The smoothing process stops in the range $3 \leq kr \leq 5$, and the $\bar{w}^e$ distribution becomes the same as the $\bar{w}^m$ distribution.

3. The radiator length weakly affects the time behavior of the instantaneous energy densities w^e and w^m. Both of them vary with time by a double-frequency harmonic law. In the immediate vicinity of the radiator, the oscillations of w^e lag in phase from the oscillations of w^m by $\pi/4$; that is, when w^e reaches a maximum, w^m is a minimum, and vice versa. The phase difference decreases with distance from the radiator, and for $kr > 3$ it can be assumed that w^e and w^m vary synchronously; that is, the field behaves as a traveling wave.

4. In the immediate vicinity of a dipole, there is an energy exchange between the electrical and the magnetic field. In the case of a dipole of resonance length, the energy exchange is accompanied by energy motions in space. This causes the phase difference between the oscillations of w^e and w^m to decrease with increasing kr. As a result, larger and larger amounts of electrical and magnetic energies start oscillating in phase; that is, the real part of the Poynting vector flux through the cylindrical surface of radius kr becomes prevailing. If a dipole is short, in the electrical energy stored the adjoining cylindrical layer is greater than the stored magnetic energy; that is, not all energy participates in the exchange and, hence, in the formation of the radiated wave. Besides, the localization of electrical and magnetic energies near a short radiator is less pronounced. Therefore, the phase difference between the oscillations of w^e and w^m decreases with increasing kr slower than in the case of a dipole of resonance length; that is, the radiated wave is formed in a larger spatial domain.

5. The difference between the electrical and magnetic energies not participating in the energy exchange repeatedly moves from the surrounding space in and out of the generator, and this is the main cause of the decrease of the pass band of a short radiator.

Similar conclusions are valid for magnetic-type radiators producing electromagnetic pulses in which magnetic energy prevails over electrical energy. In this case, the space-time distribution of the magnetic energy density is similar to that of the electrical energy of an electrical-type radiator, and vice versa.

7.3.2 The Quality Factor of a Linear Radiator

The storage of energy in the near-field region of a radiator allows us to consider it as a reactive two-terminal network with quality factor Q_a by which we mean the ratio of the total stored energy W to the energy W_s lost during an oscillation period T multiplied by 2π:

$$Q_a = 2\pi \frac{W}{W_s} = \frac{2\pi W}{T P_s} = \frac{\omega W}{P_s} \tag{7.32}$$

where ω is the oscillation frequency and P_s is the average lost power.

The notion of the quality factor of an antenna is commonly assigned with the same physical meaning as the quality factor of a lossy oscillatory circuit. A formula for the calculation of the quality factor of a linear standing-wave radiator operating near series resonance ($L = \lambda/4$) was derived by Kessenikh [9]. It reads

$$Q_a = \frac{\omega \dfrac{\partial X_a}{\partial \omega} + |X_a|}{R_a} \tag{7.33}$$

where R_a and X_a are the real and the imaginary part of the input impedance of the antenna. This formula also allows one to calculate Q_a at frequencies below the series resonance frequency, but yields incorrect results for near-parallel-resonance frequencies ($L = \lambda/2$). Besides, this expression is not suitable to calculate the quality factor of a traveling-wave radiator with $X_a \to 0$, as the simulation of a radiator by a two-terminal network does not consider the fact that the radiation process (conversion of the energy of alternating currents into the energy of free electromagnetic waves) is inertial due to the presence of stored energy, which, in the theory

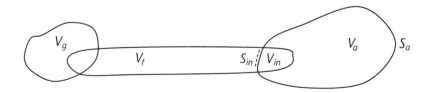

Figure 7.30 Simulation of a radiator as the transition region between free space and the feeder: V_g—region containing the voltage (current) pulse generator, V_f—region occupied by the feeder, and V_a—region, bounded by surface S_a, that contains the radiator and part of the space in which the radiation field is formed (the induction, or near-field, region); S_{in} is the surface corresponding to the radiator input and V_{in} is the transition region between the feeder and the radiator.

of antennas, is defined as the power "radiated" during a period into the region of imaginary angles (invisible region) [1].

In the general case, an antenna can be considered as a transition region between the feeder and free space [10], and the radiation source can be simulated as three overlapping regions (Figure 7.30).

Based on the Poynting theorem, without taking into account the losses in the radiator, we can write

$$\frac{1}{2}\int\limits_{S_{in}} [\mathbf{E} \times \mathbf{H}^*]\,d\mathbf{s} = \frac{1}{2}\int\limits_{S_a} [\mathbf{E} \times \mathbf{H}^*]\,d\mathbf{s} + i2\omega\int\limits_{V_a} \left(\frac{\mu}{4}|\mathbf{H}|^2 - \frac{\varepsilon}{4}|\mathbf{E}|^2\right)dv \quad (7.34)$$

The integral in the left-hand part of (7.34) describes the complex power P_{in} at the antenna input, which can also be represented as a product: $P_{in} = |I_0|^2 Z_a/2$, where I_0 is the current amplitude at the radiator input. If the surface S_a is off the radiator at a distance where the field behaves as a free wave, we have

$$\frac{1}{2}\mathrm{Im}\int\limits_{S_a} [\mathbf{E} \times \mathbf{H}^*]\,d\mathbf{s} \to 0 \quad (7.35)$$

Analysis of the energy relations for the near-field region of a radiator allows the treatment of the physical processes underlying the radiation of any antenna according to which the total energy of the antenna field contains the following components:

1. the energy of the free space wave formed in the near-field region of the antenna, W_Σ, that is, the energy "being radiated," which is determined by the power radiated during the period T or by the real part of the power flux through the cross-section S_{in} corresponding to the input of the antenna:

$$W_\Sigma = T\,\mathrm{Re}\int\limits_{S_{in}} [\mathbf{E} \times \mathbf{H}^*]\,d\mathbf{s} = R_a|I_0|^2 T \quad (7.36)$$

where $\mathbf{E}$ and $\mathbf{H}$ are the electrical and magnetic field vectors, R_a is the real part of the input impedance of the antenna, and I_0 is the current at the antenna input. This energy prevails at large distances from the antenna;

2. the reactive energy W_r equal to the difference between the magnetic and the electrical energy stored in the antenna near-field region V_a:

$$W_r = \int\limits_{V_a} \left(\frac{\mu}{4}|\mathbf{H}|^2 - \frac{\varepsilon}{4}|\mathbf{E}|^2\right)dv \quad (7.37)$$

This energy does not participate in the formation of the radiated wave and repeatedly moves from the near-field region in and out of the generator; it determines the imaginary part of the flux of the Poynting vector $\mathbf{S}$ through the cross-section S_{in} as

$$2\pi W_r = \frac{1}{2}\text{Im} \int_{S_{in}} \mathbf{S} \cdot d\mathbf{s} = X_a |I_0|^2 T \qquad (7.38)$$

This period-average energy flux through S_{in} is equal to zero, and its amplitude determines the reactive power of the antenna and the reactive component of its input impedance. The reactive energy density is a maximum near the antenna and rapidly decreases with distance from it;

3. the bound energy participating in the exchange between the electrical and the magnetic energy in the near-field region. This energy does not contribute to the energy flux through S_{in} and moves from the region of electrical energy localization in and out of the region of magnetic energy localization. At great distances from the antenna, the bound energy density tends to zero.

According to the above considerations, the energy stored in the near-field region of a radiator is equal to the sum of the reactive energy W_r and the bound energy W_c. In the case of a linear radiator, the bound energy consists of the energy stored in the current wave propagating along the radiator, $W_{c\tau}$, and the energy "radiated" into the invisible region, $W_{c\Sigma}$.

In view of the foregoing, we have for the antenna quality factor

$$Q_a = \frac{2\pi(W_r + W_{c\Sigma} + W_{c\tau})}{(P_\Sigma + P_\sigma)T} \qquad (7.39)$$

where P_Σ is the average radiated power, P_σ is the average power lost in the conductors and dielectrics of the radiator. Disregarding the losses in the radiator and in the surrounding medium ($P_\sigma \to 0$), we have $P_\Sigma = W_\Sigma/T$, and the antenna quality factor can be represented as the sum of partial quality factors [11, 12]:

$$Q_a = 2\pi \frac{W_r}{W_\Sigma} + 2\pi \frac{W_{c\tau}}{W_\Sigma} + 2\pi \frac{W_{c\Sigma}}{W_\Sigma} = Q_r + Q_{c\tau} + Q_{c\Sigma} \qquad (7.40)$$

The stored reactive energy can be expressed in terms of the antenna reactance as $W_r = X_a |I_0|^2 T/2\pi$, and then, we have

$$Q_r = \frac{X_a}{R_a} \qquad (7.41)$$

The quantity $W_{c\tau}$ is determined based on the following reasoning: a current wave propagating with velocity v_τ transfers a power P_τ through a plane orthogonal to the linear radiator. Hence, the energy stored per unit length is given by $W' = P_\tau/v_\tau$. At the same time, the lost power P' referred to a unit length is defined as $dP_\tau/dz = 2\text{Im}(\gamma)P_\tau$, where γ is the current wave propagation constant. For a radiator of length L, we have $W_{c\tau} = W'L$, and the energy lost by radiation is determined as $P_\Sigma T = 2\text{Im}(\gamma)P_\tau LT$. In view of $v_\tau = \omega/\text{Re}(\gamma)$, we have $W_{c\tau} = \text{Re}(\gamma)R_a|I_0|^2 T/4\pi\text{Im}(\gamma)$, whence

$$Q_{c\tau} = \frac{\mathrm{Re}(\gamma)}{2\,\mathrm{Im}(\gamma)} \tag{7.42}$$

For the quantitative estimation of $W_{c\Sigma}$, the quality factor $Q_{c\Sigma}$ is used that is defined as the ratio of the power "radiated" into the invisible region to the power radiated into the region of real angles (visible region) [1]:

$$Q_{c\Sigma} = \frac{\int_{-\infty}^{-k} F^2(\xi)\,d\xi + \int_{k}^{\infty} F^2(\xi)\,d\xi}{\int_{-k}^{k} F^2(\xi)\,d\xi} = \frac{\int_{-\infty}^{\infty} F^2(\xi)\,d\xi}{\int_{-k}^{k} F^2(\xi)\,d\xi} - 1 \tag{7.43}$$

where $F(\xi)$ is the normalized pattern function of the antenna and $\xi = k\cos\theta$. This quality factor is conventionally termed a radiation quality factor [13–15].

Thus, the quality factor of a linear radiator can be expressed in terms of its characteristics (input impedance, wave propagation constant, and pattern function) as

$$Q_a = \frac{|X_a|}{R_a} + \frac{\mathrm{Re}(\gamma)}{2\,\mathrm{Im}(\gamma)} + \left(\frac{\int_{-\infty}^{\infty} |F^2(\xi)|^2\,d\xi}{\int_{-k}^{k} |F^2(\xi)|^2\,d\xi} - 1 \right) \tag{7.44}$$

This equation indicates that the quality factor of an antenna can be represented by the sum of partial quality factors differently depending on the parameters of the radiator, making possible to reveal the radiator parameters that most substantially affect its match bandwidth. In the case of a traveling-wave radiator, the first term is small, as $R_a \approx const$ and $X_a \to 0$. The second term substantially depends on the radiator characteristic impedance ρ_a, that is, on the radiator length-to-width ratio, and monotonically decreases with increasing radiator electric length. The third term depends on the current distribution along the radiator and also monotonically decreases with increasing radiator length.

For a hypothetical antenna with uniform in-phase current distribution (the phase velocity of a traveling current wave tending to infinity, that is, $\mathrm{Re}(\gamma) \to 0$), the main contribution to the antenna quality factor is made by $Q_{c\Sigma}$. This antenna can be treated as a minimum-Q antenna. For a standing-wave radiator, Q_r turns to zero only if the radiator has resonant dimensions and sharply increases with decreasing the antenna longitudinal or traverse dimension, as $X_a = \mathrm{Im}[-i\rho_a\cot(\gamma L)]$. The second and third terms ($Q_{c\tau}$ and $Q_{c\Sigma}$) have qualitatively the same dependence on antenna dimensions as for a radiator with uniform in-phase current distribution, but they are noticeably smaller in value. For an antenna of length $2L$ with a standing current wave less than $\lambda/2$, the main contribution to the antenna quality factor is made by Q_r. Figure 7.31 presents plots of the frequency dependences of $Q_{c\Sigma}$, $Q_{c\tau}$, and Q_r (curves 1, 2, and 3, respectively) for a dipole of length $2L$ with characteristic impedance $\rho_a = 500\ \Omega$ (solid lines) and $200\ \Omega$ (dashed lines).

For a hypothetical antenna whose stored reactive energy is equal to zero (minimal reactance antenna), the quality factor is determined only by the portion of

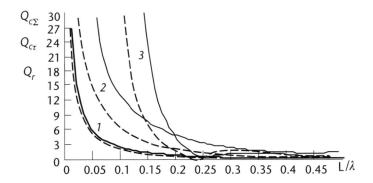

Figure 7.31 The partial quality factors of a dipole with characteristic impedance $\rho_a = 500\ \Omega$ (solid lines) and $200\ \Omega$ (dashed lines) versus dipole length. Curves 1, 2, and 3 correspond to the partial quality factors $Q_{c\Sigma}$, $Q_{c\tau}$, and Q_r, respectively.

the stored bound energy that can be treated (using the notions of the theory of antenna synthesis) as the energy radiated during one oscillation period into the invisible region. Then, by the antenna, quality factor $Q_{a\min} = Q_{c\Sigma}$ should be meant. The dependence of $Q_{a\min}$ on the antenna length for different current distributions is illustrated by Figure 7.32. The dashed line depicts the radiation quality factor calculated by the formula [13]

$$Q_\Sigma = \frac{1}{k^3 L^3} + \frac{1}{kL} \tag{7.45}$$

To estimate the maximum (theoretically) achievable match bandwidth of an antenna, we take advantage of the Fano theorem [16] from which it follows that when the antenna complex impedance $Z_a = R_a + iX_a$ is matched (using a lossless matching device) with a real impedance (the input resistance of the feeder) the wave

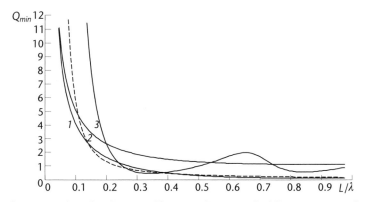

Figure 7.32 The minimal quality factors of linear radiators with different current distributions: curve 1 corresponds to a uniform in-phase distribution, curve 2 to a TWA operating in an axial radiation mode, and curve 3 to a dipole with a sinusoidal current distribution. The dashed line depicts the quality factor calculated by (7.45).

reflection coefficient R in a given frequency range cannot be less than a certain value, and, on the contrary, a given match level can be achieved within a frequency bandwidth which is not over some maximum value $2\Delta\omega_{max}$. As applied to antennas, this theorem is considered in [17]. The reflection coefficient, the match bandwidth, and the quality factor of an antenna are related by the formula

$$|R_0| = \exp\left(\frac{-2\pi Q_a \Delta\omega}{\omega_0}\right) \tag{7.46}$$

where $2\Delta\omega_{max} = |\omega_1 - \omega_2|$, $\omega_0 = (\omega_1\omega_2)^{1/2}$, ω_1 and ω_2 are the cutoff frequencies of the match band.

From this equation, it follows that the quality factor Q_a of an antenna with the relative match bandwidth $2\Delta\omega/\omega_0 \geq 1$ at a maximum admissible reflection coefficient $|R_{max}|$ should not be greater than

$$Q_a = \frac{\pi}{\ln\left(\frac{1}{|R_{max}|}\right)} \tag{7.47}$$

The quality factor Q of the load, the relative match bandwidth, and the maximum achievable voltage standing wave ratio $K_{Vmax} = (1 + |R_{max}|)/(1 - |R_{max}|)$ are related by the formula

$$Q = \frac{\pi\sqrt{\omega_1\omega_2}}{(\omega_2 - \omega_1)\left[\ln(K_{Vmax} + 1) - \ln(K_{Vmax} - 1)\right]} \tag{7.48}$$

Figure 7.33 presents plots of the dependences of the quality factor on the bandwidth ratio $b = \omega_2/\omega_1$ for several values of K_{Vmax}. For instance, for the match bandwidth of an antenna to be greater than an octave ($b > 2$) at $K_{Vmax} \leq 2$, the quality factor should be less than 4.3.

The plots in Figure 7.32 suggest that, theoretically, a balanced antenna with minimal reactance (a radiator with uniform in-phase current distribution) can be matched over a wide frequency range if its dimension, $2L$, is no less than 1/8 of the maximum operating wave length.

With increasing the antenna dimensions, not only the stored bound energy decreases, but also the conditions for the reactive energy to be a minimum become less stringent. Therefore, the frequency range in which a minimal reactance antenna can be matched with the feeder is theoretically not limited from above. Restriction of a pass band from above is generally caused by a change of the pattern shape, which, for instance, in the case of an SWA with collinear arms, "collapses" if the radiator length is greater than the radiation wave length. Thus, theoretically, the limiting accessible pass band of a minimal reactance radiator makes three octaves. This pass band suffices for a radiator to radiate no less than 90% of the energy of a bipolar exciting pulse.

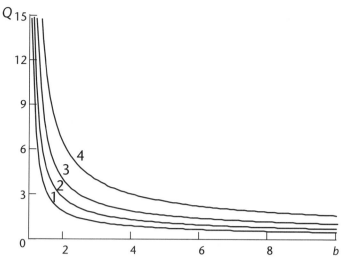

Figure 7.33 The quality factor versus match bandwidth ratio for different admissible values of the VSWR: $K_{V\max} = 1.2$ (curve 1), 1.5 (curve 2), 2.0 (curve 3), and 3.0 (curve 4).

7.3.3 The Pass Band of a Combined Radiator

Short antennas are known to be sources of spherical waves, that is, they have a pronounced phase center, and their pattern functions weakly depend on frequency. However, in most cases, antennas whose dimensions are less than the radiation wave length have a rather strong frequency dependence of the input impedance (which is the stronger, the smaller the antenna dimensions) which impedes their matching in a wide frequency range. Therefore, the lower limit of the pass band of a short antenna is uniquely determined by the lower limit of the match band. The upper limit of the pass band can be related to poor matching of the antenna and to distortion of its pattern functions. For instance, the pattern of an electric dipole is substantially distorted if the dipole dimensions are greater than the radiation wave length.

Using energy equations, we write down the conditions for an antenna to be well matched to the feeder:

$$\left| 2\,\mathrm{Re} \int_{S_{in}} \mathbf{S}\,d\mathbf{s} - \rho_f \left| I_{01} \right|^2 \right| \leq \upsilon \tag{7.49}$$

$$\left| \mathrm{Im} \int_{S_{in}} \mathbf{S}\,d\mathbf{s} \right| = \left| 2\omega \int_{V_a} (\overline{w}^m - \overline{w}^e)\,dV \right| \leq \delta \tag{7.50}$$

where I_{01} is the complex amplitude of the current at the antenna input (cross-section S_{in}), ρ_f is the characteristic impedance of the feeder, υ and δ are small positive quantities determined by the admissible value of the VSWR in the feeder.

Equations (7.49) and (7.50) indicate that to widen the match band, it is necessary, first of all, to minimize the stored reactive energy.

The well-known simplest way to reduce the reactive energy in the region V_a is to increase the transverse dimensions of the radiator, that is, to eliminate some spatial domain in which the reactive energy density is a maximum from the region V_a. This widely used method allows one to noticeably reduce the reactive energy stored in the near-field region of a radiator; however, it cannot be completely eliminated. Besides, the real part of the antenna impedance remains frequency-dependent, as the lower the frequency, the lower the bound energy density due to which the radiated energy flow is formed.

A more efficient method for widening the match band of an antenna is to make coincident the near-field regions (V_a) of two radiators having a common input but dissimilar reactive energies (if electrical energy prevails in the near-field region of one radiator, magnetic energy should prevail in the near-field region of the other). In this case, condition (7.50) is fulfilled for any frequency if the stored reactive energies of the radiators are equal and have identical frequency dependences. Another advantage of this method is that the stored reactive energy is converted into bound energy at a certain orientation of the radiators relative to each other. In this case, a decrease in frequency results in an increase in bound energy density and, hence, in radiation intensity. Therefore, the radiation resistance relatively increases and the frequency dependence of the real part of the antenna impedance becomes less pronounced, providing for the fulfillment of condition (7.49) in a wider frequency range.

It was demonstrated [6, 7] that combining electrical- and magnetic-type radiators not only minimizes reactive energy, but also ensures reasonable stability of the radiation power in a wide frequency range.

Let a balanced electric dipole of length $2L$ and two magnetic radiators be located in a Cartesian coordinate system. The axes of the radiators are oriented parallel to the x-axis and their centers lie in the plane $x = 0$ at distances d and h from the y- and the z-axis, respectively (Figure 7.34).

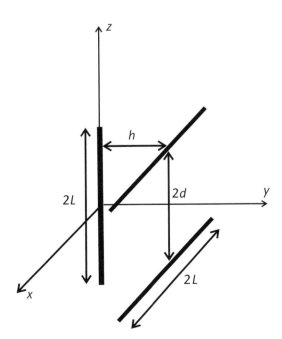

Figure 7.34 Sketch of the model of a combined antenna. (With permission from Pleiades Publishing, Ltd.)

Assume that the electrical and magnetic current distributions in the radiators are sinusoidal with complex amplitudes I_0^e, I_{01}^m, and I_{02}^m and that

$$\frac{I_{01}^m}{Z_0 I_0^e} = m_1 \left(\frac{L}{\lambda}\right)^{-1} \exp i(\varphi_1 + \Delta\varphi_1), \quad \frac{I_{02}^m}{Z_0 I_0^e} = m_2 \left(\frac{L}{\lambda}\right)^{-1} \exp i(\varphi_2 + \Delta\varphi_2) \quad (7.51)$$

where m_1 and m_2 are constants, φ_1 and φ_2 are the initial phase shifts of the currents in the magnetic radiators relative to the current in the electric dipole, and $\Delta\varphi_1 = \Delta\varphi_2 = \beta\sqrt{(kh)^2 + (kd)^2}$, where β has the meaning of slowness factor.

The factor $(L/\lambda)^{-1}$ entering into (7.51) reflects the frequency dependence of the current ratio that takes place when electrical and magnetic radiators are connected in parallel to a common feeder, and the value of β determines the slowing-down of a current wave in the feeder. Using the method of vector potentials [1], we obtain the following expressions for the components of the electromagnetic field in the far-field region ($r \to \infty$) in spherical coordinates r, θ, and φ:

$$E_\theta = \frac{i I_0^e Z_0}{2\pi \sin kL} \frac{e^{-ikr}}{r} \left\{ \frac{\cos(kL\cos\theta) - \cos kL}{\sin\theta} + \sin\varphi \frac{\cos(kL\sin\theta\cos\varphi) - \cos kL}{1 - \sin^2\theta\cos^2\varphi} \exp(ikh\sin\theta\sin\varphi) \right.$$

$$\left. \times \left[\frac{I_{01}^m}{I_0^e Z_0} \exp(ikd\cos\theta) + \frac{I_{02}^m}{I_0^e Z_0} \exp(-ikd\cos\theta) \right] \right\}$$

$$(7.52)$$

$$E_\varphi = \frac{i I_0^e Z_0}{2\pi \sin kL} \frac{e^{-ikr}}{r} \cos\theta\cos\varphi \frac{\cos(kL\sin\theta\cos\varphi) - \cos kL}{1 - \sin^2\theta\cos^2\varphi} \exp(ikh\sin\theta\sin\varphi)$$

$$\times \left[\frac{I_{01}^m}{I_0^e Z_0} \exp(ikd\cos\theta) + \frac{I_{02}^m}{I_0^e Z_0} \exp(-ikd\cos\theta) \right], \quad H_\varphi = \frac{E_\theta}{Z_0}, \quad H_\theta = -\frac{E_\varphi}{Z_0}$$

$$(7.53)$$

The power radiated by a set of radiators (combined antenna) can be found by the Poynting vector method [1]. In this case, the radiation power P_Σ divided by $|I_0^e|^2 Z_0 / \sin^2 kL$ is defined as

$$P_\Sigma = P_\Sigma^e + P_\Sigma^m + P_\Sigma^{e,m} \quad (7.54)$$

where the radiation power of the electric dipole is given by

$$P_\Sigma^e = \frac{Z_0}{2\pi} \int_0^\pi \frac{[\cos(kL\cos\theta) - \cos kL]^2}{\sin\theta} d\theta \quad (7.55)$$

the additive to this power due to the radiation of the magnetic radiators by

$$P_\Sigma^m = \frac{Z_0}{4\pi^2} \int_0^\pi \int_0^{2\pi} \frac{[\cos(kL\sin\theta\cos\varphi) - \cos kL]^2}{1 - \sin^2\theta\cos^2\varphi} \sin\theta \left| \frac{I_{01}^m}{I_0^e Z_0} \exp(ikd\cos\theta) + \frac{I_{02}^m}{I_0^e Z_0} \exp(-ikd\cos\theta) \right|^2 d\theta d\varphi$$

$$(7.56)$$

and the additive due to the "combination" effect of the summation of the radiated powers by

$$P_\Sigma^{e,m} = \frac{Z_0}{2\pi^2} \int\limits_0^\pi \int\limits_0^{2\pi} [\cos(kL\cos\theta) - \cos kL] \frac{\cos(kL\sin\theta\cos\varphi) - \cos kL}{1 - \sin^2\theta\cos^2\varphi}$$

$$\times \mathrm{Re}\left\{\exp(-ikh\sin\theta\sin\varphi)\left[\frac{I_{01}^{m*}}{I_0^e Z_0}\exp(-ikd\cos\theta) + \frac{I_{02}^{m*}}{I_0^e Z_0}\exp(ikd\cos\theta)\right]\right\}\sin\varphi\,d\theta\,d\varphi$$

(7.57)

Figure 7.35 presents the powers P_Σ^e, P_Σ^m, $P_\Sigma^{e,m}$, and P_Σ as functions of the ratio L/λ for $h/L = 0.5$, $d/L = 0.8$, $m_1 = m_2 = 0.3$, $\varphi_1 = \varphi_2 = 0$, and $\beta = -1.5$.

It can be seen that with the radiators connected to a common feeder having a pure active characteristic impedance of about 140 Ω, the electric dipole can be matched on a level $K_V = 2$ in a frequency range with a cutoff frequency ratio of 8:1, and the frequency dependence of P_Σ becomes weaker compared to that of P_Σ^e. It should be noted that this effect is observed only for certain amplitude-phase equations of the currents in the radiators. If nonoptimal values of m_1 and m_2, and also of φ_1 and φ_2, are set, the frequency dependence of P_Σ may become stronger than that of P_Σ^e.

Thus, a better match between antenna and feeder can be achieved due to minimization of the reactive energy with an increased bound energy density. This can be provided by equalizing the electrical and the magnetic energy in the near-field region of the radiator. The proposed approach has been realized in combined antennas (KAs), which are composed of certainly arranged electrical and magnetic radiators with certain relations between the amplitudes and phases of the currents flowing in them.

7.4   Flat Combined Antennas

Proceeding from the above reasoning, we can formulate general design principles for wideband matched antennas:

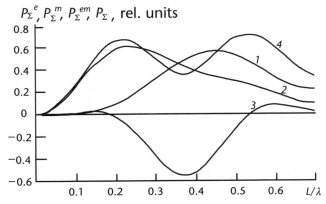

Figure 7.35   The radiated powers P_Σ^e (curve 1), P_Σ^m (curve 2), $P_\Sigma^{e,m}$ (curve 3), and P_Σ (curve 4) versus L/λ. (With permission from Pleiades Publishing, Ltd.)

- the antenna should be a combination of electrical and magnetic radiators ensuring minimization of reactive energy due to a proper proportion between the moments of electrical and magnetic currents. For electrical radiators, monopoles or dipoles can be used, whereas loops or slit radiators can serve as magnetic radiators. The phase shift between the currents feeding electrical and magnetic radiators should be close to $\pi/2$;
- the electrical and magnetic radiators should be placed in the immediate vicinity of each other, so that the distance between their phase centers would be nonzero and made no more than 1/4 of the pulse spatial length. The radiators should be so oriented relative to each other that their polarization characteristics would be identical;
- as the stored electrical and magnetic energies of actual radiators differently depend on frequency and decrease with increasing radiator transverse dimensions, it is reasonable to use electrical and magnetic radiators with increased transverse dimensions.

Combining electrical and magnetic radiators makes it possible not only to ensure matching of the antenna in a wide frequency range, but also to form a unidirectional pattern close in shape to a cardioid and thus increase the antenna power gain.

7.4.1 Unbalanced Combined Antennas

The antennas operating in the microwave range are generally excited by feeders. In this case, a combined antenna should have an unbalanced input, as with a monopole. A possible design version of an unbalanced flat KA is schematically shown in Figure 7.36. The antenna base component is an unbalanced radiator comprising a monopole and a counterpoise, both made as metal plates [Fig. 7.36(a)]. The input of the radiator is marked by solid circles. The braid of the coaxial cable is connected to the counterpoise and the inner conductor to the monopole. The pattern of this type of radiator has zeros in the direction of the monopole axis and is uniform in the plane orthogonal to the monopole. If the monopole is bent as shown in Figure 7.36(b), some portion of its current is branched, through the capacitance of the monopole–counterpoise gap, to the counterpoise, forming a closed loop carrying

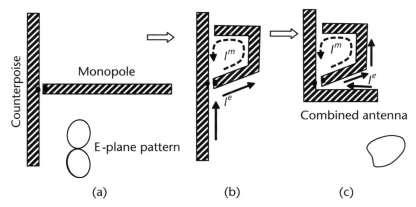

Figure 7.36 Design versions of a flat combined antenna.

a current I^m, that is, a magnetic-type radiator. The current in the lower part of the counterpoise and in the initial segment of the monopole, I^e, is not closed and endows the antenna with the properties of an electrical-type radiator. This configuration of the radiator widens the match band, but the pattern shape becomes substantially depending on frequency. To stabilize the pattern shape and reduce the radiator dimensions, the counterpoise should be shaped as shown in Figure 7.36(c).

The possibility to widen the match band of a KA with this design is supported by a numerical simulation of wire antenna models performed using the 4NEC2 code [18]. Figure 7.37 sketches a wire monopole and a wire KA model having identical overall dimensions and gives the frequency dependences of their VSWRs.

A KA of this type was realized in practice [19]. The antenna is printed on a metal-clad fiberglass FR-4 plate of dimensions $50 \times 45 \times 1$ mm. The topology of the printed board is shown in Figure 7.38(a). The antenna comprises a bent monopole (1) and a counterpoise (2). The metal-free board segment (3) is a magnetic-type radiator loaded with the capacitance of the gap (4) between the monopole end and the counterpoise. The antenna is furnished with a coaxial socket (5) connected to the monopole input via a coplanar line segment (6). The electrical-type radiator (7) is formed by the lower part of the counterpoise and the outer edge of the monopole. This design, allowing the use of one-sided foil-coated dielectric, has one drawback. The matter is that when the antenna is excited by a 50-Ω feeder, the width of the slot line exciting the electrical-type radiator is as small as 0.2–0.3 mm, so that there is a danger of electrical breakdown restricting the radiation power. Use of a bilateral foil-coated dielectric should eliminate this drawback if the counterpoise is mounted on the back side of the board and the coaxial socket is connected to the monopole input via a microstrip line segment (6), as shown in Figure 7.38. In this case, the average radiated power can reach several tens of watts.

The proposed design version of a flat KA was used as an antenna element of an UWB antenna array intended for radiotomography. The appearance of the antenna element is shown in Figure 7.39(a). The frequency dependence of its VSWR is given in Figure 7.39(b).

Figure 7.40 presents the frequency dependence of the pattern shape of the proposed design version of a flat KA.

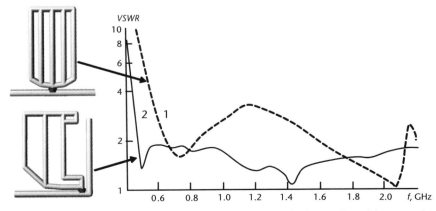

Figure 7.37 The VSWR frequency dependences for a monopole (curve 1) and for a combined antenna (curve 2).

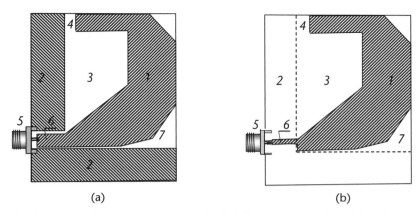

(a)   (b)

Figure 7.38 Examples of practical realization of a flat KA using a one-sided (a) and a two-sided foil-coated dielectric (b).

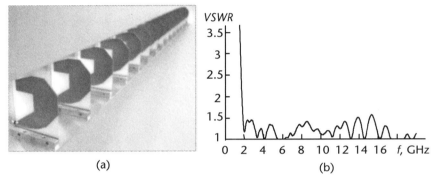

(a)   (b)

Figure 7.39 A flat combined antenna as an element of an ultrawideband antenna array (*a*) and the frequency dependence of its VSWR (b). (With permission from Science & Technology Publishing House).

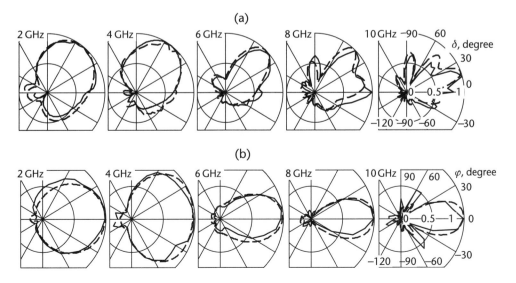

Figure 7.40 Patterns of a flat KA in the *E*-plane (a) and in the *H*-plane (b); measurement data are depicted by solid lines and simulation predictions by dashed lines. (With permission from Pleiades Publishing, Ltd.)

The peak-amplitude pattern shape and the radiated pulse waveform for the antenna excited by a bipolar voltage pulse of duration 0.2 ns are shown in Figure 7.41.

An unbalanced flat KA of similar design was used as an element of an ultra-wideband multibeam antenna array intended for operation with pulses of duration about 0.5 ns [20]. The appearance of the antenna, printed on a one-sided metal-clad fiberglass FR-4 plate of dimensions $80 \times 80 \times 1$ mm is shown in Figure 7.42(a). Figure 7.42(b) presents the frequency dependence of the reflection coefficient in a 50-Ω feeder.

The peak-power patterns of the antenna in the horizontal and vertical planes are given in Figure 7.43. The antenna was excited by a bipolar voltage pulse of duration 0.5 ns. In performing measurements, a receiving TEM antenna was used.

A flat KA of similar design printed on a metal-clad fiberglass FR-4 plate of dimensions $120 \times 170 \times 2$ mm is used as a central component in the ultrawideband

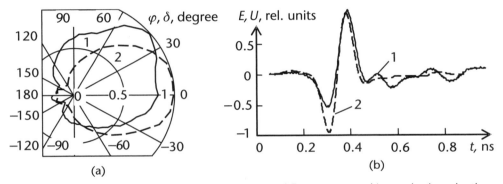

(a)   (b)

Figure 7.41   Radiation characteristics of an unbalanced flat KA operated in a pulsed mode: the peak-amplitude pattern shapes in the *E*-plane (1) and in the *H*-plane (2) (a) and the waveform of the pulse radiated in the boresight direction (1) and the voltage pulse waveform (2) (b). (With permission from Pleiades Publishing, Ltd.)

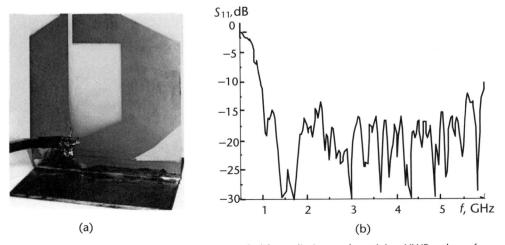

(a)   (b)

Figure 7.42   The appearance of a flat KA intended for radiating and receiving UWB pulses of duration 0.5 ns (a) and the coefficient of wave reflection from the antenna input (b). (With permission from Pleiades Publishing, Ltd.)

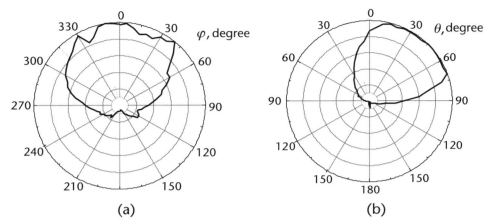

Figure 7.43 The peak-power patterns in the *H*-plane (a) and in the *E*-plane (b). (With permission from Pleiades Publishing, Ltd.)

antenna of a borehole radar [21]. The antenna was designed to operate in a medium with relative permeability $\varepsilon_r = 4$ and be excited by a bipolar pulse of duration 2 ns. The data of measurements performed on a mockup filled with humidified sand with $\varepsilon_r = 3.9$ agree with simulation predictions.

Another way of designing flat KAs was proposed and investigated by the authors of [22–24]. The antennas are intended for use in short-range UWB communication systems in the frequency range 3.1–10.6 GHz. The antennas are printed on a rectangular plate of Rogers RO3210 foil-coated dielectric. Each antenna has an SMA-type power socket mounted on the wide edge of the dielectric plate. The base geometry of the printed KA is shown in Figure 7.44. The electrical-type radiator is fabricated as a flat dipole with arms (1 and 2) whose inner edges form a profiled slot. The input (3) of the electrical radiator is connected to the back (wide) edge of the plate via a slot line (4). The magnetic-type radiator (5) is made as a hole cut out in arm *1* of the electrical radiator. The input (6) of the magnetic radiator is connected to the back edge of the plate via a slot line (7). Near the back edge, slot lines 4 and 7 form an unbalanced coplanar line (8) to which a coaxial socket (9) forming the antenna input is connected. The hole (10) cut out in arm 2 is joined by a slot with the side edge of the plate at a distance from the back edge making 1/3 of the side edge length. The characteristic impedance of line 7 is 3–4 times that of line 4. The electric length of line 4 is greater than that length of line 7 by 0.15–0.2 of the maximum working wave length.

The proposed antenna design can be considered as a combination of an electrical and a magnetic radiator. The electrical radiator is a flat dipole with arms (1 and 2) separated by a profiled slot. The radiator is excited by a slot line (4). The characteristic feature of the radiator is that at frequencies below the first resonance frequency, its impedance is capacitive in nature; hence, the stored electrical energy prevails in the radiator near-field region. The magnetic radiator is a hole (5) in arm 1 that is excited by a slot line (7). It is characterized by that at frequencies below the first resonance frequency, its impedance is inductive in nature, and the stored magnetic energy is greater than the stored electrical energy. As a signal is applied to

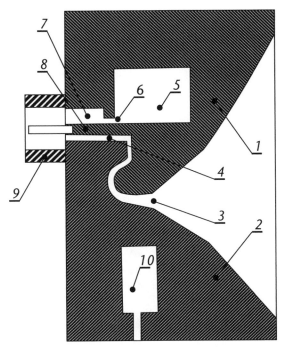

Figure 7.44 Base geometry of a printed combined UWB antenna: 1—top arm of the dipole, 2—bottom arm of the dipole, 3—input of the electrical radiator, 4—slot line of variable cross-section, 5—magnetic radiator, 6—input of the magnetic radiator, 7—slot line, 8—unbalanced coplanar line, 9—input socket, and 10—hole in the metal. (With permission from IEEE, Inc.)

the antenna input, one part of the signal excites (via a slot line) the electrical radiator and the other excites the magnetic radiator. With properly chosen dimensions of the antenna components, the proportion between the electrical and magnetic moments provides a nearly zero difference between the stored electrical and magnetic energies in the near-field region at frequencies above the lower cutoff frequency of the pass band. This minimizes the reactive component of the antenna impedance and weakens the frequency dependence of the active component, as the lower the frequency, the higher the energy density in the near-field region of the antenna and, hence, the higher the radiated power. The hole (10) in arm 2 works as a rejection filter and prevents the current flow to the back edge of the plate. This stabilizes the position of the pattern maximum and improves the match of the antenna with the feeder in the low frequency range.

The optimized geometry of two antenna versions with boards of dimensions 25×20 mm (A1) and 30×20 mm (A2) is presented in Figure 7.45. The boresight direction is shown by an arrow.

The electrode geometry of antenna A1 is presented in Figure 7.45(a). Antenna A1 is a combination of an electrical radiator fabricated as a flat dipole (with arms 1 and 2) and a magnetic radiator (3) made as a hole in the metal of arm *1*. The input (4) of the electrical radiator is connected to the back edge of the plate by a slot line of variable cross-section (5). The input of magnetic radiator 3 is connected to the back edge of the plate by a slot line (6). To increase the characteristic impedance of line 6, part of the dielectric is removed by cutting out holes 7 and 8. Near the

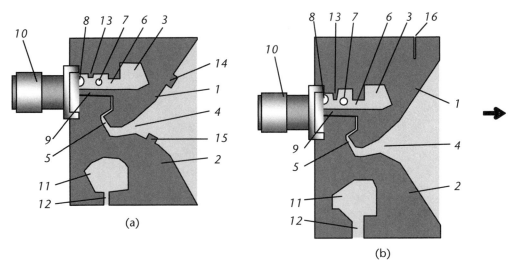

Figure 7.45 Geometry of antennas A1 (a) and A2 (b): 1—top arm of the dipole; 2—bottom arm of the dipole; 3—magnetic radiator; 4—dipole excitation region; 5—slot line of variable cross-section; 6—slot line; 7—open hole 1 mm in diameter; 8—open hole 1.6 mm in diameter; 9—unbalanced coplanar line; 10—input socket; 11—profiled hole; 12—slot; 13, 14, and 15—matching devices; 16—short-circuited slot-line segment. (With permission from IEEE, Inc.)

back edge, lines 5 and 6 form an unbalanced coplanar line (9) to which the central electrode of the coaxial socket (10) forming the antenna input is connected. In the metal of arm 2, a profiled hole (11) is made that is connected by a slot (12) with the bottom edge of the antenna. Projection 13 serves to match the antenna in the central part of the frequency band. Projections 14 and 15 are necessary to provide a demanded match level at the upper frequencies. The necessary amplitude and phase proportions between the parts of the signal exciting the electrical and the magnetic radiator are provided by properly choosing the characteristic impedances of lines 5 and 6 and the proportion between their lengths. As the distance between the central electrode and the socket leg is limited, to increase the characteristic impedance of line 6, holes 7 and 8 are cut out in the dielectric.

The electrode geometry of antenna A2 is presented in Figure 7.45(b). The purpose of electrodes 1–13 is similar to that of the respective electrodes of antenna A1. The short-circuited slot-line segment (16) serves for improving the match at the upper frequencies.

The proposed design and the chosen proportions in dimensions not only extend the match band toward the lower frequencies, but also increase the antenna directivity and keep constant the position of the boresight in the frequency band. Figure 7.46 presents the patterns of antenna A1 measured at different frequencies. The respective patterns of antenna A2 have very similar shapes.

The VSWR and gain G of the antennas were measured according to a standard procedure based on measuring complex transmission ratios with an Agilent Technologies 8719ET measurer. Plots of the VSWR versus frequency for antennas A1 and A2 are presented in Figure 7.47.

The gains of the antennas were measured by the method of two identical antennas. The frequency dependence of the gain for antennas A1 and A2 is plotted in

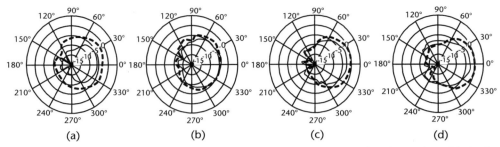

Figure 7.46   The patterns of antenna A1 in the *E*-plane (solid lines) and in the *H*-plane (dashed lines) at different frequencies: *f* = 3.1 (a), 5.1 (b), 7.1 (c), and 9.1 GHz (d). (With permission from IEEE, Inc.)

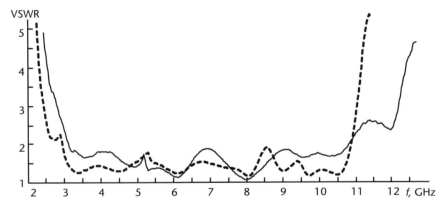

Figure 7.47   The frequency dependence of the measured VSWR for antennas A1 (solid line) and A2 (dashed line). (With permission from IEEE, Inc.)

Figure 7.48. In performing measurements, the antennas were placed 0.5 m apart from each other.

Figure 7.49(a) gives the measured modulus and argument of the transmission ratio S_{21}. The deviation of the PFR of each antenna from a linear function is shown in Figure 7.49(b).

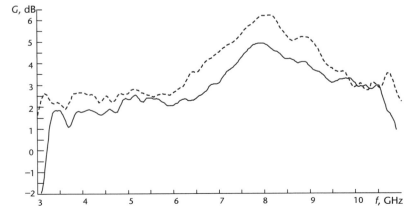

Figure 7.48   The frequency dependence of the gain for antennas A1 (solid line) and A2 (dashed line). (With permission from IEEE, Inc.)

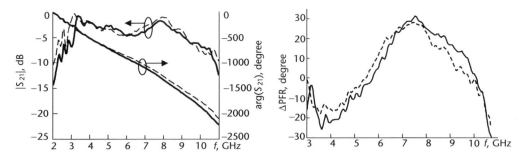

Figure 7.49 The frequency dependences of $|S_{21}|$ and $\arg(S_{21})$ (a) and the deviation of the PFR from a linear function (b) for antennas A1 (solid lines) and A2 (dashed lines). (With permission from IEEE, Inc.)

7.4.2 Balanced Combined Antennas

A rather complicated problem with unbalanced KAs is to realize a nondirectional radiation mode, such that the pattern would be isotropic at least in one plane. The reason is that electrical-type and magnetic-type radiators have different pattern shapes at the same polarization of the radiated field, or, on the contrary, when having closely similar patterns, they produce differently polarized fields. When a nondirectional operating mode is required, balanced KAs can be used. A design version of a KA based on a flat dipole was proposed by the authors of [25]. The dipole arms are flat thin metal plates located in one plane. The combined radiator is made by cutting out holes and slots in the arms of the dipole, as shown in Figure 7.50. This design can be considered a combination of electrical and magnetic radiators, as it carries both electrical (by conductive parts) and magnetic currents (in holes and slots).

The possibility of extending the match band of a KA toward the lower frequencies was demonstrated by a numerical simulation performed using the 4NEC2 code. The simulation results are presented in Figure 7.51, where the wire models used in the simulation are shown at the left.

The problem of synthesis of a fat radiator with the maximum possible match band was formulated in [26] who obtained a particular solution for the distribution of electrical and magnetic currents on a rectangular plate. Based on the above

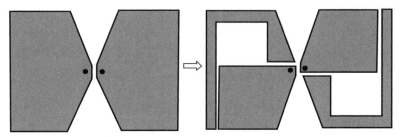

Figure 7.50 Sketch illustrating the transformation of a flat dipole into a balanced KA. (With permission from Science & Technology Publishing House, Ltd.)

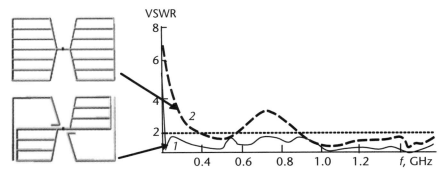

Figure 7.51   The frequency dependence of the VSWR for a balanced KA (1) and for a flat dipole (2) of the same dimensions with a source resistance of 180 Ω. (With permission from Science & Technology Publishing House, Ltd.)

principles and on numerical simulation results, a design version of a flat combined radiator with an extended match band was developed. The radiator design is sketched in Figure 7.52(a). A balanced combined radiator is printed on a square foil-coated dielectric plate (1). The radiator consists of a flat electric dipole formed by plates (2) and of two magnetic radiators (loops) (3) capacitively coupled with the dipole arms via slot gaps (4 and 5). The proportion between the moments of electrical and magnetic currents is optimized by properly choosing the length and width of the slots and the area of the loops. The maximum width of the match band is reached with the radiator excited using a balanced line with a characteristic impedance of 100 Ω. When the radiator is excited with a 50-Ω coaxial cable, a balun is used which is sketched in Figure 7.52(b).

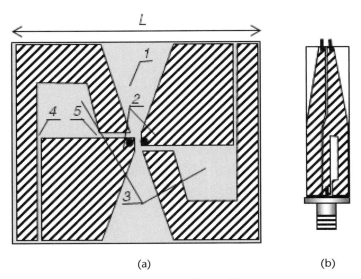

(a)   (b)

Figure 7.52   Topology of a flat KA (a) and of a balun (b).

Figure 7.53 shows a mockup of the balanced KA (a) and the frequency dependence of the VSWR in the 50-Ω duct (b), measured with a P2M-04 panoramic reflectivity measurer. The measurement data show that if the antenna dimension makes no more than 0.3 of the maximum operating wave length, the match bandwidth is over two octaves.

The frequency dependence of the pattern shape of a balanced KA in the E-plane is illustrated by Figure 7.54. The pattern shape in the H-plane differs from circular no more than by 3 dB at all frequencies of the match band.

7.5   Volumetric Combined Antennas

The research and development of three-dimensional (volumetric) combined antennas were aimed, first of all, at designing compact radiators for two-dimensional multi-element arrays of high-power UWB radiation sources. In the first KA versions [6], [27–29], the combination of an electrical monopole and a magnetic dipole made it possible to extend the pass band and realize a cardioid pattern. However, these KA versions had a number of disadvantages that constrained their further development. In particular, pattern failure limited the upper bound of the antenna pass band.

In a later proposed KA [30, 31], the pass band was extended toward the higher frequencies due to the use of a TEM horn as an electrical-type radiator. This

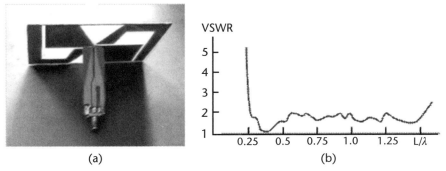

Figure 7.53 Mockup of a balanced flat KA with a balun (a) and the measured frequency dependence of the VSWR (b).

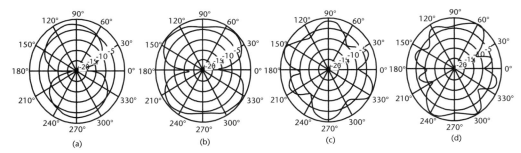

Figure 7.54 Frequency dependence of the pattern shape of a balanced flat KA: $L/\lambda = 0.3$ (a), 0.5 (b), 0.75 (c), and 1.0 (d).

substantially increased the possibilities to control the radiation characteristics by varying the proportion between the parameters of the active and passive magnetic dipoles and the TEM horn geometry. The characteristics of various KA versions are compared in detail elsewhere [12]. In the following sections, we present the results of comprehensive investigations of a KA with a TEM horn performed to reveal the factors limiting the pass band.

7.5.1   Radiation of Low-power Pulses

A combined antenna presented in [32] is intended for radiation of low-voltage (± 10 V) bipolar pulses of duration τ_p = 200 ps. These pulses are of interest in developing public-domain short-range radars [33] and low-power UWB beam communication systems. The antenna geometry was simulated using the similarity theorem. Proceeding from the dimensions of earlier developed KAs [34, 35] intended for radiation of nanosecond bipolar pulses, the antenna height (equal to its width) and length were chosen to be $h \approx 0.5\tau_p c$ and $L \geq h$, respectively.

The antenna parameters were calculated using the 4NEC2 code. The appearance of the antenna wire model is sketched in Figure 7.55(a). The model consists of 734 wires (1923 segments) of diameter 0.4 mm. The EMF source of the model is mounted on the wire located between the back wall of the antenna and the beginning of the upper lobe of the TEM horn. The wire carrying the EMF source is connected to eight fan-out back-wall wires, each carrying a lumped load of impedance 400 Ω. With this design, the total load of the wire model was 50 Ω. The model dimensions were L = 34 mm and h = 32 mm.

To determine the working bandwidth of the antenna model, its VSWR, AFR and pattern were investigated. The frequency dependence of the model VSWR is plotted in Figure 7.56(a) (curve 1). As can be seen from the plot, we have $K_V \leq 2$ in the frequency range 1.63–10 GHz, except for the gap 6.5–7.3 GHz where K_V is a little greater than two.

To determine the AFR of the model, the vertically polarized component of the electric field strength in the far-field region was calculated at fixed frequencies. The

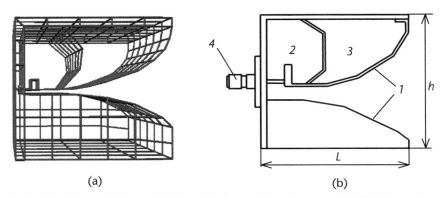

(a)   (b)

Figure 7.55   The wire model (a) and design (b) of a combined antenna: 1—TEM horn, 2—active magnetic dipole, 3—passive magnetic dipole, and 4—SMA-type input socket. (With permission from Pleiades Publishing, Ltd.)

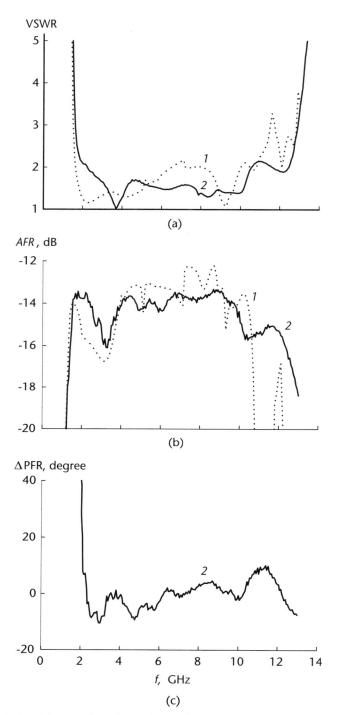

Figure 7.56 Calculated (curves 1) and experimental (curves 2) VSWR (a), AFR (b), and ΔPFR (c) of a combined antenna. (With permission from Pleiades Publishing, Ltd.)

power lost in the load (8 by 400 Ω) was assumed to depend on frequency. To provide the best fit of the calculations to the physics of electromagnetic radiation, the input power was corrected to balance the power lost in the load. In addition, the input power level was controlled according to the VSWR of the model. The input power level was reduced by the power that would be reflected in the feeder in the case of an actual antenna. Thus, for the frequencies corresponding to $K_V = 3$, the power reflected from the antenna input makes 0.25 of the input power, and for the model, the input power level was set equal to 0.75 of the input power. The frequency dependence of the model AFR is given in Figure 7.56(b) (curve 1). The AFR variations about mean in the boresight direction ($\varphi, \delta = 0$), where φ is the azimuthal angle and δ is the elevation angle are no greater than ±2 dB in the frequency range 1.3–10.6 GHz, except for the low frequency dip in the range 2.7–3.5 GHz where the maximum deviation corresponds to −2.56 dB.

The power patterns of the wire model plotted in the H-plane and E-plane, $F^2(\varphi)$ and $F^2(\delta)$, respectively, for frequencies of 2, 5, 8, and 10 GHz are given in Figure 7.57 (dashed lines). As can be seen from these plots, the boresight of the wire model has a fixed direction for the frequencies 2–8 GHz and slightly declines from this direction toward the positive δ at 10 GHz. At this frequency, side lobes appear in the H-plane pattern; their power level is higher than that in the main lobe.

The pass band of the antenna wire model could not be determined for lack of data about its phase-frequency response, which cannot be calculated using the 4NEC2 code. However, the obtained data on the model VSWR, AFR, and pattern pointed to the feasibility of an antenna with plausible characteristics in the frequency range 2–10 GHz.

Figure 7.55(b) presents the design of a KA fabricated to fit the geometry of the wire model. The antenna can be considered as a combination of an electrical radiator made as a TEM horn (1) and magnetic radiators made as an active (2) and a passive magnetic dipole (3). The antenna is connected to a 50-Ω feeder through an SMA RF connector. This KA differs from the earlier developed antennas radiating nanosecond bipolar pulses by the absence of the bottom passive magnetic dipole. The bottom part of the TEM horn is a solid metal electrode to enhance the mechanical strength of the antenna and provide a possibility of its duplicating with high reproducibility of geometric dimensions.

For measuring the antenna VSWR, AFR, and phase-frequency deviation from a linear function (ΔPFR), an Agilent 8719ET measurer of complex transmission ratios with a bandwidth of 0.05–13.5 GHz was used. To investigate the antenna characteristics in time domain, a receiving TEM antenna [36] and a TMR 8112 stroboscopic oscilloscope with a bandwidth of 12 GHz were used.

The frequency dependence of the VSWR of the combined antenna is plotted in Figure 7.56(a) (curve 2). The antenna match bandwidth on the level $K_V \leq 2$ is 2.23–12.5 GHz, except for the gap 7.5–11.6 GHz where K_V is slightly greater than two.

The variations of the antenna ARF about mean are within the limits ±1.5 dB for the boresight direction in the frequency range 1.4–12.4 GHz [Figure 7.56(b), curve 2]. The deviation of the phase-frequency response from a linear function for the same direction of observation makes less than ±π/16 for the frequencies from 2.1 GHz to, at least, 13 GHz [Figure 7.56(c), curve 2].

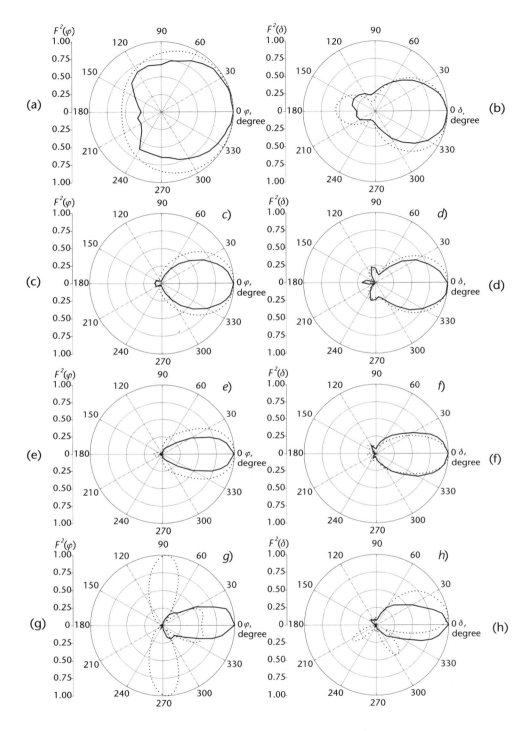

Figure 7.57 The calculated (dashed lines) and experimental (solid lines) power patterns of the model antenna in the *H*-plane (a, c, e, and g) and in the *E*-plane (b, d, f, and h) for the frequencies 2 (a, b), 5 (c, d), 8 (e, f), and 10 GHz (g, h). (With permission from Pleiades Publishing, Ltd.)

The calculated (dashed lines) and experimental (solid lines) power patterns of the KA in the main planes for frequencies of 2, 5, 8, and 10 GHz are given in Figure 7.57. The experimental patterns were plotted by the data obtained in pulsed measurements: The squared amplitude spectra of the pulses radiated by the antenna were found at specified frequencies in two main planes with a step varied from 5° to 15°. The low spectral amplitudes of the radiated pulses at frequencies above 10 GHz (with the antenna excited by bipolar pulses of duration 200 ps) gave no way of constructing plausible patterns in this frequency range. The voltage at the output of the receiving TEM antenna was assumed to be proportional to the radiation electric field strength.

Comparing the patterns, we see good agreement between calculations and measurements in the frequency range 2–8 GHz. At the same time, the experimental patterns indicate that the position of the pattern principal maximum remains stable up to 10 GHz. The discrepancy between the calculated and experimental results in the range of high frequencies is possibly related to that the thin wire approximation fails, first of all, because the condition of smallness of the wire segment length is not fulfilled at the given wave lengths.

Thus, from the measurements of the VSWR, AFR, and ΔPFR it follows that the KA has a pass band in the boresight direction in the frequency range 2.23–12.4 GHz; that is, the frequency band ratio is 5.6:1. The lower bound of the pass band is determined by the lower bound of the match band and the upper bound by the AFR of the antenna. The ratio of the antenna length to the lower-bound wave length of the match band, L/λ_L, is 0.25.

The time–domain measurements of the antenna characteristics were performed with the use of a generator of bipolar voltage pulses of duration 200 ps and amplitude 13 V. The voltage pulse waveform is given in Figure 7.58(a). The receiving TEM antenna was at a distance $r = 0.9$ m from the transmitting antenna in the boresight direction, which corresponded to the radiation far-field region for all frequencies of the pass band. In Figure 7.58(b), the waveform of the pulse radiated by the KA in the boresight direction is presented. The peak field strength efficiency of the KA, k_E, was 0.9 and its energy efficiency k_w was 0.92.

The peak-power pattern functions of the antenna, $F^2(\varphi)$ and $F^2(\delta)$, in the H-plane and in the E-plane, respectively, are given in Figure 7.59. The half-peak-power width of the pattern in the H-plane is 87° and that in the E-plane is 103°. The antenna boresight directivity D_0 equals to four. The radiation of the antenna is linearly polarized in the vertical plane. The procedures of determining k_E, k_w, and D_0 are described elsewhere [31] (Section 2.4.2).

For comparison, we briefly discuss the results of a numerical simulation performed for a bicone antenna optimized for the radiation of low-voltage bipolar pulses of duration $\tau_p = 150$ ps [37]. The antenna geometry with cone angle $2\theta_0 = 120°$ and flat end faces is given in Figure 2.15. The cone generator length L is 60 mm. For the time–domain simulation, the finite-difference method was used. The code based on a finite-difference scheme for two-dimensional Maxwell equations in a cylindrical coordinate system (r,z) possessing rotational symmetry is described elsewhere [38].

Calculations of the antenna VSWR, AFR, and ΔPFR have shown that the lower bound of the antenna pass band in the boresight direction is determined by the lower

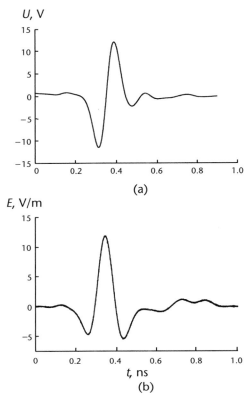

Figure 7.58 Waveforms of the voltage pulse at the antenna input (a) and of the radiated pulse (b). (With permission from Pleiades Publishing, Ltd.)

bound of the AFR and its upper bound is determined by the ΔPFR. The frequency ratio is 5.6:1, and the ratios of the bicone generator length to the lower and upper bound wave lengths of the pass band are, respectively, $L/\lambda_L = 0.7$ and $L/\lambda_H = 3.9$. Note that the VSWR does not restrict the pass band of the bicone antenna. The lower bound of the bicone match band on the level $K_V = 2$ corresponded to $L/\lambda_L = 0.35$ and was higher than that for the KA ($L/\lambda_L = 0.25$), and the upper bound lied far in the high-frequency range and could not be determined because of the limited calculation accuracy. Thus, the match band of a bicone antenna can be substantially wider than its pass band.

The above investigation results indicate that the KA radiating bipolar pulses of duration 150–200 ps has substantially smaller dimensions and higher pattern directivity than the bicone antenna radiating similar pulses and having a pass band with a close frequency ratio. To increase the pattern directivity, the bicone antenna was modernized by using a short-circuited loop [39]. As a result, the lower bound of the match band $L/\lambda_L = 0.39$ was achieved. This correlates with the calculated value for a bicone antenna ($L/\lambda_L = 0.35$) and is half again that for a KA ($L/\lambda_L = 0.25$). This antenna can also be treated as a combination of an electrical radiator (bicone) and a magnetic radiator (loop). A KA of similar design was developed and investigated [40]. A novel X-range flat antenna combining an electrical and a magnetic radiator

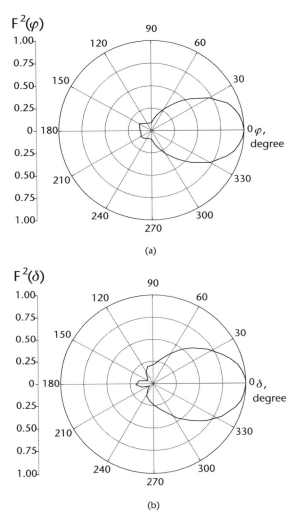

Figure 7.59 Peak-power patterns of the antenna in the *H*-plane (a) and in the *E*-plane (b). (With permission from Pleiades Publishing, Ltd.)

is described elsewhere [41]. Also, noteworthy is an unbalanced loopback TEM horn [42] with a wide match band (100:1 frequency ratio), which is intended for use in low-power wideband receiving-transmitting systems.

7.5.2   Antennas Intended for Radiation of High-power Pulses

In this section, we consider the characteristics of KAs with an extended pass band intended for radiation of high-power pulses. The distinctive feature of these antennas is in the design of the lead-in that ensures high electric strength at a bipolar pulse amplitude of up to 200 kV. In first experiments [30, 31, 12], a KA with nonoptimal geometry was used (Figure 8.10). The optimization procedure consists in changing the antenna geometry based on results of numerical simulations [4, 43, 47]. Below, only two KAs are considered for which the exciting bipolar voltage pulses differ in duration by an order of magnitude. The characteristics of the antennas were measured with the use of low-voltage bipolar pulses.

We first consider a KA excited by a bipolar pulse of duration τ_p = 3 ns. Antennas of this type were used in UWB radiation sources based on a single antenna [48] and on a four-element array [49] and differed only by the design of the high-voltage bushing. The antenna had the transverse dimension $h = 0.5c\tau_p$ = 45 cm and the longitudinal dimension L = 47 cm. The antenna characteristics were optimized using the 4NEC2 code. The appearance of the antenna that was used as an array element is shown in Figure 7.60. Below, the measurement data for this KA are given.

Figure 7.61 presents the antenna VSWR as a function of the frequency of the wave propagating in the 50-Ω feeder. It can be seen that in the frequency range 0.13–1.1 GHz, we have $K_V \leq 2$. For a system of two antennas, one operated to transmit and the other to receive radiation pulses, measurements of the AFR and PFR were performed. The AFR variations about mean (Figure 7.62, curve 1) do not fall outside the limits ±1.5 dB for the principal direction ($\varphi, \delta = 0$) in the frequency range 0.14–0.85 GHz. The deviation of the phase-frequency response from a linear function, ΔPFR, for the same observation direction is no more than ±π/16 for the frequencies 0.14–0.9 GHz (Figure 7.62, curve 2). The relative pass band, determined by three criteria satisfied simultaneously (VSWR ≤ 2, ΔAFR ≤ ±1.5 dB, and ΔPFR ≤ ±π/16), is 6.1:1 (0.14–0.85 GHz) for the boresight direction. It follows that the lower bound of the pass band for this KA is determined by the AFR and PFR and the upper bound by the AFR. In this case, the lower bound of the match band corresponds to L/λ_L = 0.2.

Figure 7.60 Appearance of the antenna excited by a bipolar pulse of duration 3 ns: 1—TEM horn, 2—active magnetic dipole, and 3—passive magnetic dipoles. (With permission from Pleiades Publishing, Ltd.)

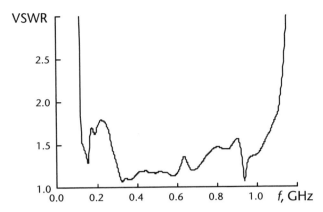

Figure 7.61   The frequency dependence of the antenna VSWR. (With permission from Pleiades Publishing, Ltd.)

Figure 7.63 presents the antenna peak-power patterns in the H-plane and in the E-plane. The half-peak-power widths of the patterns are approximately identical and make 80°.

The antenna peak-power directivity was estimated, based on the low-voltage measurements, to be $D_0 = 5$. The peak-field-strength efficiency and the energy efficiency of the KA were estimated as $k_E = 2$ and $k_w = 0.93$, respectively.

The results of investigations of the characteristics of a KA used in a high-power UWB radiation source excited by a bipolar pulse of duration $\tau_p \approx 200$ ps are discussed in [50]. The appearance of the KA is shown in Figure 7.64(a). The antenna length L is 4.3 cm and its both transverse dimensions, h, are 4 cm.

The antenna characteristics were optimized using the 4NEC2 code. The appearance of the antenna wire model is presented in Figure 7.64(b). The simulation was aimed at minimizing the antenna VSWR. The calculation results are given in Figure 7.65 (curve 1).

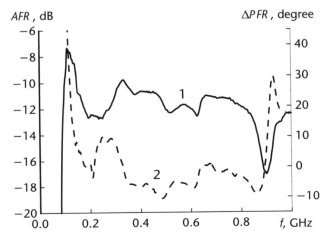

Figure 7.62   The measured antenna characteristics: the AFR (curve 1) and the deviation of the PFR from a linear function (curve 2). (With permission from Pleiades Publishing, Ltd.)

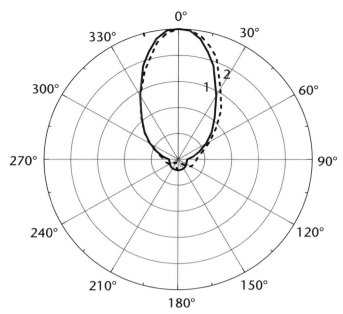

Figure 7.63 The antenna patterns in the *H*-plane (curve 1) and in the *E*-plane (curve 2). (With permission from Pleiades Publishing, Ltd.)

The measured match bandwidth on the level $K_V \leq 2$ is 1.4–9.8 GHz (Figure 7.65, curve 2). The AFR variations about mean (Figure 7.66, curve 1) do not fall outside the limits ±1.5 dB for the boresight direction ($\varphi, \delta = 0°$) in the frequency range 1.1–8.1 GHz. The deviation of the phase-frequency response from a linear function, ΔPFR, for the same observation direction is no more than ±π/16 for the frequencies 1.7–9.2 GHz (Figure 7.66, curve 2).

The relative pass band determined by the simultaneously satisfied three criteria makes 4.8 : 1 (1.7–8.1 GHz) for the boresight direction. It follows that for this KA, the lower and the upper bound of the pass band are determined by the PFR and

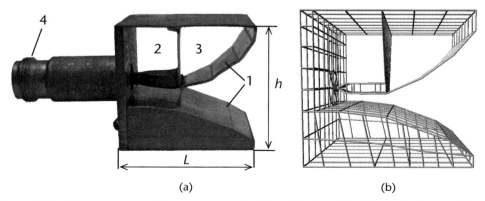

Figure 7.64 The appearance of an antenna excited by a bipolar pulse of duration 200 ps (a) and its wire model (b): 1—TEM horn, 2—active magnetic dipole, 3—passive magnetic dipole, and 4—N-type input connector. (With permission from Pleiades Publishing, Ltd.)

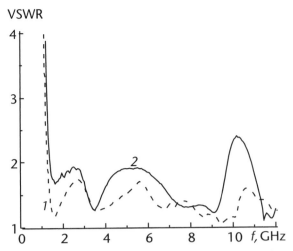

Figure 7.65 The calculated (curve 1) and measured VSWR (curve 2) of the antenna. (With permission from Pleiades Publishing, Ltd.)

AFR, respectively. In this case, the lower bound of the match band corresponds to $L/\lambda_L = 0.2$.

The peak-power antenna patterns in the H-plane and in the E-plane are given in Figure 7.67. The half-peak-power pattern width is 90° in the H-plane and 100° the E-plane. The peak-power pattern maxima in the two planes, like for the above antenna, correspond to $\varphi = \delta = 0$.

Based on the low-voltage measurements, the antenna directivity at the pattern peak was estimated to be $D_0 = 4$. The peak-field-strength efficiency and the energy efficiency were, respectively, $k_E = 0.9$ and $k_w = 0.92$.

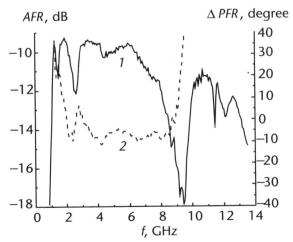

Figure 7.66 The measured antenna characteristics: the AFR (curve 1) and the deviation of the PFR from a linear function (curve 2). (With permission from Pleiades Publishing, Ltd.)

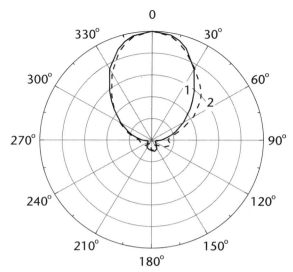

Figure 7.67 The antenna peak-power patterns in the *H*-plane (curve 1) and in the *E*-plane (curve 2). (With permission from Pleiades Publishing, Ltd.)

Conclusion

The basic requirements for the antennas intended for radiation of UWB electromagnetic pulses have been formulated. The concept of the transfer function of an antenna operating in the transmission mode has been introduced, and it has been shown that the waveform of the radiated pulse is always different from that of the voltage pulse exciting the antenna. For linear radiators, expressions for the current distribution functions have been derived and the factors have been revealed that affect the distortion of the radiated pulse waveform. In addition, the calculation equations have been derived for the transfer functions of monopoles, collinear and noncollinear dipoles, and loop antennas operating in the transmission mode.

It has been revealed that the main causes of the distortion of the radiated pulse waveform are

- the strong frequency dependence of the input impedance of the radiator if its dimensions are less than half the spatial length of the pulse, and
- the frequency dependence of the pattern shape for dipoles and loops if the dipole length and the loop perimeter are greater than the spatial length of the pulse.

It has been shown that the pass band of a short radiator can be widened due to the extension of the match band toward the lower frequencies that can be provided by using combined antennas comprising electrical- and magnetic-type radiators. Conditions have been formulated under which the frequency dependence of the input impedance of a combined radiator can be substantially reduced.

Design versions of flat KAs have been discussed that feature low cost and simplicity of manufacture and are suitable for radiation of pulses with peak powers of tens of watts. Volumetric KAs intended for radiation of UWB pulses with peak powers of hundreds of megawatts have been developed and investigated. It has been demonstrated experimentally that the KA pass band, determined by three characteristics, such as the VSWR, the AFR, and the deviation of the PFR from a linear function (ΔPFR) is narrower than the match band determined by the frequency dependence of the antenna VSWR.

Problems

7.1   What is meant as the transfer function of a transmitting antenna?

7.2   Which parameters of a transmitting antenna determine its pass band?

7.3   What are the reasons why the waveform of the radiated pulse differs from that of the voltage pulse exciting the antenna?

7.4   What determines the lower bound of the pass band of an antenna?

7.5   Why the match band of a combined antenna containing radiators of electrical and magnetic types can be wider than the match band of each of the radiators?

7.6   Why the dimensions of a volumetric KA can be less than the dimensions of a flat KA having the same pass band?

References

[1]   Markov, G. T., and D. M. Sazonov, *Antennas*, Moscow: Energia, 1975 (in Russian).

[2]   Buyanov, Yu. I., V. I. Koshelev, and V. V. Plisko, "Radiation of a Long Conductor Excited by a Short Pulse," *Proc. VII Inter. Conf. Mathematical Methods in Electromagnetic Theory*, Kharkov, June 2–5, 1998, Vol. 1, pp. 312–314.

[3]   Efimova, N. A., and V. A. Kaloshin, "On Matching of Symmetric TEM Horns," *J. Commun. Technol. Electron.*, Vol. 59, No. 1, 2014, pp. 54–60.

[4]   Cadilhon, B., et al., "Ultra Wideband Antennas for High Pulsed Power Applications," *Ultra Wideband Communications: Novel Trends—Antennas and Propagation*, 2011, Dr. Mohammad Matin (ed.), ISBN: 978-953-307-452-8, InTech, DOI: 7.5772/20305. Available from: http://www.intechopen.com/books/ultra-wideband-communications-novel-trends-antennas-and-propagation/ultra-wideband-antennas-for-high-pulsed-power-applications.

[5]   Litvinenko, O. N., and V. I. Soshnikov, *Theory of Nonuniform Lines and Their Use in Radio Engineering*, Moscow: Sov. Radio, 1964 (in Russian).

[6]   Koshelev, V. I., et al., "High-Power Ultrawideband Electromagnetic Pulse Radiation," *Proc. SPIE*, Vol. 3158, 1997, pp. 209–219.

[7]   Belichenko, V. P., et al., "On the Possibility of Extending the Passband of Small-Size Radiators," *J. Commun. Technol. Electron.*, Vol. 44, No. 2, 1999, pp. 167–172.

[8]   Wait, J. R., *Electromagnetic Radiation From Cylindrical Structures*, New York: Pergamon Press, 1959.

[9]   Kessenikh, V. N., *Propagation of Radio Waves*, Moscow: GITTL, 1953 (in Russian).

[10]   Walter, C. H., *Traveling Wave Antennas*, New York: McGraw-Hill, 1965.

[11]   Buyanov, Yu. I., "The Quality Factor and Pass Band of Linear Radiators," *Proc. All-Russian Conf. Physics of Radio Waves*, Tomsk, Sept. 23–28, 2002, pp. VII20–23 (in Russian).

[12]   Andreev, Yu. A., Yu. I. Buyanov, and V. I. Koshelev, "Small-Sized Ultrawideband Antennas Radiating High-Power Electromagnetic Pulses," *Zh. Radioelektron.*, No. 4, 2006. Available from: http://jre.cplire.ru/mac/apr06/1/text.html (in Russian).

[13] McLean, J. S., "A Re-examination of the Fundamental Limits on the Radiation Q of Small Antennas," *IEEE Trans. Antennas Propagat.*, Vol. 44, No. 5, 1996, pp. 672–676.

[14] Geyi, W., P. Jarmuszewski, and Y. Qi, "The Foster Reactance Theorem for Antennas and Radiation Q," *IEEE Trans. Antennas Propagat.*, Vol. 48, No. 3, 2000, pp. 401–408.

[15] Kwon, D. H., "On the Radiation Q and Gain of Crossed Electric and Magnetic Dipole Moments," *IEEE Trans. Antennas Propagat.*, Vol. 53, No. 5, 2005, pp. 1681–1687.

[16] Fano, R. M., "Theoretical Limitations on the Broadband Matching of Arbitrary Impedances," *J. Franklin Inst.*, Vol. 249, January and February 1950, pp.57–83 and 139–154.

[17] Vershkov, M. V., and O. B. Mirotvorsky, *Marine Antennas*, Leningrad: Sudostroenie, 1990 (in Russian).

[18] *NEC Based Antenna Modeler and Optimizer.* Available from: http://www.qsl.net/4nec2.

[19] Balzovsky, E. V., and Yu. I. Buyanov, "An Ultrawideband Antenna Element for a Synthesized Aperture," *Izv. Vyssh. Uchebn. Zaved., Fiz.*, Vol. 55, No. 8/2, 2012, pp. 60–61.

[20] Buyanov, Yu. I., V. I. Koshelev, and P. F. Shvadlenko, "Investigation of the Characteristics of the Elements of an Ultrawideband Multibeam Antenna Array," *Izv. Vyssh. Uchebn. Zaved., Fiz.*, Vol. 55, No. 9/2, 2012, pp. 22–26.

[21] Balzovsky, E. V., et al., "Directional Ultrawideband Combined Antenna for a Borehole Radar," *Proc. 1st All-Russian Microwave Conf.*, Moscow, Nov. 27–29, 2013, pp. 312–316 (in Russian).

[22] Kwon, D.-H., et al., "Small Printed Combined Electric-Magnetic Type Ultrawideband Antenna with Directive Radiation Characteristics," *IEEE Trans. Antennas Propagat.*, Vol. 56, No. 1, 2008, pp. 237–241.

[23] Kwon, D.-H., et al., "Small Printed Ultra-Wideband Antennas Combining Electric- and Magnetic-type Radiators," In *Ultra-Wideband, Short-Pulse Electromagnetics 9*, pp. 425–431, F. Sabath, et al. (eds.), New York: Springer, 2010.

[24] Balzovsky, E. V., Yu. I. Buyanov, and V. I. Koshelev, "Small-Sized Flat Antenna as an Element of an Ultrawideband Double Polarized Array," *Proc. 3rd All-Union Conf. Radar and Radio Communications*, Moscow, Oct. 26–30, 2009, pp. 77–82 (in Russian).

[25] Belichenko, V. P., Yu. I. Buyanov, and S. N. Litvinov, "Combined Radiators with an Extended Match Band," *Izv. Vyssh. Uchebn. Zaved., Fiz.*, Vol. 49, No. 9, Suppl., 2006, pp. 23–27.

[26] Belichenko, V. P., et al., "Synthesis of Ultrawideband Small-Sized Radiators Based on Minimization of Reactive Energy," *Proc. 2nd Intern. Conf. Acoustooptical and Radar Methods for Measuring and Data Processing*, Suzdal, Sept. 25–27, 2007, pp. 32–35 (in Russian).

[27] Andreev, Yu. A., et al., "A High-Power Ultrawideband Electromagnetic Pulse Generator," *Instrum. Exp. Tech.*, Vol. 40, No. 5, 1997, pp. 651–655.

[28] Andreev, Yu. A., et al., "An Element of the Scanning Antenna Array for Radiation of High-Power Ultrawideband Electromagnetic Pulses," *J. Commun. Technol. Electron.*, Vol. 44, No. 5, 1999, pp. 492–498.

[29] Andreev, Yu. A., et al., "Multichannel Antenna System for Radiation of High-Power Ultrawideband Pulses," In *Ultra-Wideband, Short-Pulse Electromagnetics 4*, pp. 181–186, E. Heyman, B. Mandelbaum, and J. Shiloh (eds.), New York: Plenum Press, 1999.

[30] Koshelev, V. I., et al., "Ultrawideband Radiators of High-Power Pulses," *Proc. IEEE Pulsed Power Plasma Science Conf.*, Las Vegas, June 17–22, 2001, Vol. 2, pp. 1661–1664.

[31] Andreev, Yu. A., Yu. I. Buyanov., and V. I. Koshelev, "A Combined Antenna with Extended Bandwidth," *J. Commun. Technol. Electron.*, Vol. 50, No. 5, 2005, pp. 535–543.

[32] Andreev, Yu. A., V. I. Koshelev, and V. V. Plisko, "Combined Antenna and Linear Arrays for Radiation of Low-Power Picosecond Pulses," *J. Commun. Technol. Electron.*, Vol. 56, No. 7, 2011, pp. 812–823.

[33] Federal Communication Commission USA (FCC) 02-48, ET Docket 98-153, First Report and Order, April 2002.

[34] Gubanov, V. P., et al., "Sources of High-Power Ultrawideband Radiation Pulses with a Single Antenna and a Multielement Array," *Instrum. Exp. Tech.*, Vol. 48, No. 3, 2005, pp. 312–320.

[35] Efremov, A. M., et al., "Generation and Radiation of High-Power Ultrawideband Nanosecond Pulses," *J. Commun. Technol. Electron.*, Vol. 52, No. 7, 2007, pp. 756–764.

[36] Andreev, Yu. I., V. I. Koshelev, V. V. Plisko, "Characteristics of Receiving-Transmitting TEM Antennas," *Proc. 5th Scientific and Technical Conf. Radar and Radio Communications*, Moscow, Nov. 21–25, 2011, pp. 77–82 (in Russian).

[37] Koshelev, V. I., et al., "Frequency and Time Characteristics of Conical TEM Antennas," *Proc. 4th All-Union Conf. Radar and Radio Communications*, Moscow, Nov. 29–Dec. 3, 2010, pp. 336–340 (in Russian).

[38] Koshelev, V. I., A. A. Petkun, and S. Liu, "Numerical Simulation of Axially Symmetric Ultrawideband Radiators," *Russ. Phys. J.*, Vol. 49, No. 9, 2006, pp. 970–975.

[39] Desrumaux, L., et al., "On Original Antenna for Transient High Power UWB Arrays: The Shark Antenna," *IEEE Trans. Antennas Propagat.*, Vol. 58, No. 8, 2010, pp. 2515–2522.

[40] Kim, J. S., Y. J. Yoon, and J. Ryu, "A Compact High-Power Antenna for Ultrawideband Nanosecond Bipolar Pulse," *Microwave Opt. Technol. Lett.*, Vol. 57, No. 6, 2015, pp. 1296–1301.

[41] Haider, N., et al., "Directive Electric-Magnetic Antenna for Ultra-Wideband Applications," *IET Microwave Antennas Propagat.*, Vol. 7, No. 5, 2013, pp. 381–390.

[42] Biryukov, V. L., et al., "Investigation of an Ultrawideband Ring Antenna Array," Zh. Radioelektron., No. 1, 2013. Available from: http://jre.cplire.ru/jre/jan13/20/text.pdf.

[43] Andreev, Yu. A., V. I. Koshelev, and V. V. Plisko, "Extension of the Pass Band of a Combined Antenna," *Proc. 4th All-Union Conf. Radar and Radio Communications*, Moscow, Nov. 29–Dec. 3, 2010, pp. 331–335 (in Russian).

[44] Andreev, Yu. A., et al., "Generation and Radiation of High-Power Ultrawideband Pulses with Controlled Spectrum," *J. Commun. Technol. Electron.*, Vol. 58, No. 4, 2013, pp. 297–306.

[45] Godard, A., et al., "A Transient UWB Antenna Array Used With Complex Impedance Surface," *Inter. J. Antennas Propagat.*, Vol. 2010, Article ID 243145, 8 pages, doi: 7.1155/2010/243145.

[46] Mehrdadian, A., and K. Forooraghi, "Design of a UWB Combined Antenna and an Array of Miniaturized Elements with and without Lens," *Progress Electromagn. Res. C.*, Vol. 39, 2013, pp. 37–48.

[47] Mehrdadian, A., and K. Forooraghi, "Design and Fabrication of a Novel Ultrawideband Combined Antenna," *IEEE Antennas Wireless Propagat. Lett.*, Vol. 13, No. 1, 2014, pp. 95–98.

[48] Koshelev, V. I., et al., "Study on Stability and Efficiency of High-Power Ultrawideband Radiation Source," *J. Energy Power Eng.*, Vol. 5, No. 6, 2012, pp. 771–776.

[49] Andreev, Yu. A., et al., "A High-Performance Source of High-Power Nanosecond Ultrawideband Radiation Pulses," *Instrum. Exp. Tech.*, Vol. 54, No. 6, 2011, pp. 794–802.

[50] Andreev, Yu. A., et al., "Generation and Emission of High-Power Ultrabroadband Picosecond Pulses," *J. Commun. Technol. Electron.*, Vol. 56, No. 12, 2011, pp. 1429–1439.

CHAPTER 8

Antenna Arrays

Introduction

The useful working range of multipurpose radio systems can be extended by using antenna arrays [1–5]. Multielement arrays make it possible to reduce the pattern width and, accordingly, increase the power density radiated at the boresight and to improve the spatial resolution of radar objects. Besides, antenna arrays are necessary to perform electron scanning of objects by a wave beam and to design multibeam and adaptive antennas. Antenna arrays are classed as transmitting, receiving, transmitting–receiving, and also as passive and active. In active arrays, the active device (amplifier) is integrated into an array element and has a significant impact on its performance. Structurally, three main parts of an array are distinguished: the radiating (receiving) system, the control devices, and the distribution system.

The studies of antenna arrays were largely aimed at their use in narrowband radio systems. The radio pulse duration was assumed to be substantially greater than the array aperture fill time, $L\sin\theta/c$, where L is the linear dimension of the array, θ is the pulse incidence angle, counted from the normal to the array, and c is the velocity of light in free space. The radiation spectrum width was estimated as $\Delta f/f = \lambda_0/L(\sin\theta)$, where λ_0 is the center wavelength. In this case, on summation of the array element signals, the waveform of the output pulse should be the same for all pattern angles and represent a quasi-harmonic oscillation, except for the pulse beginning and end. It seems that Harmuth [6] was the first who showed, by the example of a Hertz dipole linear array, that for short UWB pulses, the radiated pulse waveform depends on direction.

In high-power ($\geq$100 MW) UWB arrays, TEM antennas [7, 8] and combined antennas (KAs) [9, 10] are widely used. Tapered slot antennas (TSAs) [11, 12] have found wide application as elements of dual-polarized arrays in radio systems of low peak power. Note that transmitting–receiving arrays of this type operate mainly with long radio pulses whose center frequency f_0 is tunable in a wide range with the frequency ratio reaching 12:1 (time–frequency modulation). High spatial resolution of probed objects is achieved in the processing of reflected signals [13].

In high-power radio systems and short-range radars, transmitting and receiving arrays are separated apart. In the first case, this is due to the difficulties in ensuring electrical insulation between transmitter and receiver, and in the second due to small distance ($\sim$ 1 m) to the probed object. In these systems, various antenna types can be used as elements of transmitting and receiving antenna arrays. Among receiving arrays intended for detecting short UWB radiation pulses, arrays based on electric dipoles [14] and on magnetic loops [15] can be mentioned.

This chapter is focused on the transmitting and receiving arrays designed to operate with short UWB radiation pulses. In some cases, for comparison, data are given for arrays operating with long radio pulses whose central frequency is tunable in a wide range.

8.1   Directional Properties of Antenna Arrays

8.1.1   Numerical Calculations

Consider the space and time characteristics of short electromagnetic pulses generated by a planar array of radiating elements using the results of numerical calculations [16]. Let an array of radiators, each having a cardioid pattern like in the case of a KA [17, 18], be located in the yOz plane. The strength of the far electromagnetic field radiated by an array element in the xOy plane is determined, not taking into account its interaction with other elements, by the equation

$$E_{mn}\left(t, r_{mn}\right) = \frac{A}{r_{mn}} \frac{dI_{mn}}{dt} (1 + \cos\varphi) \qquad (8.1)$$

where A is a dimensional constant, r_{mn} is the distance from the array element to the observation point, $I_{mn} = I(t - r_{mn}/c)$ is the current of the mnth radiator taken at the time t with a delay of r_{mn}/c, and φ is the angle specifying the direction toward the observation point in spherical coordinates at $\theta = 90°$. The total field is the superposition of the fields radiated by the array elements.

Generally, to provide a desired space and time localization of the radiated electromagnetic pulse, each array element is excited by a current pulse with an appropriate time shift

$$\Delta t_{mn} = \frac{\left(y_n \sin\theta_0 \sin\varphi_0 + z_m \cos\theta_0\right)}{c} \qquad (8.2)$$

where y_n, z_m are the coordinates of the mnth array element and θ_0, φ_0 are the angles specifying the localization direction of the radiated field in the spherical coordinate system.

The space–time characteristics of the radiated electromagnetic pulse were investigated for an 8×8 square array whose elements were excited by monopolar and bipolar current pulses and by pulses containing a larger number of current oscillation periods. The waveform of the current pulse in the radiator was set as a combination of gaussoids: $I(t) = \exp[-(10t/3\tau_p - 2)^2]$ for a monopolar pulse, $I(t) = \exp[-(5t/\tau_p - 2)^2] - \exp[-(5t/\tau_p - 4)^2]$ for a bipolar pulse, and $I(t) = \sum_{i=1}^{p}(-1)^i \exp[-(4t/T - 2i)^2]$ for a pulse with two and more periods. Here, τ_p is the pulse duration at a 0.1 amplitude level, T is the pulse period, and p is the number of half-periods in the pulse.

As the waveform of the electromagnetic pulse radiated by an array excited by short current pulses depends on direction (Figure 8.1), the notion of the pattern function for this case is ambiguous. For a narrowband signal, by the pattern function is meant the amplitude and angular dependence of the radiated field intensity at a fixed distance in the far-field region. However, for short radiation pulses, it should be borne in mind that the time at which the signal reaches the observation point and its shape are different for different directions. In this case, when measuring the field at a fixed time at different distances from the array (Figure 8.1), we will obtain different pattern functions.

Therefore, it was proposed to plot the patterns for UWB radiation pulses based on peak field strength E_p or peak power (power density) E_p^2. The values of E_p are taken as the maximum radiation field strengths for a given direction (θ,φ) in a certain spherical layer whose thickness is equal to the maximum duration of the radiated pulse (Section 2.4). In practice, these pattern functions are measured as follows: the waveforms of pulses radiated in different directions (θ,φ) are recorded at a specified distance r in the far-field region, and E_p values corresponding to different times are captured.

The differences between the radiation pattern for a narrowband signal and the peak field pattern are illustrated by Figure 8.2, where the corresponding curves are given for a square array excited by a bipolar current pulse for the spacing between the centers of the elements radiating in the vertical and in the horizontal direction $d_v = d_h = d = 0.5\tau_p c$. Curve 1 represents the narrowband signal pattern at the same center frequency radiated from the same array for the distance corresponding to r_2 in Figure 8.1. Curves 2 and 3 depict, respectively, the peak field pattern and the pattern plotted by the energy transported by the radiated pulse in specified directions. Note that on the decibel scale, the peak field pattern coincides with the peak power pattern. As the obtained results indicate that the peak field (peak power) pattern function for a short UWB pulse has no zeros, it should be spoken of side radiation rather than of sidelobe radiation.

A numerical calculations of the field radiated by an antenna array excited by monopolar and bipolar current pulses have shown that the best way to reduce side radiation is to use an array with uniform current amplitude distribution. This is

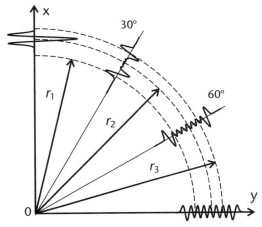

Figure 8.1 Waveforms of electromagnetic pulses radiated at different angles by an array whose elements, spaced by $0.5\tau_p c$, are synchronously excited by a bipolar current pulse.

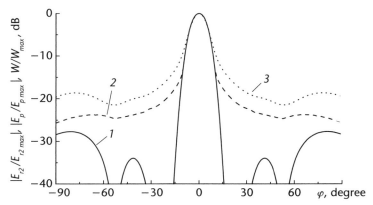

Figure 8.2 Pattern functions of an array with the elements synchronously excited by a bipolar current pulse: narrowband signal pattern (curve 1), peak field pattern (curve 2), and energy pattern (curve 3).

its essential difference from arrays excited by long radio pulses for which a current amplitude distribution tapered at the array edges is the best choice [3].

Figure 8.3 presents radiation patterns obtained for an array excited by a bipolar current pulse. The pattern width decreases with increasing the array aperture, which is the greater, the greater the element spacing [Figure 8.3(a, b)], and the side radiation becomes more intense. In this case, with the spacing between the element centers increased to $d = \tau_p c$ and the wave-beam scan angle $\varphi_0 = 45°$, the grating lobes are not pronounced [Figure 8.3(b)], although the side radiation intensity and the pattern width are increased. This is also an essential distinction of UWB arrays from narrowband ones, for which grating lobes arise if

$$\frac{d}{\lambda_0} \geq \frac{1}{(1 + \sin\varphi_0)} \tag{8.3}$$

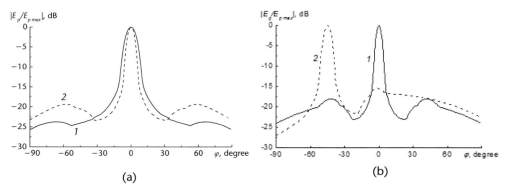

Figure 8.3 Peak field patterns of an array with elements spaced by $d = 0.5\tau_p c$ (curve 1) and $0.75\tau_p c$ (curve 2), synchronously excited by a bipolar current pulse (a), and of an array with elements spaced by $d = \tau_p c$, synchronously excited (curve 1) and wave-beam scanned within 45° (curve 2) (b).

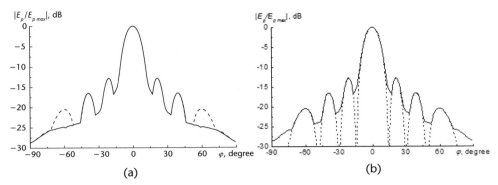

Figure 8.4 Peak field patterns of an array with elements ($d = 0.5Tc$) synchronously excited by current pulses with the number of periods equal to 3 (solid line) and 4 (dashed line) (a) and by a 10-period current pulse (solid line) and a harmonic current (dashed line) (b).

Note that for a bipolar pulse we have $\lambda_0 = \tau_p c$.

When the spatial length of the exciting pulse is no greater than the array dimension, the number of the pattern side maxima is determined by the number of oscillation periods in the current pulse. Figure 8.4(a) shows the patterns for an array excited by a three-period (solid line) and a four-period current pulse (dashed line) for the element spacing $d = 0.5Tc$. The patterns of the same array excited by a 10-period current pulse (solid line) and by a harmonic current (dashed line) are shown in Figure 8.4(b). It can be seen that for the array excited by pulses of finite duration, there are no zeros in the patters, and the pattern width and the side radiation intensity are greater than the respective characteristics for the array with the same element spacing excited by a bipolar pulse.

8.1.2 Experimental Investigations

Detailed experimental investigations were performed for rectangular KA arrays [18] with the number of elements equal to 2×2, 4×4, and 8×8 that were excited by bipolar pulses of duration $\tau_p = 0.2–3$ ns [10], [19–24]. The KAs, except for those used in the experiment [23], had transverse dimensions $h = 0.5\tau_p c$. The spacing between the element centers was $d_{v,h}/\tau_p c = 0.5–0.6$ or $d_{v,h}/h = 1–1.2$. The elements of the radiating systems were fastened on a dielectric or metal plate and were excited synchronously using a distribution system.

We first discuss the results of investigations of an antenna array with a 4×4 radiating system excited by a bipolar pulse of duration 0.5 ns [21] and then summarize the data for other arrays. The appearance of the array is shown in Figure 8.5. The elements of the square-radiating system are spaced by $d/\tau_p c = 0.6$ in the horizontal and vertical directions. The voltage pulse (Figure 8.6) produced by a generator with a characteristic impedance of 50 Ω arrives at the input of a wave impedance transformer (50 Ω /3.125 Ω) and then, having passed through a 16-channel power divider, is distributed over 50-Ω coaxial cables and synchronously excites the array elements. The waveform of the pulse radiated by the array is given in Figure 8.7.

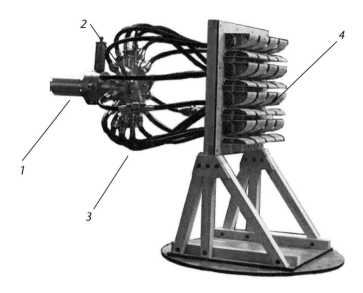

Figure 8.5 A 16-element array excited by a 0.5-ns bipolar voltage pulse: 1—wave transformer, 2—power divider, 3—polyethylene-insulated coaxial cables, and 4—array elements. (With permission from Pleiades Publishing, Ltd.)

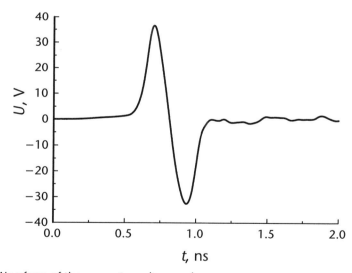

Figure 8.6 Waveform of the generator voltage pulse.

Let us compare the directional characteristics of the antenna array with those of an isolated antenna element. Figure 8.8 presents the peak power patterns of the isolated antenna in the H- and E-planes. Note that the angles in the E-plane are elevation angles counted as $\delta = 90° - \theta$. In this case, to the direction perpendicular

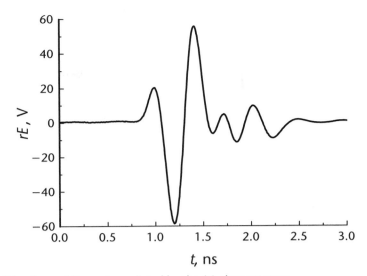

Figure 8.7 Waveform of the pulse radiated by the 16-element array.

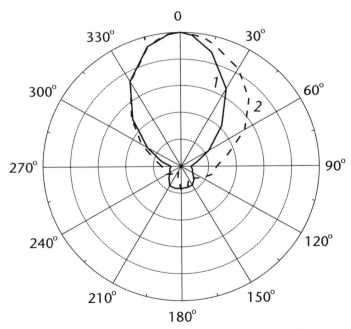

Figure 8.8 The *H*-plane (curve 1) and the *E*-plane peak power pattern (curve 2) of an isolated antenna excited by a 0.5-ns bipolar pulse. (With permission from Pleiades Publishing, Ltd.)

to the radiator plane ($\theta = 90°$) there correspond the angles $\varphi = 0$ and $\delta = 0$. This simplifies the graphical representation of patterns. For the given isolated antenna, the peak power directivity at $\varphi = \delta = 0$, D_0, is equal to four. The technique of measuring D_0 is described elsewhere [18] (Section 2.4).

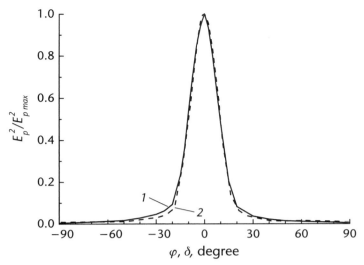

Figure 8.9 The *H*-plane (curve 1) and the *E*-plane peak power pattern (curve 2) of a 16-element array excited by a bipolar pulse of duration 0.5 ns. (With permission from Pleiades Publishing, Ltd.)

It is well known that the pattern function of an array is determined by the product of the pattern function of an array element by the array factor, which is the pattern function of an array of omnidirectional radiators. The pattern of an array element can be different from that of an isolated antenna due to the interaction between array elements. Figure 8.9 presents peak power patterns of the 16-element array in two planes. Note that the patterns of the array are symmetrical in contrast to the isolated antenna pattern in the *E*-plane (Figure 8.8). The half-peak-power widths of the array patterns in two planes are approximately equal to each other and are about one fourth of the widths of the isolated antenna patterns.

The measured peak power gain of the array in the main-lobe direction ($\varphi = \delta = 0$), $D_0 \approx 54$, is close to its maximum estimate corresponding to the product of the directivity of an isolated antenna by the number of the array elements, obtained without regard for the element interaction, $D_0 = 64$. Similar results were obtained for a 16-element array excited by a bipolar voltage pulse of duration 200 ps [23].

A way of symmetrizing the *E*-plane pattern of a linear array consisting of KAs, each having an asymmetric *E*-plane pattern, was investigated [24]. The combined antenna used in the experiment (Figure 8.10) was optimized to radiate bipolar pulses of duration 1 ns. This was the first version of a KA with an extended bandwidth [18, 25]. The antenna VSWR was not above three in the frequency range from 350 to 2000 MHz. The antenna patterns are given in Figure 8.11. The half-peak-power pattern width was 90° for the *H*-plane (curve 1) and 100° for the *E*-plane (curve 2). The *E*-plane pattern maximum corresponds to the elevation angle $\delta = 15°$, which is due to the significant antenna asymmetry in this plane.

The peak power patterns were numerically calculated for different arrays. Two methods were used: (i) direct summation of pulses radiated by an isolated antenna in view of the pulse delays and waveforms varying with angle and (ii) multiplication of the pattern function of an isolated antenna by the array factor calculated for the wavelength corresponding to the center frequency of the boresight radiation pulse.

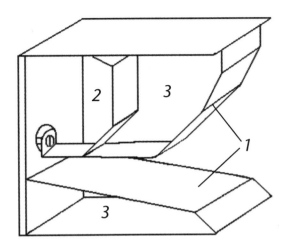

Figure 8.10   Sketch of a combined antenna excited by a 1-ns bipolar voltage pulse: 1—TEM horn, 2—active magnetic dipole, and 3—passive magnetic dipoles. (With permission from Science & Technology Publishing House, Ltd.)

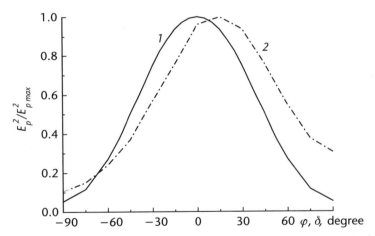

Figure 8.11   The *H*-plane (curve 1) and the *E*-plane peak power pattern (curve 2) of the combined antenna excited by a 1-ns bipolar pulse. (With permission from Science & Technology Publishing House, Ltd.)

Figure 8.12 presents the *E*-plane peak power patterns of a vertical 2×1 array $(d_v = h)$ calculated by the pulse summation method (curve 1) and by using the array factor (curve 2), and also the measured array pattern (curve 3) and, for comparison, the pattern of the isolated antenna (curve 4). It can be seen that the maximum of the *E*-plane pattern of the array (curve 3) is within 0–2.5°, which corresponds to the angular step of the measurements. The calculations indicate that the array pattern maximum is within 0–1°. Note that the maximum of the isolated antenna pattern corresponds to $\delta = 15°$.

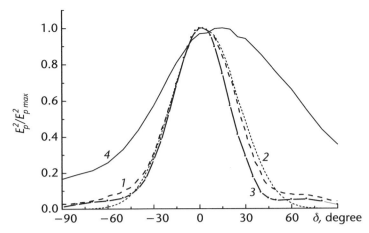

Figure 8.12 The calculated (curves 1 and 2) and the measured *E*-plane peak power pattern (curve 3) of a vertical 2 × 1 array excited by a 1-ns bipolar pulse, and the pattern of an isolated antenna (curve 4). (With permission from Science & Technology Publishing House, Ltd.)

Figure 8.13 presents the *E*-plane peak power patterns of a vertical 4 × 1 array (d_v = *h*) calculated by the pulse summation method (curve 1) and by using the array factor (curve 2), and the measured pattern (curve 3). There is good agreement between the calculated patterns and the measured pattern, and the pattern maximum corresponds to $\delta = 0$. This indicates that the degree of pattern symmetry increases with the number of array elements. The closeness of the patterns calculated by the two methods suggests that the main factor affecting the symmetrizing of the pattern is the array factor. It should be noted that the asymmetry of the *E*-plane pattern of the isolated antenna has the result that the peak field strength at the pattern maximum in the main direction ($\varphi = 0, \delta = 0$), E_{pN}, for an array with a number of elements N_t is less than the product of the peak field strength at the pattern maximum, E_{p1}, for the isolated antenna by the number of array element ($E_{pN} < N_t E_{p1}$).

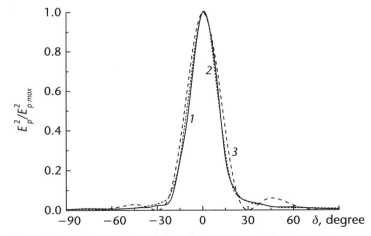

Figure 8.13 The calculated (curves 1 and 2) and the measured *E*-plane peak power pattern (curve 3) of a vertical 4 × 1 array excited by a 1-ns bipolar pulse. (With permission from Science & Technology Publishing House, Ltd.)

An important characteristic of single antennas and antenna arrays radiating UWB pulses is the dependence of the pulse waveform on the observation angle. The variation of the pulse waveform was estimated as the RMSD of the pulse waveform at the output of the receiving antenna at an arbitrary radiation angle, $U_a(t)$, from the pulse waveform at the antenna output on boresight, $V_a(t)$,

$$\sigma = \sqrt{\frac{\int_T \left[u(t) - v(t)\right]^2 dt}{\int_T v^2(t)\, dt}} \qquad (8.4)$$

where $u(t) = U_a(t)/|U_{amax}|$ and $v(t) = V_a(t)/|V_{amax}|$ are the respective normalized functions, and T is the integration time window.

Let us consider the results of measuring the radiated pulse waveform for the square 16-element array (Figure 8.5) excited by a 0.5-ns bipolar pulse as an example [21]. For comparison, the angular dependence of the RMSD of the H-plane (curve 1) and E-plane (curve 2) waveforms of the pulse radiated by the single antenna is shown. A visible difference in the pulse waveforms shows up for $\sigma > 0.1$. The angular dependence of the RMSD of the H-plane (curve 1) and E-plane (curve 2) waveforms of the pulse radiated by the 4×4 array indicate that the pulse waveform starts varying at smaller angles compared to that obtained for the single antenna.

This is the reason why the peak power efficiency of an array is lower compared to the k_p of a single antenna [18]. Thus, in the study [21], the values of k_p for a single antenna and a 4×4 array were, respectively, 0.8 and 0.36. It follows that with the same parameters of the bipolar voltage pulse, the peak radiation power of an array is less than that of a single antenna, whereas the effective radiation potential of an array source is greater than the rE_p of a single antenna source. This is a distinctive feature of UWB radiation sources.

Common to all arrays is $\sigma \approx 0.2$ for the UWB radiation pulse waveform at half the peak power. This implies that the radiation pulse waveform can be considered invariable within the boundaries of the irradiated object, which is important in object recognition research.

8.2   Energy Characteristics of Antenna Arrays

8.2.1   Distribution Systems

To excite the elements in UWB radiation sources comprising one voltage pulse generator and an antenna array, distribution systems are used. This increases the energy loss compared to the UWB sources in which the number of generators, N_g, equals the number of radiating elements, N_t. For the case of identical generators and identical array elements, the peak electric field strength of the array, assuming no interaction of the radiators, is given by $E_{pN} = N_t E_{p1}$ where E_{p1} is the peak electric field strength of an isolated antenna element at the pattern maximum corresponding to the angles φ and δ both equal to zero. For the case of one generator with characteristic impedance ρ_g equal to the characteristic impedance of the element

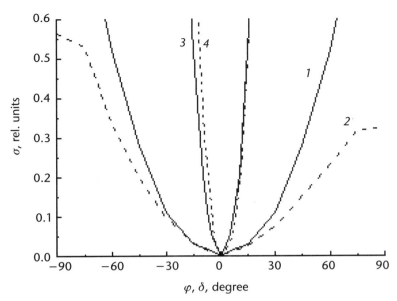

Figure 8.14 RMSD of the *H*-plane (curves 1 and 3) and *E*-plane waveforms (curves 2 and 4) of pulses radiated by a single antenna (curves 1 and 2) and by a 4 × 4 array (curves 3 and 4), both excited by a 0.5-ns bipolar voltage pulse. (With permission from Pleiades Publishing, Ltd.)

feeder, ρ_f, the peak electric field strength of the array with no energy loss in the distribution system and no interaction of the radiators is given by $E_{pN} = \sqrt{N_t}E_{p1}$.

Let us consider the distribution systems for rectangular 4 × 4 arrays excited by bipolar pulses of duration 0.2–2 ns [10, 19, 21, 23] and for a 8 × 8 array excited by a 1-ns bipolar pulse [20]. The distribution systems of the 4 × 4 arrays are made up according to the same circuit. A simplified design of the distribution system of the array excited by a 0.5-ns bipolar pulse (Figure 8.5) is sketched in Figure 8.15. The bipolar voltage pulse arrives, via a coaxial waveguide (1) and a bushing insulator (2), at the input of a transformer (3) whose output (4) is connected to a 16-channel power divider (5). Using 50-Ω coaxial cables (6), the bipolar voltage pulse is transferred to the array elements. The exponential-line wave transformer serves for matching the output impedance of the bipolar pulse generator (50 Ω) and the total characteristic impedance of the antenna array feeder (3.125 Ω).

To increase the electric strength of the insulation in distribution systems driven by bipolar pulses of duration from 0.5 to 2 ns, transformer oil is used that is pumped through the gaps between the polyethylene insulator and the coaxial cable cores and braids [10, 19, 21]. In this case, the peak power (energy) lost in the distribution system can make up 20% of the input pulse peak power [21]. When the pulse waveforms at the input and at the output of the distribution system are close to each other, the peak power loss corresponds to the energy loss. Note that for the bipolar pulse duration equal to 2 ns [10] and the pulse repetition frequency equal to 100 Hz, breakdown traces were detected after one-hour operation, which limited the

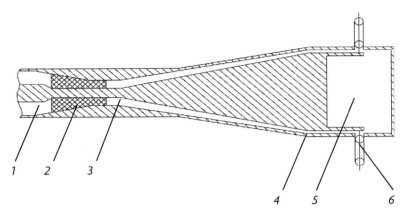

Figure 8.15 Sketch of the distribution system of a 16-element array excited by a 0.5-ns bipolar pulse: 1—output of the bipolar pulse generator, 2—bushing insulator, 3—input of the transformer, 4—output of the transformer, 5—16-channel power divider, and 6—coaxial cables. (With permission from Pleiades Publishing, Ltd.)

utility of the radiation source. Breakdowns in the cables did not occur when the bipolar pulse duration was reduced to 1 ns [19] and to 0.5 ns [21].

With bipolar pulses of shorter duration (200 ps), cord-insulated cables were used in the array distribution system (Figure 8.16), and the system was filled with SF$_6$ gas to a pressure of 5 atm. This reduced the distortion of the voltage pulse waveform and ensured high electric strength and low lost energy, which made up 30% of the input pulse energy [23]. In this case, the energy loss was primarily due to the too narrow bandwidth of the distribution system at both low and high frequencies. This resulted in the increase of the voltage pulse duration at the system output to 260 ps and in the occurrence of an additional time lobe. Figure 8.17 presents the normalized voltage pulse waveforms at the distribution system input (curve 1) and output (curve 2).

The length of the wave transformer was chosen with regard to two conflicting requirements. On the one hand, the length must be small lest the size of the radiation source be increased. On the other hand, it is necessary to increase the length of the transformer to improve its efficiency by shifting the lower bound of the pass band toward the lower frequencies. The optimum electric length of the transformer was chosen as about one and a half of the voltage pulse duration [10, 21, 23]. As the transformer electric length was doubled compared to the optimum length, the energy lost in the distribution system with a 1-ns bipolar pulse [19] made up 40% of the input energy, and the energy was lost mainly in the transformer oil.

The general trend is to replace the oil and solid insulation by gaseous insulation, which provides high electric strength for the bipolar voltage pulse duration increased to 3 ns at a pulse repetition frequency of 100 Hz [22]. This is due to the replacement of coaxial cables with solid polyethylene insulation by cord-insulated cables, whose design allows one to pump SF$_6$ gas into them under high pressure.

The distribution system discussed earlier has a limitation on the number of elements in the array. This is due to the restriction on the diameter of the power

Figure 8.16 A 16-element antenna array excited by a 200-ps bipolar voltage pulse: 1—wave transformer, 2—16-channel power divider, 3—cord-insulated cables, and 4—array elements. (With permission from Pleiades Publishing, Ltd.)

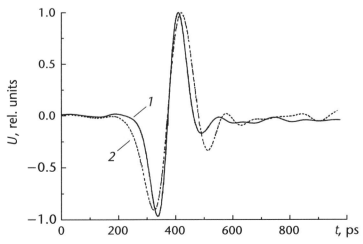

Figure 8.17 Pulse waveforms at the input (curve 1) and output (curve 2) of the distribution system of a 16-element array excited by a 200-ps bipolar voltage pulse. (With permission from Pleiades Publishing, Ltd.)

divider (Figure 8.15) aimed to prevent the excitation of higher modes that can lead to a distortion of the pulse waveform and to a nonuniform amplitude distribution over the channels. For the center wavelength of the bipolar pulse, $\lambda_0 = \tau_p c$, about which most of the energy is concentrated, provided that the diameter of the inner electrode is close to that of the outer electrode, D, the criterion for the absence of higher modes [26] can be written as

$$\frac{D}{\tau_p c} < \frac{1}{\pi\left(\sqrt{\varepsilon_r}\right)} \tag{8.5}$$

where ε_r is the relative permittivity of the insulation.

This criterion was satisfied only for the power divider with a bipolar pulse of duration $\tau_p = 2$ ns [10]. For the power dividers with bipolar pulses of duration 1 ns [19] and 0.5 ns [21], the value of $D/\tau_p c$ was twice the criterion level $1/\pi\left(\sqrt{\varepsilon_r}\right)$, and for $\tau_p = 0.2$ ns [23], it was above this level by a factor of 3.5. In this case, the difference in pulse amplitudes at the output of the divider was not above 6% [23], which is acceptable for practical use. It should be noted that the criterion is strictly valid for a coaxial line of unlimited length. The longitudinal dimensions of the power divider are small compared to the pulse spatial length, and this is why it can operate under conditions where the criterion for the excitation of higher modes does not hold.

Additional factors limiting the development of this type of distribution system are the difficulties associated with reducing the characteristic impedance of the bipolar pulse generator and increasing the impedance transformation ratio.

To increase the number of array elements, a distribution system was developed which, unlike the previous system, possessed the properties of both transformer and power divider. The system was tested with an 8×8 antenna array [20]. Structurally, the 64-channel power divider consists of three series-connected stages of four-channel power dividers (Figure 8.18). The characteristic impedance at the first-stage input is equal to that of the bipolar pulse generator (12.5 Ω). The characteristic impedance at the beginning of each arm of the divider first stage is 12.5 $\times 4 = 50$ Ω. The total impedance of the feeders of the 16-element array, which is the load of each arm of the first stage, is 50/16 = 3.125 Ω, and the total impedance of the feeders of the 64 elements is 50/64 $\approx$ 0.78 Ω. To minimize reflections during impedance transformation, a compensated exponential adapter was used whose impedance was calculated by the formula [27]

$$\rho(x) = \rho(0)\exp\left\{\ln\frac{\rho(l)}{\rho(0)}\left[\frac{x}{l} - 0.133\sin 2\pi\frac{x}{l}\right]\right\} \tag{8.6}$$

where $\rho(0)$ and $\rho(l)$ are the impedance values at the beginning and at the end of the adapter, respectively.

The appearance of the 64-channel power divider is shown in Figure 8.19. The transmission lines are filled with transformer oil. The total length of the divider arm is 120 cm. To match the input impedance of the divider with the characteristic impedance of the test low-voltage generator of 1-ns bipolar pulses, a 31.5-cm long wave transformer matching 50 Ω to 12.5 Ω was used.

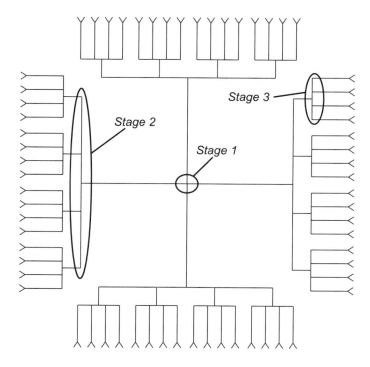

Figure 8.18 Sketch of a 64-channel power divider.

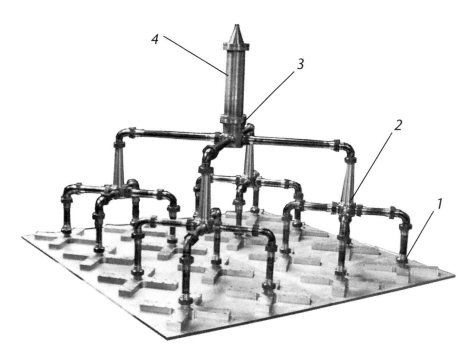

Figure 8.19 A 64-channel power divider: 1, 2, and 3—four-channel dividers, and 4—wave transformer intended for test measurements. (With permission from Springer.)

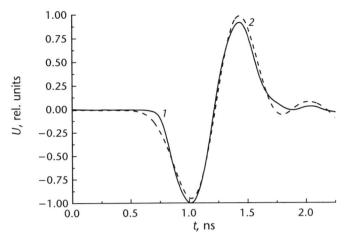

Figure 8.20   Normalized voltage pulse waveforms at the input (curve 1) and output (curve 2) of the 64-channel power divider insulated with transformer oil. (With permission from Springer.)

Figure 8.20 presents the normalized voltage pulse waveforms at the input (curve 1) and output (curve 2) of one of the divider channels. It can be seen that the pulse waveform is distorted slightly. The amplitude jitter at the divider outputs is ~4%. Estimates show that the peak power lost in the 64-channel power divider reaches 50% of the input peak power, and it is lost mainly in the transformer oil.

According to investigations [28] performed in the frequency range 0.6–2 GHz, the power loss in vacuum oil is significantly smaller than that in transformer oil. Therefore, in the 64-channel power divider, transformer oil was replaced by vacuum oil. The measurements [29] showed that as a result, the output pulse amplitude in the channels increased by about a factor of 1.3. The peak power loss in the divider decreased to about 30% of the input peak power. Figure 8.21 shows the voltage

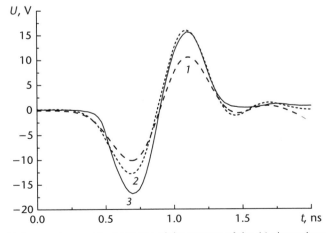

Figure 8.21   Voltage pulse waveforms at one of the outputs of the 64-channel power divider using transformer oil (curve 1) and vacuum oil (curve 2) and the waveform corresponding to the no-loss case (3). (Reprinted with permission Cambridge University Press.)

pulse waveform at the output of the feeder system of the 64-element array using transformer oil (curve 1) and vacuum oil (curve 2). For comparison, the voltage pulse waveform for the no-loss case is given (curve 3). This distribution system allows increasing the number of elements in the array as a multiple of four and the effective potential of the radiation source as a multiple of two.

8.2.2 Structure of the Radiating System

Before considering KA-based radiating systems, which is the main objective here, we briefly discuss the results of theoretical and experimental studies of TSA-based arrays [11, 12, 30–32]. The characteristics under investigation were the coefficient of voltage reflection from the inputs of the array elements and the VSWR, which determine the energy efficiency of the radiating system. Linear and planar arrays with a linearly polarized field and with two orthogonally polarized field components, respectively, were studied. In the E-plane, the antennas are connected galvanically to form a row. In a linearly polarized planar array, the spacing between the rows in the H-plane can vary. A dual-polarized planar array has a mesh structure with the mesh size equal to the height of the antenna.

An important result of numerical simulations and experimental studies of the arrays is the conclusion that the lower bound of the match band, estimated for K_V = 2, shifts toward the lower frequencies with increasing the number of elements for a linear array, and the shift is more pronounced for a planar array. The shift of the lower frequency bound is a maximum for an array of unlimited dimension. As follows from the simulations, a 50×50 planar array should be similar in performance to an array of unlimited dimension. The theoretical and experimental data indicate that linear arrays consisting of 30 elements and 30×30 planar arrays are also similar in performance to arrays of unlimited dimension. The shift of the lower frequency bound of an array element relative to that of an isolated antenna is accounted for by an increased aperture dimension. The radiation of the elements nearest to the test element induces a current in the test element and suppresses the reflections at lower frequencies.

To reduce the backscattered radiation, which may affect the electronics of the transceiver, the array is mounted on a ground plate. The presence of the metal plate leads to the reflection of the backscattered field and to a narrower match band. To avoid this, it was proposed to use absorbing plates of ferrite material [33, 34].

For planar KA arrays, two types of structure can be distinguished. In first-type arrays, the elements of transverse dimension h are equally spaced in the vertical and horizontal directions ($d_v = d_h = 1.2h$) and are fastened on a dielectric plate (Figure 8.5) [19, 21] or on a metal plate (Figure 8.22) [20]. In second-type arrays, the elements are galvanically connected in the vertical direction ($d_v = h$) and spaced by $d_h = 1.2h$ in the horizontal direction (Figure 8.16) [10, 23] and are fastened on a dielectric plate.

Consider the results of comparative measurements of K_V for an isolated KA and for a 16-element array, both optimized for excitation by a 2-ns bipolar pulse [10]. Figure 8.23 shows plots of the VSWR versus frequency ($K_V(f)$) for an isolated antenna (curve 1) and for an inner element of the array (curve 2). The measurements indicate that the K_V of the element decreased in the frequency ranges 200–650 MHz and f

Figure 8.22   Appearance of a 64-element array. (With permission from Springer.)

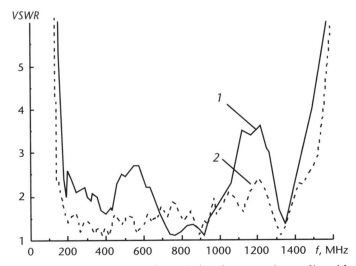

Figure 8.23   The VSWR versus frequency for an isolated antenna (curve 1) and for an array inner element (curve 2), both optimized for excitation by a 2-ns bipolar pulse. (With permission from Pleiades Publishing, Ltd.)

> 1000 MHz, and the lower frequency bound for the level $K_V = 2$ shifted to lower frequencies. A difference in VSWR between the edge and the inner elements of the array was observed, which was negligible at low frequencies ($f < 400$ MHz) and significant at high frequencies ($f > 1300$ MHz). Similar results for the low-frequency range were obtained for the 16-element arrays excited by bipolar pulses of duration 1 ns [19] and 200 ps [23]. These results suggest that in planar arrays mounted on dielectric plates, regardless of the structure type, the K_V of the elements decrease in the low-frequency range, and the lower frequency bound is shifted toward the lower frequencies compared to that of the isolated antenna.

Investigations of a 2×2 array ($d_v = d_h = 1.2h$) mounted on a metal plate have shown that the galvanic connection of the output apertures of vertical elements increases the peak field strength by 3% [20], which is accounted for by the inhibition of resonant fields in the gaps between the elements. This array is a 1/16 part of the 64-element array (Figure 8.22) whose structure was dictated by the design of the divider third stage (Figure 8.19) and by the need to change the radiation polarization by rotating the elements, not the array, by 90° to measure the array pattern in the E-plane.

Reducing the K_V of the array elements compared to that of the isolated antenna increases the efficiency of the radiating system. This makes it possible to partially compensate for the energy loss in the distribution system. In particular, for the radiation sources comprising a single antenna and a 16-element array, both excited by 2-ns bipolar pulses, the effective radiation potentials $rE_p = 440$ kV and 1670 kV were obtained, respectively [10]. The ratio of these values is 3.8, which is close, to within 5%, to $\sqrt{N_t} = 4$ corresponding to a perfect array free of energy loss.

Consider next the results of studies [24] aimed at optimizing the position of the KA in four-element linear and rectangular arrays. Figure 8.24 presents schematic configurations of a horizontal (a) and a vertical (b) linear array. The arrows show the polarization plane of the vector **E** of the radiated pulse. Figure 8.25 shows configuration versions of a rectangular 2×2 array. The elements are fastened on a dielectric plate. In the case of $d_{v,h} = h$, the array elements are galvanically connected.

For an element of the test arrays, the antenna shown in Figure 8.10 was used. The array antennas were excited synchronously by 1-ns bipolar pulses supplied

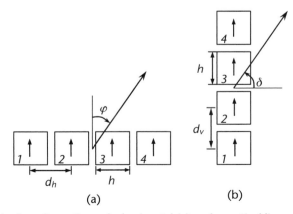

Figure 8.24 Sketched configurations of a horizontal (a) and a vertical linear array (b). (With permission from Science & Technology Publishing House, Ltd.)

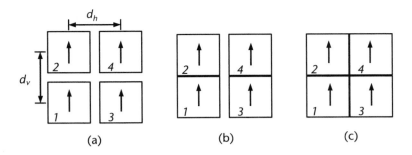

Figure 8.25 Sketched configurations of rectangular arrays with $d_v = d_h = 1.2h$ (a), $d_v = h$, $d_h = 1.2h$ (b), and $d_v = d_h = h$ (c). (With permission from Science & Technology Publishing House, Ltd.)

through a four-channel power divider from a low-voltage pulse generator. In the experiments, the optimization parameters were the linear dimensions of the array, the reflected energy, and the peak field strength corresponding to the pattern maximum. The first two parameters should be minimal, and the third one should be maximal.

To directly measure the reflected energy, a bipolar voltage pulse was applied, through a matched coupler, at the input of the test element, and matched loads were connected to the remaining elements. In addition, the reflected energy was estimated, using the $K_V(f)$ of the element and the spectrum of the excitation voltage pulse, by the equation

$$\frac{W_{\text{ref}}}{W_g} = \frac{\int U_g^2(f)\left(\frac{K_V(f)-1}{K_V(f)+1}\right)^2 df}{\int U_g^2(f)\,df} \tag{8.7}$$

where W_{ref} is the reflected energy, W_g is the generator pulse energy, and $U_g(f)$ is the spectral function of the generator voltage pulse. The estimated reflected energy for the isolated antenna, W_{ref1}, made up 13% of the energy of the generator pulse, which is close to the results of direct measurements (11 %). Note that the energy efficiency of an antenna is defined as $k_w = 1 - W_{\text{ref}}/W_g$.

The energy reflected from array elements was measured and estimated for different array configurations. Figure 8.26(a) gives the reflected energy values for the elements of a vertical array [Figure 8.24(b)]. Curves 1 and 2 represent experimental data obtained for an open ($d_v = 1.2h$) and a closed array ($d_v = h$), respectively, and curves 3 and 4 represent estimates for the respective arrays. The difference between the reflected energies measured in time domain and those estimated using the $K_V(f)$ and the pulse spectrum in frequency domain seems to be due to the short time ($\Delta t = 8$ ns) used to measure the reflected pulse characteristics.

For the vertical array, the measured total reflected energy for the version with $d_v = h$ is lower than that for the version with $d_v = 1.2h$ by 17%. The calculated total reflected energy was about the same for the open and the closed array. From the measurements, it also follows that the boresight peak field strength is nearly the same for the open and the closed array. This implies that the optimum structure

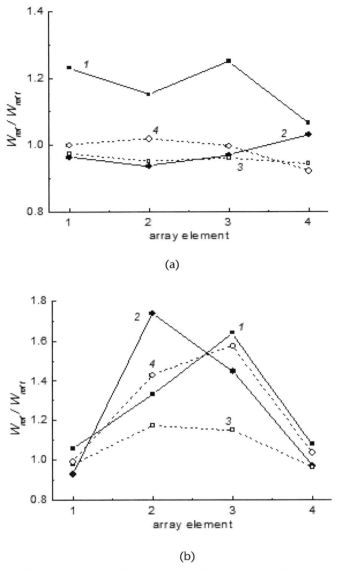

Figure 8.26 The ratio of the measured (curves 1 and 2) and calculated energies (curves 3 and 4) reflected from the elements of an open (curves 1 and 3) and closed (curves 2 and 4) vertical (a) and horizontal array (b) to the energy reflected from an isolated antenna element. (With permission from Science & Technology Publishing House, Ltd.)

of a vertical array is one with $d_v = h$ to which there corresponds the minimum linear dimension.

Plots of the reflected energy for the horizontal array elements [Figure 8.24(a)] are given in Figure 8.26(b). Curves 1 and 2 were obtained in experiments with an open ($d_h = 1.2h$) and a closed array ($d_h = h$), respectively, and curves 3 and 4 represent estimates for the respective arrays. It can be seen that the energy reflected from the inner elements increased significantly for both array versions, which indicates the interaction of the elements in the H-plane.

For the horizontal array, the measured total reflected energy is nearly the same for the open and the closed array versions, whereas the estimated total reflected energy for the open array is less than that for the closed one by 3%. In addition, the peak field strength corresponding to the pattern maximum for the open array is greater than that for the closed one by 7%. It follows that the optimum structure for a horizontal array is one with $d_h = 1.2h$, to which there corresponds the maximum peak field strength with a slightly increased linear dimension.

The values of the reflected energy for elements of the 2×2 rectangular arrays (Figure 8.25) are given in Figure 8.27. Curves 1, 2, and 3 were obtained in experiments with arrays a, b, and c (Figure 8.25), respectively. Curves 4, 5, and 6 represent estimates for the respective arrays. The measured total reflected energy for the array version given in Figure 8.25(b) ($d_v = h$, $d_h = 1.2h$) is lower than that for the versions given in Figure 8.25(a) ($d_v = d_h = 1.2h$) and Figure 8.25(c) ($d_v = d_h = h$) by 4.5% and 10%, respectively. The estimated total reflected energy for the array version shown in Figure 8.25(b) is lower than that for the versions shown in Figures 8.25(a) and 8.25(c) by 2% and 10%, respectively. The peak field strength corresponding to the pattern maximum for the array of Figure 8.25(b) is greater than that for the arrays of Figures 8.25(a) and 8.25(c) by 2% and 5%, respectively. Thus, the structure of a rectangular KA array that is optimal in energy efficiency and in peak field efficiency is one in which the elements are galvanically connected in the E-plane ($d_v = h$) and spaced in the H-plane ($d_h = 1.2h$).

Subsequent, more detailed, investigations [35] of the effective radiation potential in the H-plane versus element spacing for a linear [Figure 8.24(a)] and a rectangular array [Figure 8.25(b)] have shown that the maximum value of rE_p corresponds to $d_h = 1.5h$ for both arrays. Reducing the element spacing reduces the effective radiation potential by 2–4% compared to its maximum value. With galvanically connected KAs in these arrays, the effective radiation potential decreases by about 10%.

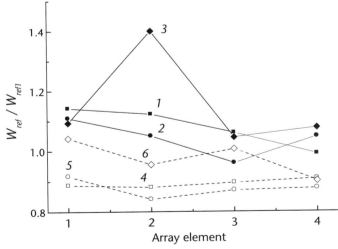

Figure 8.27 The ratios of the measured (curves 1, 2, and 3) and calculated energies (curves 4, 5, and 6) reflected from the elements of the arrays shown in Figure 8.25(a) (curves 1 and 4), Figure 8.25(b) (curves 2 and 5), and Figure 8.25(c) (curves 3 and 6) to the energy reflected from an isolated antenna element. (With permission from Science & Technology Publishing House, Ltd.)

Galvanic connection of antennas in the *E*-plane and antenna spacing in the *H*-plane were also employed in modified KA arrays [36, 37]. With KAs galvanically connected in both planes [38], the backscattered radiation was reduced.

In high-power UWB radiation sources with semiconductor generators connected so that $N_g = N_t$, TEM horn antennas are most often used [8, 39]. To improve the efficiency of rectangular arrays at low frequencies, it is recommended [40] to connect galvanically the output apertures of the TEM antennas in both the *E*- and the *H*-plane.

8.3   Antenna Arrays Radiating Orthogonally Polarized Pulses

The antenna arrays [19] consist of 16 elements not connected galvanically that are fastened on a dielectric plate at equal spacings $d_v = d_h = 1.2h = 18$ cm. Each radiating element is a KA excited by a bipolar voltage pulse of duration 1 ns. The array elements are different in design from the KA shown in Figure 8.10. With the used arrangement of the elements, the array can radiate vertically polarized pulses [Figure 8.28(a)] with synchronously excited elements and pulses with mutually perpendicular polarization vectors [Figure 8.28(b, c)] with the perpendicularly oriented elements excited by pulses shifted in time by 2 ns. The direction of the electric field polarization plane is shown by arrows. Let us label the arrays sketched in Figures 8.28(a–c) as AA1–AA3, respectively. The time shift between the exciting pulses is provided by different lengths of the cable feeders. When an array is broken into two subarrays, the elements of one of which being turned by 90° relative to the elements of the other, two orthogonally polarized pulses are radiated successively. The data for a linearly polarized array are given here for ease of comparison with orthogonally polarized arrays.

Figure 8.29 (curves 1–3) presents the VSWRs for two elements of the test arrays, which are labeled by the Figures 1 and 2 in Figure 8.28. For comparison, the VSWRs of these elements operating as single radiating antennas rather than as array elements are also given (curve 4). One can see a difference in their VSWRs, which is associated with the precision of fabrication of the antennas. Note that near the low-frequency bound, the K_V of the elements of array AA1 is less than the K_V of

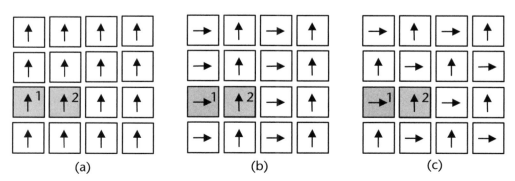

Figure 8.28 Options of the element arrangement in antenna arrays. (With permission from Pleiades Publishing, Ltd.)

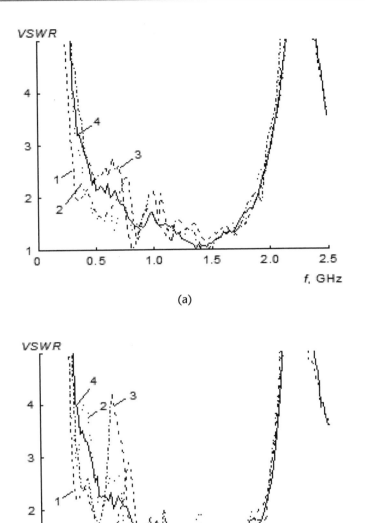

(a)

(b)

Figure 8.29   The VSWR of elements *1* (a) and *2* (b) in AA1–AA3 arrays (curves 1–3) corresponding to Figures 8.28(a–c) and of the same elements operating as isolated antennas (curve 4). (With permission from Pleiades Publishing, Ltd.)

the single antenna. The VSWR of an antenna element depends on its location in the array [Figure 8.29(a, b)] and on the array configuration (curves 1–3). The average value of K_V for the inner array elements at frequencies of 0.3–1.5 GHz is somewhat greater than that for the edge elements. This is confirmed by the measurements of the energy reflected from the element inputs. Figure 8.30(a–c) presents the reflected energy as a percentage of the energy of the incident bipolar pulse of duration 1 ns for arrays AA1–AA3, respectively (Figure 8.28). In this case, individual elements reflect, on the average, 9% of the energy of the bipolar pulse. The average reflected

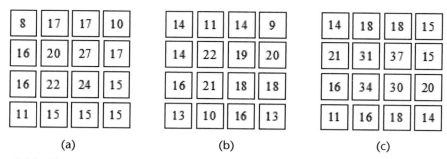

(a)   (b)   (c)

Figure 8.30 The energy reflected from the array element inputs as a percentage of the incident energy. (With permission from Pleiades Publishing, Ltd.)

energy for arrays AA1–AA3 makes up, respectively, 17%, 16%, and 21% of the incident bipolar pulse energy. The difference in the reflected energies for arrays AA1 and AA2 is within the measurement error. The increase in reflected energy and the variation in VSWR are due to the mutual influence of the array elements.

Figure 8.31 presents the quantity rE_p versus distance r between the receiving antenna and the array for different array configurations. The data are normalized to the maximum value of the rE_p for a subarray of a vertically polarized AA3-type array. For $r > 7$ m, the curves vary slightly with r for all arrays; that is, the field varies in proportion to $1/r$, which is indicative of the far-field region and is consistent with the far-field region boundary estimated by (2.72) ($r = 6.3$ m). In addition, as can be seen from Figure 8.31, the value of rE_p for array AA3 (curve 3) is less than that for array AA2 (curve 2), although the number of radiating elements in the subarrays is the same. This is accounted for by a significant influence of elements on each other in array AA3. Note that for AA2 and AA3 arrays, the relations $rE_p(r)$ for the horizontal and vertical polarizations are the same within the measurement error.

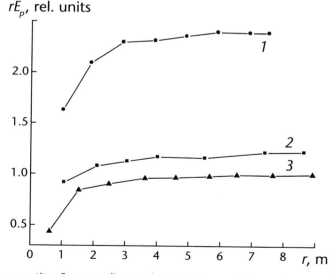

Figure 8.31 The quantity rE_p versus distance between the receiving antenna and the array for arrays AA1–AA3 (curves 1–3) corresponding to Figure 8.28(a–c). (With permission from Pleiades Publishing, Ltd.)

The E-plane and H-plane peak power patterns $E_p^2/E_{p\max}^2$ of the arrays are shown in Figure 8.32(a, b, d) for the vertically polarized electric field and in Figure 8.32(c) for the horizontally polarized electric field. The E-plane and H-planes patterns are symmetrical [Figure 8.32(a, b, d)] and their width at half the peak power is ~20° for all test arrays. For the vertically polarized subarray of array AA2, the boresight of the H-plane pattern is offset by $\Delta\varphi \approx 2.5°$ [Figure 8.32(b)]. This offset is also caused by the element interaction in the array. Figure 8.33 gives configuration options of array AA2 shown in Figure 8.28(b). The corresponding off-boresight angle in the H-plane pattern of the vertically polarized array is 2.5° for the configurations shown in Figure 8.33(a, b) and −2.5° for the configuration shown in Figure 8.33(c).

Note that for array AA2, the side radiation intensity increases significantly for both polarizations [Figure 8.32(b, c)], making the array difficult to use. Array A3 is characterized by low side radiation intensity [Figure 8.32(d)] and by the coincidence of the pattern maxima in the two planes for both polarizations, which makes this type of array more suitable for the formation of orthogonally polarized wave beams separated in time. A disadvantage of this array is rather low energy efficiency.

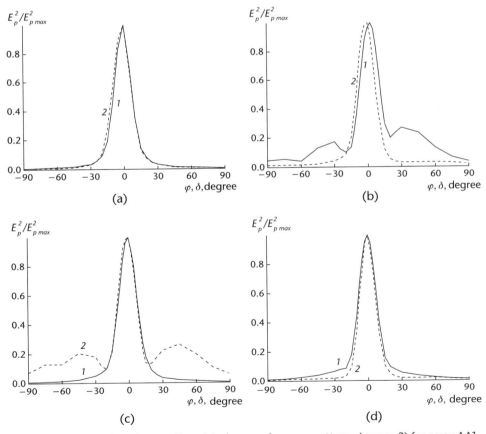

Figure 8.32 The H-plane (curves 1) and E-plane peak power patterns (curves 2) for array AA1 with vertical polarization (a), array AA2 with vertical polarization (b), array AA2 with horizontal polarization (c), and array AA3 with vertical polarization (d). (With permission from Pleiades Publishing, Ltd.)

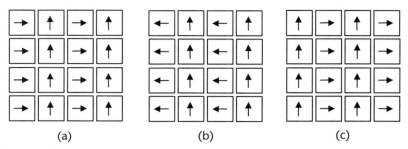

Figure 8.33 Configuration options of array AA2 presented in Figure 8.28(b). (With permission from Pleiades Publishing, Ltd.)

8.4 Characteristics of Wave-beam-scanning Linear Antenna Arrays

8.4.1 Nanosecond Pulse Excitation of the Arrays

Consider first the wave-beam scanning in linear KA arrays designed to radiate high-power bipolar pulses of duration 1 ns [41]. For an array element, the antenna shown in Figure 8.10 was used in two modifications differing by the perimeter of the active magnetic dipole (reference 2 in Figure 8.10) and, accordingly, by the H-plane pattern width at half the peak power, $\Delta\varphi$ [18]. The H-plane patterns for the two antennas are given in Figure 8.34 (hereinafter, $F^2(\varphi,\delta)$ corresponds to $E_p^2(\varphi,\delta)$). For the first antenna (A1), we have $\Delta\varphi = 90°$ (curve 1) and for the second one (A2) $\Delta\varphi = 140°$. Antenna A2 was used in experiments to find the maximum scan angle in the H-plane.

The configurations of the linear arrays are shown in Figure 8.24. A four-element horizontal (a) and vertical array (b) with the element spacing $d_h = 1.2h$ and $d_v = h$, respectively, were investigated. The excitation voltage pulse was applied to the input of a four-channel power divider and then passed via cables to the antennas. The scan mode was realized by selecting the lengths of the cables transmitting the pulses exciting the array elements to provide the desired pulse delay.

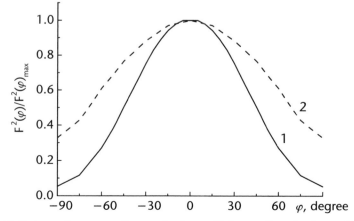

Figure 8.34 *H*-plane peak power patterns of antenna A1 (curve 1) and antenna A2 (curve 2). (With permission from Science & Technology Publishing House, Ltd.)

For an array operating in the scan mode, the variation of the location of the peak power pattern maximum was investigated. To calculate the array pattern in the scan mode, a code was developed to simulate the summation of the radiation pulses taking into account the pattern of an isolated antenna element and the delay in excitation of the pulses in the array. Initially, for an array element, antenna A1 was used. The experimental (solid lines) and calculated (dashed lines) H-plane patterns of the horizontal array are presented in Figure 8.35 for specified scan angles φ_0 = 0, 15°, 30°, 45°, and 60°, to which there correspond curves 1–5. All patterns were normalized to the pattern maximum $F^2(\varphi = 0, \delta = 0)$ corresponding to no-scanning operation of the array.

The results indicate that the electromagnetic field strength corresponding to the pattern maximum decreases and the pattern width increases with increasing scan angle. In addition, there is a difference between a specified and a measured scan angle. This difference is not over 5° for specified scan angles up to 45°. However, for the specified angle φ_0 = 60°, the experimental scan angle of the array was 43.5°. As can be seen from Figure 8.35, the calculated and experimental patterns agree well with the exception of the pattern obtained for φ_0 = 60°.

Figure 8.36 presents the experimental (solid lines) and calculated E-plane peak power patterns (dashed lines) of a vertical array corresponding to the scanning in the range of positive elevation angles δ. All patterns were normalized to the array pattern maximum $F^2(\varphi = 0, \delta = 0)$ corresponding to no-scanning operation. In the investigations of the vertical array, the same set of feeders as for the horizontal array was used. The spacing of the vertical array was smaller than that of the horizontal one. Therefore, the specified scan angles δ_0 were increased. Instead of the angles 0, 15°, 30°, 45°, and 60°, the specified angles were 0, 17°, 34.5°, 53°, and 79°. To these angles there correspond curves 1–5 in Figure 8.36. The difference between the specified scan angle and the measured one was not over 5° for the specified scan angles up to 53°.

It should be noted that during wave-beam scanning, the waveform of the radiated pulse varies with respect to that of the pulse radiated on synchronous excitation of

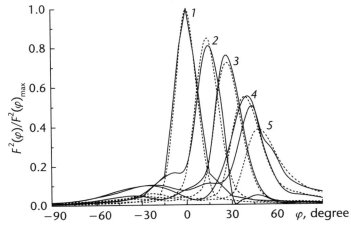

Figure 8.35 Calculated (dashed lines) and experimental H-plane peak power patterns (solid lines) of the horizontal array of antennas A1 for the scan angles equal to 0, 15°, 30°, 45°, and 60° (curves 1–5). (With permission from Science & Technology Publishing House, Ltd.)

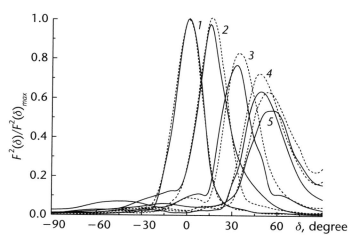

Figure 8.36 Calculated (dashed lines) and experimental *E*-plane peak power patterns (solid lines) of the vertical array of antennas A1 for the scan angles equal to 0, 17°, 34.5°, 53°, and 79° (curves 1–5). (With permission from Science & Technology Publishing House, Ltd.)

the array elements ($\varphi = \delta = 0$). For the wave-beam scanning within ±45°, the RMSD of the radiated pulse waveform reached 0.4 for both arrays of antennas A1 [42].

To investigate the possibility of increasing the scan angle in the *H*-plane, antenna A2 was used as a horizontal array element. Figure 8.37 presents the experimental (curve 1) and calculated *H*-plane pattern (curve 2) for the specified scan angle equal to 60°; herein, the experimental pattern for an array of antennas A1 is given (curve 3). As can be seen from the figure, although the pattern maximum angle was closer to the 60° specified angle, the radiation amplitude decreased significantly. This was due to that the pattern of an element in the array changed due to its interaction with neighboring elements.

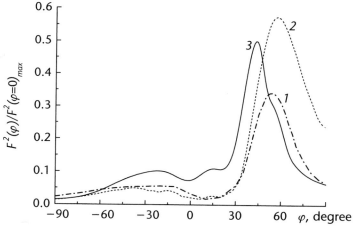

Figure 8.37 The experimental (curve 1) and the calculated *H*-plane peak power pattern (curve 2) of an array of antennas A2 for a specified scan angle of 60°, and the experimental *H*-plane peak power pattern (curve 3) of a horizontal array of antennas A1 for the same specified scan angle. (With permission from Science & Technology Publishing House, Ltd.)

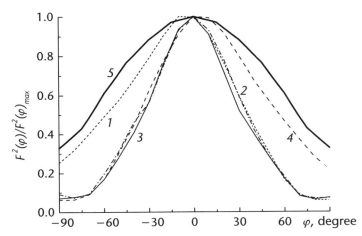

Figure 8.38   The *H*-plane peak power experimental patterns of antenna A2 as an element of a horizontal array (curves 1–4) and of antenna A2 operating individually (curve 5). (With permission from Science & Technology Publishing House, Ltd.)

Figure 8.38 presents the *H*-plane patterns of antenna A2 used as an array element (curves 1–4) and of antenna A2 operating as an isolated one (curve 5). The numbering of the curves corresponds to that of the array elements in Figure 8.24(a). In measuring the pattern of a test element, the remaining elements are loaded by matched loads. It can be seen that the patterns of the array inner elements are symmetrically narrowed almost twice and those of the edge elements are narrowed by a factor of 1.5, becoming nonsymmetrical. Using the patterns of the array elements (Figure 8.38, curves 1–4), we calculate the pattern of the array (Figure 8.39, curve 2) that is much better consistent with the experimental pattern (Figure 8.39, curve 1).

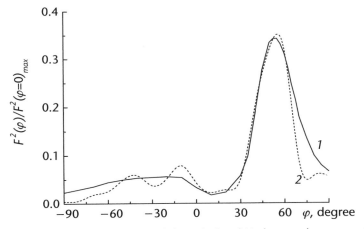

Figure 8.39   The experimental (curve 1) and the calculated *H*-plane peak power pattern (curve 2) of a horizontal array of antenna A2 elements for a specified scan angle of 60°. (With permission from Science & Technology Publishing House, Ltd.)

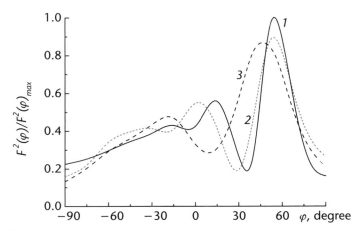

Figure 8.40 The peak power patterns of two-element horizontal arrays with the spacing between the element centers equal to 54 (curve 1), 36 (curve 2), and 18 cm (curve 3) for a specified scan angle of 60°. (With Permission from Science & Technology Publishing House, Ltd.)

The pattern of an array element varies due to its interaction with adjacent elements. Figure 8.40 presents the patterns of two-element horizontal arrays with different element spacings for a specified scan angle of 60°. It can be seen that the pattern maximum decreases and shifts toward lower angles with decreasing element spacing. For the element spacing equal to 54 cm, the maximum scan angle was 55°.

The arrays of combined antennas A1 provide for wave-beam scanning of UWB radiation within ±43° for a horizontal array and within +56/−48° for a vertical array. Increasing the width of the H-plane pattern of antenna A2 by a factor of 1.5 compared to that of antenna A1 increased the scan angle in a horizontal four-element array with $d_h = 18$ cm ($d_h = 1.2h$) only slightly. With large element spacings ($d_h \gg h$) in an array of antennas A2, the scan angle can be increased to 50–60°.

8.4.2 Picosecond Pulse Excitation of Antenna Arrays

The combined antennas excited by low-voltage (~10 V) bipolar pulses of duration $\tau_p = 200$ ps were also used as elements of linear arrays (Figure 8.24) intended for UWB radiation wave-beam scanning [43]. In the horizontal array, d_h was 35 mm ($\approx 1.1h$). In the vertical array with $d_v = h = 32$ mm, antennas of length $L = 34$ mm were galvanically connected.

In the experiment, the receiving TEM antenna was in the far-field region at a distance of 1.6 m from the array. The generator pulse was applied to a four-channel voltage divider. The wave-beam scanning was realized by properly setting the delays of arrival of the input pulse at the array elements. The delay for each element was determined by the length of the cable feeders between the voltage divider and the antenna input.

The calculated (dashed lines) and experimental H-plane peak power patterns (solid lines) of the horizontal array (for negative scan angles φ_0) are presented in

Figure 8.41. All patterns are normalized to the maximum $F^2(\varphi = 0, \delta = 0)$ corresponding to no-scanning operation of the array. The data for the horizontal array are summarized in Table 8.1. The specified scan angles φ_0 in the table correspond to the pattern maxima for the array with selected feeder sets for the case that the array elements are omnidirectional radiators in the H-plane.

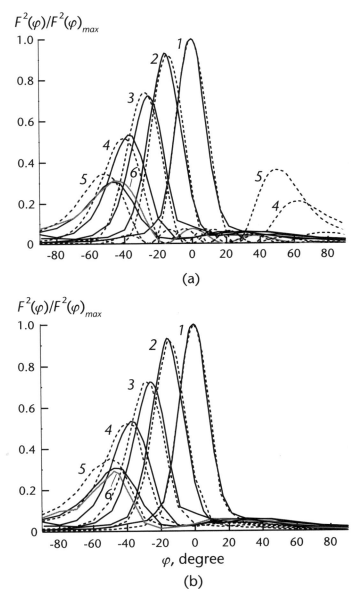

Figure 8.41   Calculated (dashed lines) *H*-plane peak power patterns of an isolated antenna obtained for the frequency $f_0 = 5$ GHz (a) and by summation of pulses radiated by the antenna (b) and experimental (solid lines) *H*-plane peak power patterns of the horizontal array for the specified scan angles equal to 0, −15, −30°, −45°, −60° (curves 1–5). Curve 6 in (a) was calculated for the average pattern and $f_0 = 3.6$ GHz; curve 6 in (b) was calculated by summing up the pulses radiated by the array elements. (With permission from Pleiades Publishing, Ltd.)

Table 8.1 Characteristics of the *H*-plane Patterns of the Horizontal Array. (With permission from Pleiades Publishing, Ltd.)

Specified scan angle φ_0, deg	Calculated scan angle φ_0, deg		Experimental scan angle φ_0, deg	Pattern maximum normalized to its value with no scanning			Half-maximum width of the pattern main lobe, deg		
	$f_0 = 5$ GHz	Σ		Calculation f_0	Calculation Σ	Experiment	Calculation f_0	Calculation Σ	Experiment
0	0	0	0	1	1	1	22	22	21
±15	±14	±15	±15.5	0.92	0.92	0.93	22.5	22.5	23
±30	±28	±28	±26	0.74	0.72	0.79	24	25	23
±45	±40	±40	±37	0.52	0.52	0.53	28	30	27
±60	±53	±50	±47	0.38	0.35	0.31	33.5	36	35
	(±46)	(±47.5)		(0.31)	(0.29)		(37)	(30)	

The calculated patterns [Figure 8.41(a)] were plotted by using the conventional formula for the harmonic radiation of a linear equispaced, equal-amplitude, linear-phase array

$$F_N^2(\varphi) = F_A^2(\varphi) \left(\left| \frac{\sin \Psi}{N \sin \dfrac{\Psi}{N}} \right| \right)^2 \tag{8.8}$$

where $F_A^2(\varphi)$ is the H-plane peak power pattern function of a single combined antenna. The second multiplier in (8.8) is the squared array factor of an N-element array, and Ψ is a generalized angle variable, which, for a tilted radiation mode, is determined as

$$\Psi = \frac{Nkd}{2}(\sin \varphi - \xi) \tag{8.9}$$

where φ is the angle counted from boresight [Figure 8.24(a)].

To calculate the wave number $k = 2\pi/\lambda_0$, the wavelength λ_0 was chosen as $\tau_p c =$ 6 cm, which corresponds to the maximum frequency ($f_0 = 5$ GHz) of the amplitude spectrum of the combine antenna radiation in the main-beam direction.

The slowness factor, defined for narrowband arrays as

$$\xi = \frac{\Delta \Phi}{kd} \tag{8.10}$$

where $\Delta \Phi$ is the phase difference between any two neighboring radiators, was calculated by the equation

$$\xi = \frac{\Delta \tau c}{d} \tag{8.11}$$

where $\Delta \tau$ is the difference in pulse travel time between the neighboring and c is the velocity of light in free space. The calculated patterns were plotted with a varied step of 1–5° and the experimental ones were plots of measurements performed with a varied step of 1–15°.

As can be seen from Figure 8.41(a) and Table 8.1, the calculated ($f_0 = 5$ GHz) and the experimental main-lobe patterns of the horizontal array are in good agreement up to the calculated maximal scan angles equal to ±40° (corresponding to experimental angles of ±37°). In the experiment, as the scan angle was further increased, the pattern maximum amplitude decreased significantly and the measured scan angle was significantly different from the calculated one. Also, for the calculated scan angles $\varphi_0 = \pm 40°$ and greater, a first grating lobe maximum appeared in the real angle range in the direction opposite to the main lobe direction [Figure 8.41(a)].

To improve the accuracy of the calculations for large scan angles, it is necessary to take into account the variation of the waveform of the radiated UWB pulse corresponding to these angles. The amplitude spectrum maximum of the radiation pulse produced by a KA in the horizontal plane for $\varphi_0 = \pm 47°$ corresponds to a frequency of 3.6 GHz. To find the wave number k, we choose the wavelength λ_0 corresponding to this frequency. In addition, we take into account the variation of the pattern of the combined antenna used as an array element. Comparative measurements of peak power patterns were performed for a single KA excited by one of the power divider channels (with matched load connected to the remaining channels) and for the same antenna used in a linear horizontal array as an edge element [references 1 and 4 in Figure 8.24(a)] and as an inner element [references 2 and 3 in Figure 8.24(a)]. In measuring the pattern of an array element, matched loads were connected to the inputs of the remaining antennas in the array.

The measurement results are presented in Figure 8.42. Curve 1 corresponds to the pattern of the single KA excited through a four-channel voltage divider that coincides with the pattern of the same KA excited directly by the generator of bipolar pulses. Curve 2 corresponds to the pattern of the antenna installed at the array edge [reference 4 in Figure 8.24(a)]. The rotation axis of the array passed through a partial phase center (radiation center) of the test element; a positive rotation angle φ corresponded to clockwise rotation. Curve 3 corresponds to the pattern of an array inner element [reference 3 in Figure 8.24(a)]. Curve 4 represents the average of the four pattern functions of the array elements. This pattern function was substituted in formula (8.8) to calculate the peak power pattern for the linear array operating in a wave-beam scanning mode at a specified angle $\varphi_0 = -60°$. The calculation results are presented in Figure 8.41(a) (curve 6) and in Table 8.1 (data in parentheses). It can be seen that the position of the maximum and the amplitude are almost the same for the experimental and the calculated pattern, and the estimated pattern half-maximum width is different from the measured one by 2°.

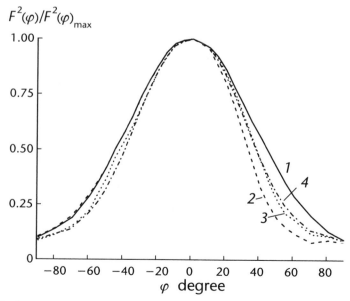

Figure 8.42 Peak power patterns of an isolated antenna element (curve 1) and of the edge (curve 2) and inner elements (curve 3) of a horizontal array and the average of the patterns of the array four elements (curve 4). (With permission from Pleiades Publishing, Ltd.)

The pattern of a UWB antenna array can be calculated and plotted using a more laborious method: direct pulse summation. For this purpose, a computer code was used that provided for summation of arbitrarily time-sampled pulses with specified delay $\Delta\tau$. In investigating the radiation of the single antenna, the step in angle φ was 5°–10°. If it was necessary to calculate a radiation pulse for an intermediate observation angle, the data on two neighboring pulses were interpolated. Four identical pulses radiated by an isolated antenna element in the direction specified by a fixed angle φ were summed up taking into account the spatial path difference $d_h\sin\varphi$ and the time delay $\Delta\tau$ for any neighboring array elements. For the integrated pulse, the maximum deviation from the horizontal axis corresponding to the peak radiation field strength E_p of the horizontal array was found for a specified angle φ. The data of calculations using pulse summation are presented in Figure 8.41(b) and in Table 8.1, where they are indicated by the symbol Σ.

As can be seen from Figure 8.41(b) and Table 8.1, for the pulse summation there is good agreement between the calculated and experimental patterns of the horizontal array in the measurement range for the numerically found scan angles from 0 to ±40° (corresponding to the experimental $\varphi_0 = \pm37°$). In this calculation variant, in contrast to the calculation for a fixed frequency, $f_0 = 5\text{GHz}$, there are no grating lobes in the array pattern at large scan angles in the actual angle range [Figure 8.41(b)].

As shown above, the pattern of an antenna varies slightly when it is installed as an element in an array and depends on its location in the array. Hence, the pulses radiated by an isolated antenna and by an antenna element of an array must be different in waveform. This is due to that the VSWR of an antenna changes on installing it in an array. This is illustrated by Figure 8.43 that presents the pulses radiated in the main direction ($\varphi,\delta = 0$) by an edge (curve 3) and an inner element

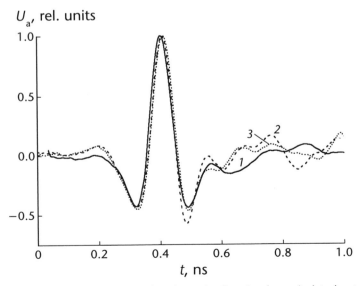

Figure 8.43 Waveforms of pulses radiated in the main direction by an isolated antenna (curve 1) and by an inner (curve 2) and an edge element of a horizontal array (curve 3). (With permission from Pleiades Publishing, Ltd.)

(curve 2) of a horizontal array. Herein, the pulse radiated by an isolated KA (curve 1) is given for the same observation direction. In this case, the KA was excited by a bipolar voltage pulse via a four-channel voltage divider.

Thus, to increase the accuracy of the calculation of the array pattern at large scan angles, it is necessary to sum up the pulses radiated by each element in a fixed direction φ. The radiation pattern thus constructed for the specified scan angle φ_0 = −60° is shown in Figure 8.41(b) (curve 6). The characteristics of this pattern are also presented in Table 8.1 (data in parentheses). It can be seen that the maxima of the experimental and calculated patterns almost coincide in position and have nearly the same values (0.31 and 0.29, respectively). The difference between the calculated and the measured pattern width at half maximum is 5°.

Figure 8.44(a) presents the waveforms of the pulses radiated by a horizontal array at the pattern maxima for different scan angles. As can be seen from the figure, the first two time lobes of the radiated pulses are close in waveform for all

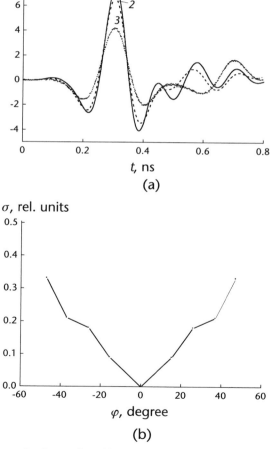

Figure 8.44 Waveforms of pulses radiated by a horizontal array at the pattern maxima for the scan angle equal to 0 (curve 1), 26° (curve 2), and 47° (curve 3) (a) and the pulse waveform RMSD versus scan angle (b). (With permission from Pleiades Publishing, Ltd.)

scan angles, and only the third lobe shows noticeable differences. Quantitatively, the difference in the waveforms of pulses radiated in different directions from the waveform of the pulse radiated in the main direction is characterized by the RMSD σ [Figure 8.44(b)] that was determined by formula (8.4).

Figure 8.45 presents the experimental (solid lines) and calculated (dashed lines) E-plane peak power patterns of a vertical array. Figure 8.45(a) shows the patterns corresponding to the scanning in the range of negative angles δ (calculation for $f_0 = 5$ GHz) and Figure 8.45(b) shows the patterns corresponding to the scanning in the range of positive angles δ (calculated by summing up the pulses radiated by an isolated antenna). All patterns were normalized to the maximum $F^2(\varphi = 0, \delta = 0)$ for the array operating in a no-scanning mode. The data for a vertical array are

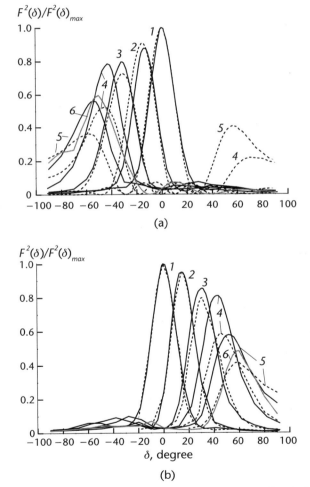

Figure 8.45 Calculated (dashed lines) E-plane peak power patterns of an isolated antenna obtained for the frequency $f_0 = 5$ GHz (a) and by direct summation of the pulses radiated by the antenna (b) and experimental (solid lines) E-plane peak power patterns of a vertical array for specified scan angles of 0, ±16.4°, ±33°, ±50.6°, and ±71.3° (curves 1–5). Curve 6 in Figure 8.45(a) was calculated for an averaged pattern and $f_0 = 3$ GHz; curve 6 in Fig. 8.45(b) was calculated by summing up the pulses radiated by the array elements. (With permission from Pleiades Publishing, Ltd.)

given in Table 8.2. In the investigations of the vertical antenna array, the same set of feeders was used as that for the horizontal array. However, the element spacing in the vertical array was smaller than that in the horizontal one, resulting in increased specified scan angles δ_0 (Table 8.2).

As can be seen from Figure 8.45(a) and Table 8.2, the calculated (f_0 = 5 GHz) and experimental main-lobe patterns of the vertical array are in good agreement for the array operating in a no-scanning mode and at a minimal deviation of the pattern maximum direction from the normal to the array (the experimental scan angles are $-14°$ and $+15°$). For the specified scan angles $\delta_0 = \pm31°$ (corresponding to the experimental angles $\delta_0 = \pm31°$), the difference between the calculated and measured pattern maxima is 9% for negative δ and 14% for positive δ. However, the positions of the pattern maxima and the half-maximum widths of the main lobes for these scan angles are in good agreement.

For larger scan angles, the calculated deviation of the pattern maximum direction from the normal to the array is greater than the experimental one, as in the case of the horizontal array. However, for the vertical array, the calculated pattern maxima are considerably lower than the experimental ones. Thus, for the specified scan angles $\delta_0 = \pm50.6°$ (experimental $\delta_0 = \pm43°$), the calculated pattern maxima make up about 68% of their experimental values, and for the specified $\delta_0 = \pm71.3°$, they make up 65%.

This might be due to the distortion of the pattern of the combined antenna on installing it in a vertical array. To verify this supposition, additional investigations were performed. The peak power radiation patterns of vertical array elements are shown in Figure 8.46. Curves 1–4 correspond to the numbered array elements shown in Figure 8.24(b). Curve 5 is the pattern of the isolated antenna and curve 6 is the average of the patterns of the four array elements. This average pattern was used in the calculation aimed at refining the pattern of the vertical array.

The calculated patterns of the array for the specified scan angles equal to 0, $\pm16.5°$, and $\pm33°$ obtained by substituting the average pattern function in formula (8.8) are insignificantly different from those given in Figure 8.45(a) and in Table 8.2. This is accounted for by that in the angle range from $-35°$ to $+40°$, the average pattern is close to the pattern of the isolated KA and the maxima of the amplitude spectra of the pulses radiated in these directions correspond to the frequency f_0 = 5 GHz.

Figure 8.45(a) shows the refined pattern of the array for the $-71.3°$ calculated scan angle (curve 6). For this calculation, the frequency f_0 was set, according to the measurements, equal to 3 GHz. The refined calculation data for a fixed frequency (f_0 = 3 GHz) are given in Table 8.2 in brackets (for the calculated scan angles $\delta_0 = \pm50.6°$ and $\delta_0 = \pm71.3°$). As can be seen from Figure 8.45(a) and Table 8.2, the calculated patterns of the vertical array differ significantly from the experimental ones, even when corrected by using the average array element pattern and the refined value of f_0.

For the vertical array, the pattern calculations were also performed using direct pulse summation [Figure 8.45(b)]. The characteristics of the vertical array patterns obtained by summing up the pulses radiated by the isolated KA are also presented in Table 8.2 (indicated by the symbol Σ). Here, as in the case of the calculation for f_0 = 5GHz, there is a significant difference between the calculated and the experimental patterns for the specified scan angles $\delta_0 = \pm50.6°$ and $\delta_0 = \pm71.3°$.

Table 8.2 Characteristics of the E-plane Patterns of a Vertical Array. (With permission from Pleiades Publishing, Ltd.)

Specified scan angle δ_0, deg	Calculated scan angle δ_0, deg		Experimental scan angle δ_0, deg	Pattern maximum normalized to its value with no scanning			Half-maximum width of the pattern main lobe, deg		
	$f_0 = 5$ GHz	Σ		calculation f_0	calculation Σ	experiment	calculation f_0	calculation Σ	experiment
0	0	0	0	1	1	1	24	22	23
−16.4	−15	−15	−14	0.91	0.87	0.88	25	23.5	23
+16.4	+15.5	+15.5	+15	0.93	0.96	0.95	24.5	23.5	23
−33	−31	−31	−31	0.73	0.67	0.8	28	27	28
+33	+31	+31	+31	0.74	0.8	0.86	28	27.5	26
−50.6	−45	−45	−43	0.53	0.48	0.79	34.5	34	29
	(−48)	(−47)		(0.71)	(0.7)		(41)	(28)	
+50.6	+46	+45	+43	0.56	0.59	0.81	34	35	31
	(50)	(48)		(0.62)	(0.62)		(49)	(31)	
−71.3	−57	−58	−54	0.36	0.33	0.56	>47	>48	36
	(−51)	(−55)		(0.6)	(0.47)		(35)	(41)	
+71.3	+57	+58	+52	0.38	0.42	0.58	>47	>48	38
	(56)	(60)		(0.55)	(0.48)		(43)	(32)	

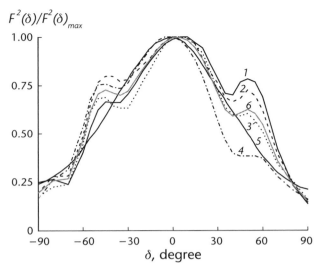

Figure 8.46 Peak power patterns of the elements of the vertical array shown in Figure 8.24(b) (curves 1–4) and of the isolated antenna (curve 5), and the average of the element patterns (curve 6). (With permission from Pleiades Publishing, Ltd.)

The waveforms of the pulses radiated by the elements of the vertical array in the main direction are shown in Figure 8.47. It can be seen that, in contrast to the horizontal array (Figure 8.43), the waveforms of the pulses radiated by the vertical array elements differ only slightly. Figure 8.45(b) shows the pattern of the array for the 71.3° calculated scan angle (curve 6), obtained by summing up the pulses radiated by the array elements. The data for the calculated angles $\delta_0 = \pm 50.6°$ and $\delta_0 = \pm 71.3°$ are given in brackets in Table 8.2. As can be seen from Figure 8.45(b) and Table 8.2, the calculated patterns of the vertical array obtained by direct summation of the pulses radiated by the array elements also differ significantly from the

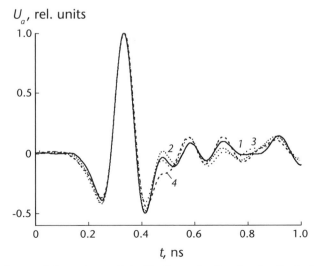

Figure 8.47 Waveforms of the pulses radiated in the main direction by the bottom (curve 1), the-second-from-the bottom (curve 2), the third-from-the-bottom (curve 3), and the top element of the vertical array (curve 4). (With permission from Pleiades Publishing, Ltd.)

experimental patterns. Note that the different calculated up and down scan angles for the vertical array (Table 8.2) are due to the asymmetry of the E-plane peak power pattern of the isolated antenna that was used in the calculation by formula (8.8).

Figure 8.48(a) presents the waveforms of the pulses radiated in the pattern maximum directions by the vertical array (when scanning is performed toward negative δ). As can be seen from the figure, the first two time lobes of the radiated pulses are close in waveform, and only the third lobes are noticeably different. Quantitatively, the difference in waveform between the pulses radiated in different directions and the pulse radiated on boresight is characterized by the RMSD σ [Figure 8.48(b)].

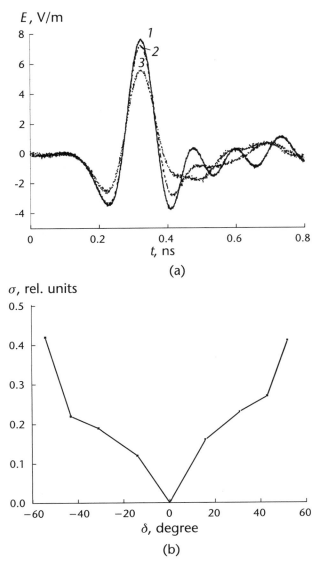

Figure 8.48 Waveforms of the pulses radiated in the pattern maximum directions by a vertical array for the scan angles equal to 0 (curve 1), −31° (curve 2), and −54° (curve 3) (a), and the RMSD of the pulse waveforms versus scan angle (b). (With permission from Pleiades Publishing, Ltd.)

The linear arrays of KAs excited by a bipolar pulse of duration 200 ps can provide half-peak-power scanning within ±50° in the *E*-plane and within ±40° in the *H*-plane. For the wave-beam scanning within ±40° in both planes, the RMSD of the pulse waveforms from the waveform realized in a no-scanning mode is not over 25%. This allows application of arrays of this type in UWB radar systems where the radiated pulse waveform is used for object recognition.

Note that the UWB TEM-antenna arrays provide smaller wave-beam-scan angles, ±15° [8] and ±20° [44], which is due to their narrower patterns. The characteristics of TSA-based arrays for scan angles of ±45° are discussed elsewhere [45]. The study was performed in the frequency range 500–1000 MHz and was restricted to the characteristics of the phase-shifting device. In the studies [8, 44, 45], the wave-beam scanning was carried out only in the *H*-plane.

8.5   Active Receiving Antenna Arrays

8.5.1   A Dual-polarized Planar Array

8.5.1.1   A 2 × 2 Antenna Module

To construct a dual-polarized receiving antenna array consisting of $2n \times 2m$ elements, where n and m are integers, it is proposed to use a modular design based on an array of four crossed dipoles located in the corners of a square, forming a 2 × 2 antenna module (AM) [14]. Each antenna array element consisting of crossed dipoles is intended to detect the orthogonal vector components of the electric field of UWB pulses of subnanosecond and nanosecond duration. The appearance and topology of the conductors of the array element are presented in Figure 8.49. The antenna was fabricated by printing technology on two 1-mm-thick double-sided fiberglass foil plates with dimensions 48 × 48 mm. Each dipole arm (1) is loaded by the input of one of the four identical active elements (AEs) (2) having a coaxial cable output (3). In the middle of each arm, a transverse cut is made in the metal in which a resistor (4) is mounted. Power supply voltage is applied to the AE via an individual cable.

To investigate the directional response and the detected pulse waveform and effective length of a UWB array as functions of element spacing, an AM with equally spaced elements, such that the element spacing could be varied. The AM (Figure 8.50) consists of four active crossed dipoles (1), a 160 × 160 mm dielectric base (2), and feeder paths coated with bulk absorber (3). The distance from the dipole arms to the dielectric base is 140 mm. The spacing between the dipole centers, $d_v = d_h = d$, was varied in the range 48–100 mm.

The block diagram of the AM is shown in Figure 8.51. Initially, the signals from the analogous arms of the active dipoles are summed in phase, and then, the antiphase component is separated. The dipoles are conditionally subdivided into vertical and horizontal. The vertical channel dipoles have four upper and four lower arms. The signals of the upper and lower arms of the dipoles are summed in phase in power combiners (1 and 2, respectively). To separate the antiphase component that corresponds to vertical polarization, a balun (3) is used. Similarly, power combiners (4 and 5) and a balun (6) are used for the horizontal channel.

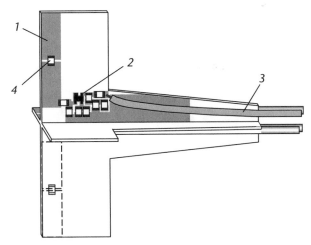

Figure 8.49 Sketch of the antenna array element: 1—dipole arm, 2—active element, 3—coaxial cable, and 4—resistor. (With permission from Pleiades Publishing, Ltd.)

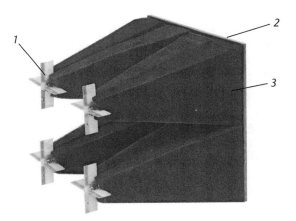

Figure 8.50 Appearance of the antenna module: 1—crossed dipoles, 2—dielectric base, and 3—absorber. (With permission from Pleiades Publishing, Ltd.)

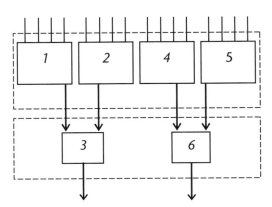

Figure 8.51 Block diagram of the antenna module: 1, 2—power combiners; 3—balun of the vertical channel; 4, 5—power combiners, and 6—balun of the horizontal channel. (With permission from Pleiades Publishing, Ltd.)

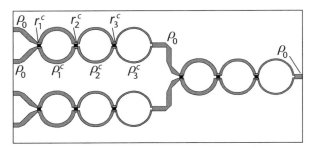

Figure 8.52   Sketch of a power combiner. (With permission from Pleiades Publishing, Ltd.)

Each power combiner (Figure 8.52) has four inputs and one output and represents two binary summation stages. Three identical three-link ring power combiners are printed on a 100×60 mm, 1-mm thick plate of FLAN foil dielectric with $\varepsilon_r = 5$. The power combiner elements were designed using a method described elsewhere [46]. The characteristic impedances of the semicircular lines have the following values: $\rho_1^c = 57\ \Omega$, $\rho_2^c = 71\ \Omega$, and $\rho_3^c = 87\ \Omega$; the length of each line corresponds to a quarter wavelength at 2 GHz. Each connecting line has characteristic impedance $\rho_0 = 50\ \Omega$. The resistances of the surface-mount resistors used, r_1^c, r_2^c and r_3^c, are, respectively, 390, 200, and 100 Ω. The match bandwidth of each power combiner is 0.3–5 GHz with the VSWR ≤ 1.5 for any of the four inputs. The input-to-output propagation ratio in this frequency band is no less than -7 dB. The four power combiners are structurally combined in a power combiner unit.

The characterization of an AM with the use of the KA [21] excited by a 0.5-ns low-voltage bipolar pulse as a UWB pulse radiation source is discussed below. The difference in waveform between the incident electromagnetic field pulse $E(t)$ and the output voltage pulse $U_a(t)$ was estimated for an isolated antenna, an AM and a multielement array using the RMSD criterion

$$\sigma = \sqrt{\frac{\int_T \left[E(t) - M U_a(t - \tau) \right]^2 dt}{\int_T E^2(t)\, dt}} \tag{8.12}$$

where T is the time interval for which the comparison is performed, M is the scaling ratio, and τ is the shift of $U_a(t)$ relative to $E(t)$ at which σ is a minimum. For a reference antenna the TEM antenna [47] (Section 6.3.1) was used; the voltage waveform at its output was assumed to be the scaled waveform of the incident field pulse. The TEM antenna was 90 cm long with the effective length $l_e^{\mathrm{TEM}} = 4$ cm. The effective length of the test antenna was determined from the ratio of the peak voltage at its matched output, U_{ap}, to the peak voltage at the matched output of the TEM antenna, U_{ap}^{TEM}, for the same distance of the antennas to the radiation source by the formula

$$l_e = l_e^{\mathrm{TEM}} \left(\frac{U_{ap}}{U_{ap}^{\mathrm{TEM}}} \right) \tag{8.13}$$

For this estimation, l_e was assumed to be a constant independent of the radiation spectrum frequency.

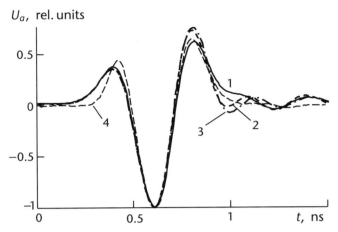

Figure 8.53 Waveforms of pulses detected by the antenna module with d = 48 (curve 1), 64 (curve 2), and 80 mm (curve 3) and by the TEM antenna (curve 4). (With permission from Pleiades Publishing, Ltd.)

The waveforms of pulses detected by one of the AM channels for d = 48, 64, and 80 mm are shown in Figure 8.53 (curves 1–3, respectively). For comparison, the waveform of a pulse detected by the TEM antenna is given (curve 4). When d was varied in the range 60–100 mm, the detected pulse waveforms and the module effective length varied inappreciably. For $48 \leq d < 60$ mm, the pulse waveform was observed to vary and l_e to increase noticeably with decreasing d (Figure 8.54), which can be accounted for by the mutual influence of neighboring dipoles. For the minimum d = 48 mm, the arms of neighboring dipoles touched one another, but there was no galvanic contact between them. The RMSD of the waveforms of the pulses detected by the TEM antenna from those detected by the AM was 0.25, 0.34, and 0.36 for d = 48, 64, and 80 mm, respectively. The polarization isolation between the AM channels was no less than 25 dB. The peak power patterns measured in two planes for the AM are shown in Figure 8.55.

Increasing the element spacing increases the pattern directivity of the AM but decreases the range of angles in which the detected pulse waveform is retained. Figure

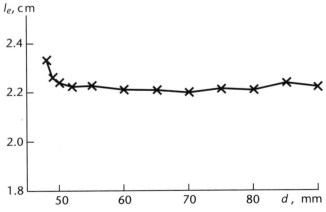

Figure 8.54 The effective length of the antenna module versus element spacing. (With permission from Pleiades Publishing, Ltd.)

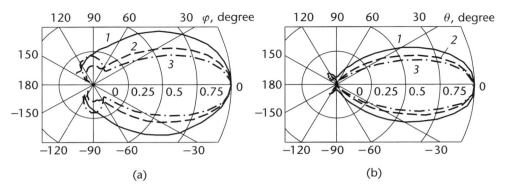

Figure 8.55 The peak power patterns of the antenna module measured in the H-plane (a) and in the E-plane (b) for d = 48 (curves 1), 64 (curves 2), and 80 mm (curves 3). (With permission from Pleiades Publishing, Ltd.)

8.56 shows the RMSD of the waveforms of pulses detected in different directions from the boresight pulse waveform as a function of the radiation incidence angle. Within the half-power pattern width, the distortions of the waveforms of pulses detected by the AM at different d are not over $\sigma = 0.2$. These distortions are additional to the waveform distortion illustrated by Figure 8.53.

As for the minimum d, the range of angles in which the detected pulse waveform is retained is a maximum, but there is a mutual influence of neighboring dipoles, for the antenna module to be used as the basis for multielement arrays, $d = 52$ mm was chosen at which the mutual influence of the dipoles is negligible.

8.5.1.2 A 4 × 4 Antenna Array

The appearance of a 4 × 4 dual-polarized antenna array [14] is shown in Figure 8.57. The array consists of four identical antenna modules (1), four power combiner units (2) of the first summation stage, a terminal power combiner unit (3), and a two-channel balun (4), whose output voltages correspond to the two orthogonal

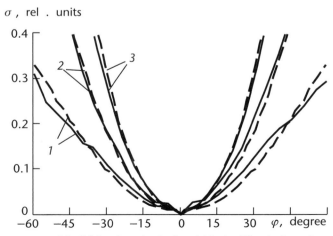

Figure 8.56 The measured (solid lines) and calculated (dashed lines) distortions of detected pulse waveforms in the *H*-plane versus radiation incidence angle for *d* = 48 (curves 1), 64 (curves 2), and 80 mm (curves 3). (With permission from Pleiades Publishing, Ltd.)

Figure 8.57 The appearance of the antenna array: 1—antenna modules, 2—first-stage power combiners, 3—terminal power combiner, and 4—balun. (With permission from Pleiades Publishing, Ltd.)

components of the electric field vector of the UWB pulses. The spacing between the centers of the array elements, d, is 52 mm. The power supply voltage of the antenna array is 3 V, the consumption current is 1.6 A, and the dimensions and mass are, respectively, $56 \times 21 \times 21$ cm and 2.6 kg.

The waveform of a pulse detected by one of the array channels at normal incidence of the UWB pulse field radiated by the KA [21] excited by a 0.5-ns low-voltage bipolar pulse is shown in Figure 8.58. The difference in waveform between the pulses detected by the array (curve 1) and by the TEM antenna (curve 2) in the main direction is no more than $\sigma = 0.2$. The effective length of each array channel, l_e, determined by formula (8.13), is 4.5 cm and the effective length of the array element is 1.2 cm. The polarization isolation between the channels is at most 25 dB. The measured H-plane pattern of the array is shown in Figure 8.59 (curve 1). Its difference from the calculated pattern (curve 2) can be due to the influence of the bulk absorber covering the feeder lines that was not considered in the calculations.

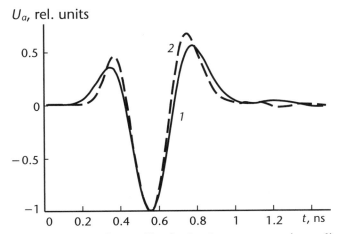

Figure 8.58 Waveforms of pulses detected by the 4×4 antenna array (curve 1) and by the TEM antenna (curve 2). (With permission from Pleiades Publishing, Ltd.)

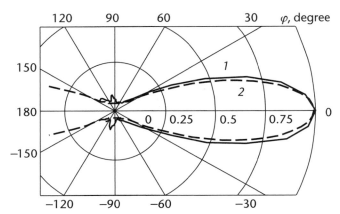

Figure 8.59 The measured (curve 1) and calculated (curve 2) *H*-plane peak power pattern of the 4 × 4 antenna array. (With permission from Pleiades Publishing, Ltd.)

The *E*-plane pattern function is different from the *H*-plane pattern function by the multiplier $\cos \theta$ (the angle θ is counted from the normal to the array plane) like for the pattern functions of a dipole antenna.

Figure 8.60 presents the RMSD of the waveforms of radiation pulses detected in different directions from the normally incident radiation waveform. These distortions are additional to the distortion of the pulses detected by the TEM antenna. The half-power width of the array patterns in both planes is about 40°. In the angular sector ±20°, additional distortions of the pulse waveforms are not above $\sigma = 0.2$.

Along with the data presented earlier, the frequency response and dynamic range of an active antenna array are of interest. The frequency response not only determines the ability of an array to detect UWB pulses with low distortion, but it has an independent value for arrays used for receiving narrowband pulses in a wide central frequency tuning range. The dynamic range of an array affects the radar line of sight.

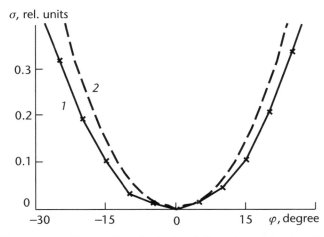

Figure 8.60 The measured (curve 1) and calculated distortion of the detected pulse waveforms in the *H*-plane versus radiation incidence angle (curve 2). (With permission from Pleiades Publishing, Ltd.)

To determine the frequency dependence of the effective length of an active receiving antenna, the following procedure was used. For a voltage pulse $U_a(t)$ detected by the test antenna, the spectrum $\dot{U}_a(f)$ was calculated. For a reference antenna, a TEM antenna was used for which the frequency dependence of the effective length, $l_e^{\text{TEM}}(f)$, is given elsewhere [48] (Section 6.3.1). Using these data, the frequency dependence of the effective length of the test antenna was calculated by the formula

$$I_e(f) = \frac{l_e^{\text{TEM}}(f)\dot{U}_a(f)}{\dot{U}_a^{\text{TEM}}(f)} \tag{8.14}$$

where $\dot{U}_a^{\text{TEM}}(f)$ is the spectrum of the voltage pulse detected by the TEM antenna at the same distance to the transmitting antenna. To determine $l_e(f)$ in the frequency range 0.2–6 GHz, for electromagnetic field sources, low-voltage generators with three KAs optimized for radiating pulses of different duration were used: the transmitting KA [19] with a generator of 1-ns bipolar voltage pulses for the frequency range 0.2–2 GHz, the transmitting KA [21] with a generator of 0.5-ns bipolar pulses for the frequency range 1–4 GHz, and the transmitting KA [23] with a generator of 0.2-ns bipolar pulses for the frequencies from 2.5 GHz and higher.

Figure 8.61 presents $|l_e(f)|$ for the array element operating as an isolated antenna (curve 1), for the AM (curve 2), and for the 4×4 array (curve 3). The three different line styles correspond to the mentioned three frequency ranges. The deviation of the phase-frequency response from a linear function (ΔPFR) for the 4×4 array is shown in Figure 8.62. Due to the linearity of the phase-frequency response at a rather strong frequency dependence of the effective length $|l_e(f)|$, the antenna array is capable to detect UWB electromagnetic pulses with low distortion.

The dynamic range of the 4×4 antenna array was evaluated experimentally using two criteria: the decrease in signal amplitude by 1 dB due to saturation of the AEs and the distortion of the signal waveform by $\sigma = 0.1$ (Section 6.3.3).

In the experiment, a generator of 0.5-ns monopolar pulses with an output voltage of about 200 V and a four-element 2×2 array of transmitting KAs [20] were used.

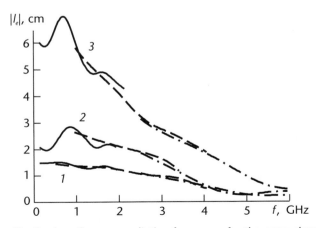

Figure 8.61 The effective length versus radiation frequency for the array element operating as an isolated antenna (curve 1), for the antenna module (curve 2), and for the 4×4 antenna array (curve 3). (With permission from Pleiades Publishing, Ltd.)

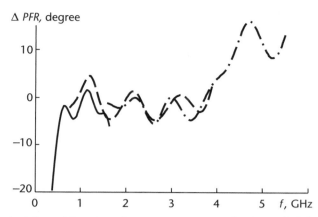

Figure 8.62 The deviation of the phase-frequency response of a channel of the 4 × 4 antenna array from a linear function. (With permission from Pleiades Publishing, Ltd.)

The field strength versus distance, measured using a TEM antenna, ranged from 30 to 250 V/m. The pulse waveform was detected by the test 4 × 4 active antenna array under the same conditions. The maximum E-field strength at which the pulse amplitude decreased by 1 dB was 100 V/m, and when estimated by the waveform distortion criterion ($\sigma = 0.1$), it was 130 V/m. The dynamic range of the array was no less than 100 dB, which is 10 dB greater than that for the isolated active antenna. The results indicate the possibility of further increasing the dynamic range of an active receiving antenna array by increasing the number of array elements.

8.5.2   A Switched Dual-Polarized Linear Antenna Array

8.5.2.1   The Antenna Array Design

The appearance of the dual-polarized antenna array [49] is shown in Figure 8.63. The antenna element (1) of the array is a pair of crossed active dipoles, one oriented vertically and the other horizontally. A detailed description of the dipole antenna design is given elsewhere [14] (Section 8.5.1.1). The spacing between the centers of the antenna array elements is 52 mm. Each dipole is encapsulated in a polystyrene foam holder to fix it mechanically and to protect the embedded electronics against static discharges. The feeder lines are coated with bulk absorber (2). The active dipoles have balanced outputs. To connect them to standard 50-Ω ports, eight broadband baluns (3) are used. Having passed through the baluns, the voltage pulses enter controlled delay lines combined into two units (4 and 5) for horizontal and vertical dipoles, respectively. Subsequently, the four signals received by the horizontal and the vertical dipoles are summed in phase by UWB summation units (6 and 7, respectively). The voltages at the outputs of summation units 6 and 7 are proportional to the strengths of horizontally and vertically polarized incident fields, respectively. The voltages are recorded by a digital oscilloscope. A microcontroller (8) is used to control the delay lines from a personal computer. is offset by in the E-plane for the horizontal dipoles and in the H-plane for the vertical ones.

The block diagram of the antenna array vertical channel is shown in Figure 8.64. The arms (1) of dipoles A1–A4 are loaded by the electronics circuits (2) that

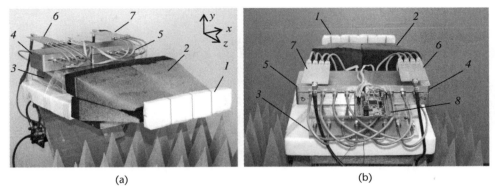

Figure 8.63 Dual-polarized receiving antenna array: 1—antenna elements, 2—absorber, 3—baluns, 4, 5—controlled delay lines, 6, 7—power combiners, and 8—microcontroller. (With permission from American Institute of Physics.)

are integrated into the dipoles to compensate the variations in frequency response due to the small dipole length. Having passed through a balanced feed line (3), one of the input signals is inverted in polarity by an inverter (4). The other signal is sent to a strip line (5) whose length is chosen to match the delay time of signal propagation through the inverter. Then the two signals are summed by a 2-stage Wilkinson power combiner (6). The inverter consists of two exponential transition couplers: the first one coupling the microstrip line to the balanced line and the second coupling the balanced line back to the microstrip line, in which the signal and the ground conductors are swapped. The design of the balun is described in detail elsewhere [14]. The pass band of the balun is 0.4–3.5 GHz.

Having passed through baluns B1–B4, the voltage pulses enter unit (7) consisting of four controlled delay lines. Subsequently, the four signals are summed in phase by a UWB power combiner (8). The four-port power combiner performs

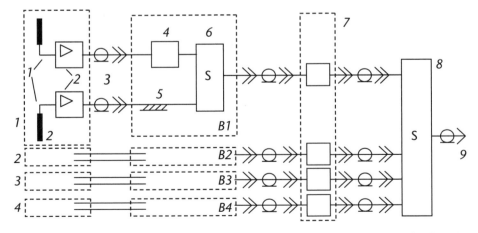

Figure 8.64 Block diagram of the antenna array vertical channel: A1–A4—active dipoles, B1–B4—UWB balanced-to-unbalanced units, 1—dipole arms, 2—amplifiers, 3—coaxial lines, 4—polarity inverter, 5—strip line, 6—power combiner, 7—four-channel unit of controlled delay lines, 8—power combiner, and 9—output port. (With permission from American Institute of Physics.)

binary summation using three-stage Wilkinson ring combiners. The match band of the power combiner is 0.4–3.2 GHz with the VSWR at each port less than 2. In this frequency band, the insert losses of a signal having passed from any input port to the output port are no less than −7 dB.

The schematic circuit diagram of the delay line core is presented in Figure 8.65. The delay line core consists of two placed mirrored RF switches of type HMC321LP4 (*Hittite Microwave*) (DA1 and DA2). Segments L1–L7 of coaxial waveguides of different length are connected between ports RF1–RF7 of the switches. The waveguides are insulated with teflon tubes of diameter 1 mm. Binary codes corresponding to the port numbers are sent to the data bus (A, B, C) of a microprocessor to control the state of the RF switches. Appropriate pairs of the switch ports (RF1–RF8, RF2–RF7, RF3–RF6, and RF4–RF5) were chosen to avoid waveguide crossings due to the mirrored placement of the RF switches. For proper connection the data bits are inverted by the elements $DD1.1$–$DD1.3$ for the switch $DA2$. Seven pairs of the RF switch outputs are put into operation. The remaining outputs are loaded by 50-Ω resistors (R) to disconnect the input port of the recording device from the antennas to protect it against high-power pulsed radiation.

The lengths of cables L1–L7 are chosen so that the scan angles of the antenna array would be 0, ±13°, ±26°, and ±40° for both polarizations. The measured insert loss for each of the delay lines varies from −3.5 dB at a frequency of 0.2 GHz to −6.5 dB at 4 GHz. In the range 0.1–8 GHz, the deviation of the phase-frequency response from a linear function is no more than ±11.5°. The VSWR in the frequency range 0.1–7.8 GHz is no greater than 2.2.

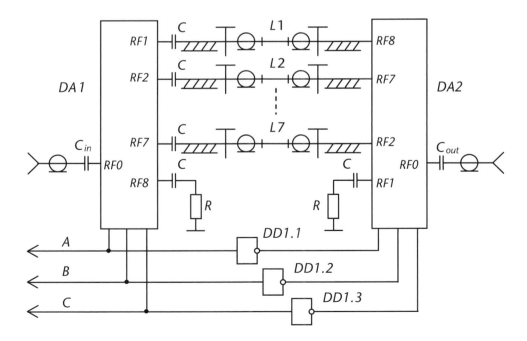

Figure 8.65 Schematic circuit diagram of the controlled delay line core. (With permission from American Institute of Physics.)

8.5.2.2 Characteristics of the Antenna Array

The directional properties of the receiving antenna array were measured in time domain. A combined antenna [21] excited by bipolar voltage pulses of duration 0.5 ns was used as a source of electromagnetic pulses. The spacing between the antennas was 3 m. The pulses at the output of the receiving antenna array were recorded by a digital sampling oscilloscope with a 12-GHz bandwidth. To measure the pattern of the antenna array, the received voltage waveforms at the antenna outputs were recorded with an angular step of 5°. The waveform peak value was measured for each angular point. As the antenna output is proportional to the field strength of the incident wave, the angular dependence of the absolute values of these peaks represents the peak field pattern of the antenna array.

Figure 8.66(a) presents the peak field pattern of the antenna array in the H-plane plotted based on the waveforms of vertically polarized radiation recorded at different scan angles θ_0. The patterns are normalized to the maximum value corresponding to $\theta_0 = 0$. Figure 8.66(b) shows the peak field pattern of the antenna array in the E-plane plotted based on the recorded waveforms of horizontally polarized radiation.

Figure 8.67 presents the difference $\Delta\theta$ between the measured and the specified angular position of the pattern maximum depending on θ_0 for the antenna array

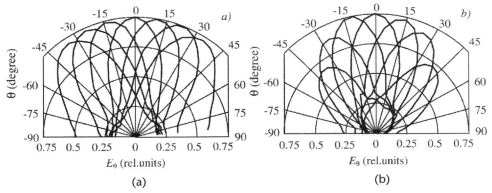

Figure 8.66 Peak field patterns of the antenna array measured for all scan angles for vertical (a) and horizontal polarization (b). (With permission from American Institute of Physics.)

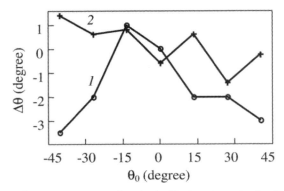

Figure 8.67 Difference between measured and specified scan angles for the vertical (curve 1) and horizontal dipoles (curve 2). (With permission from American Institute of Physics.)

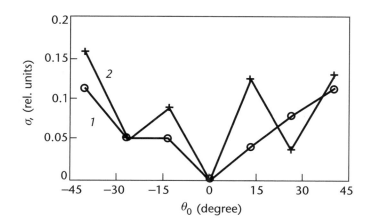

Figure 8.68 Measured waveform distortion for the vertical (curve 1) and horizontal dipoles (curve 2). (With permission from American Institute of Physics.)

vertical (curve 1) and horizontal polarization (curve 2). The error in setting the pattern maximum position is no more than 4°. For the vertical and horizontal dipoles, polarization isolation was measured for specified directions θ_0.

The recorded pulse waveforms vary slightly with scan angle. The difference between the waveforms of two received pulses, $U(t)$ and $V(t)$, was estimated quantitatively using the RMSD determined by (8.4). Figure 8.68 presents the angular dependence of the waveform distortion σ estimated for the waveform of the pulse detected at $\theta_0 = 0$ and the waveforms of pulses detected at specified scan angles.

The four-element linear antenna array was developed to measure the polarization structure of electromagnetic pulses of nano- and subnanosecond duration. The array elements are based on crossed active dipoles. The electronics circuits are integrated into the dipoles to compensate the variations in frequency response due to the small dipole length in the frequency range 0.4–4 GHz. Digital scanning was realized in seven directions ranging from –40° to +40°. The error of setting the pattern maximum is no more than 4°. The polarization isolation is over 29 dB in the direction orthogonal to the array axis and over 23 dB in other directions.

Conclusion

The fundamentals of designing multielement antenna arrays have been developed to create UWB pulse receiving-transmitting systems of both high and low peak power. Combined antennas were used as elements of transmitting arrays and crossed active dipole antennas as elements of receiving arrays.

In a transmitting array optimal in design, the elements are galvanically connected in the E-plane and are spaced a small distance from each other in the H-plane. The minimum level of side radiation relative to the boresight peak field strength is achieved with equal-amplitude excitation of the array elements. The radiation pulse waveform at the half-peak-power level is retained within $\sigma = 0.2$.

Distribution systems for equal-amplitude excitation of array elements have been developed. The amplitude jitter at the outputs of multichannel power dividers is ~4%. The energy loss in optimized distribution systems reaches 20–30%. To reduce the energy loss, gas insulation is proposed to be used.

Breaking an array into two subarrays makes it possible to build a system radiating successively two orthogonally polarized radiation pulses. It has been shown that the best choice for systems of this type is diagonal arrangement of the elements in orthogonally polarized subarrays.

The wide pattern of a combined antenna provides for the wave beam scanning at a half-peak-power level within $\pm 50°$ and $\pm 40°$ in the E-plane and in the H-plane, respectively. For scanning within $\pm 40°$, the deviation of the radiated pulse waveform corresponding to the pattern maximum from the waveform of the pulse radiated with synchronous excitation of array elements is not over $\sigma = 0.35$.

Based on a modular approach, active dual-polarized receiving arrays have been developed whose dynamic range is no less than 100 dB. The distortion of the received UWB pulse waveform within the pattern width at a half-peak-power level ($\pm 20°$) is no more than $\sigma = 0.4$. The electron scanning of the pattern of a linear dual-polarized array within $\pm 40°$ has been realized.

Problems

8.1 What are the main differences between a UWB and a narrowband transmitting antenna array?

8.2 Why to reduce side radiation, the elements of UWB arrays are excited by equal-amplitude voltage pulses, whereas the excitation of the elements in narrowband arrays is performed using voltage pulses with amplitudes falling toward the array edges?

8.3 Why in UWB arrays excited from a single voltage pulse generator, the effective radiation potential increases with the number of elements, whereas the peak radiation power decreases?

8.4 Compare the distribution systems for a 4×4 and an 8×8 array and specify their advantages and disadvantages.

8.5 Formulate criteria for optimizing the structure of an antenna array and provide a comparative analysis of linear and rectangular arrays.

8.6 What factors affect the scan angle in the H-plane and how it can be maximized?

References

[1] Vendik, O. G., and M. D. Parnes, *Antennas with Electrical Scanning (Introduction to the Theory)*, Moscow: Science-Press, 2002 (in Russian).

[2] Active Phased Antenna Arrays (ed. D. I. Voskresensky and A. I. Kanashenkov), Moscow: Radiotekhnika, 2004 (in Russian).

[3] Mailloux, R. J., *Phased Array Antenna Handbook. Second edition*, Boston: Artech House, 2005.

[4] Fenn, A. J., *Adaptive Antennas and Phased Arrays for Radar and Communications*, Boston: Artech House, 2008.

[5]   Fenn, A. J., and P. T. Hurst, *Ultrawideband Phased Array Antenna Technology for Sensing and Communications Systems*, Cambridge, Massachusetts: MIT Press, 2015.

[6]   Harmuth, H. F., *Nonsinusoidal Waves for Radar and Radio Communications*, New York: Academic Press, 1981.

[7]   Shpak, V. G., et al., "Generation of Powerful Superbroad-Band Subnanosecond Electromagnetic Pulses," *Russ. Phys. J.*, Vol. 39, No. 12, 1996, pp. 1257–1263.

[8]   Efanov, V. M., et al., "Multiunit UWB Radiator of Electromagnetic Waves With Controlled Directional Pattern," *Proc. 13th Inter. Symposium on High Current Electronics*, Tomsk, July 25–29, 2004, pp. 262–266.

[9]   Andreev, Yu. A., et al., "An Ultrawideband Gigawatt Pulse Generator," *Instrum. Exp. Tech.*, Vol. 43, No. 2, 2000, pp. 224–229.

[10] Gubanov, V. P., et al., "Sources of High-Power Ultrawideband Radiation Pulses With a Single Antenna and a Multielement Array," *Instrum. Exp. Tech.*, Vol. 48, No. 3, 2005, pp. 312–320.

[11] Holter, H., T. H. Chio, and D. H. Schaubert, "Experimental Results of 144-Element Dual-Polarized Endfire Tapered-Slot Phased Arrays," *IEEE Trans. Antennas Propagat.*, Vol. 48, No. 11, 2000, pp. 1707–1718.

[12] Kindt, R. W., and W. R. Pickles., "Ultrawideband All-Metal Flared-Notch Array Radiator," *IEEE Trans. Antennas Propagat.*, Vol. 58, No. 11, 2010, pp. 3568–3575.

[13] Chen, V. C., and H. Ling, *Time-Frequency Transforms for Radar Imaging and Signal Analysis*, London: Artech House, 2002.

[14] Balzovskii, E. V., Yu. I. Buyanov, and V. I. Koshelev, "Dual Polarization Receiving Antenna Array for Recording of Ultra-Wideband Pulses," *J. Commun. Technol. Electron.*, Vol. 55, No. 2, 2010, pp. 172–180.

[15] Yarovoy, A., et al., "Near-Field Focusing Within a UWB Antenna Array," *Proc. European Conference on Antennas and Propagation*, Nice, Nov. 6–10, 2006 (ESA SP–626, October 2006).

[16] Belichenko, V. P., et al., "Short Electromagnetic Pulse Formation by Plane Antenna Array," *Proc. of Seminar/Workshop on Direct and Inverse Problems of Electromagnetic and Acoustic Wave Theory*, Lviv, Sep. 15–17, 1997, pp. 43–46.

[17] Andreev, Yu. A., et al., "An Element of the Scanning Antenna Array for Radiation of High-Power Ultrawideband Electromagnetic Pulses," *J. Commun. Technol. Electron.*, Vol. 44, No. 5, 1999, pp. 492–498.

[18] Andreev, Yu. A., Yu. I. Buyanov, and V. I. Koshelev, "A Combined Antenna With Extended Bandwidth," *J. Commun. Technol. Electron.*, Vol. 50, No. 5, 2005, pp. 535–543.

[19] Efremov, A. M., et al., "Generation and Radiation of High-Power Ultrawideband Nanosecond Pulses," *J. Commun. Technol. Electron.*, Vol. 52, No. 7, 2007, pp. 756–764.

[20] Koshelev, V. I., V. V. Plisko, and K. N. Sukhushin, "Array Antenna for Directed Radiation of High-Power Ultra-Wideband Pulses," In *Ultra-Wideband, Short-Pulse Electromagnetics 9*, pp. 259–267, F. Sabath, et al. (eds.), New York: Springer, 2010.

[21] Efremov, A. M., et al., "High-Power Sources of Ultra-Wideband Radiation With Subnanosecond Pulse Lengths," *Instrum. Exp. Tech.*, 2011. Vol. 54, No. 1, 2011, pp. 70–76.

[22] Andreev, Yu. A., et al., "A High-Performance Source of High-Power Nanosecond Ultrawideband Radiation Pulses," *Instrum. Exp. Tech.*, Vol. 54, No. 6, 2011, pp. 794–802.

[23] Andreev, Yu. A., et al., "Generation and Emission of High-Power Ultrabroadband Picosecond Pulses," *J. Commun. Technol. Electron.*, Vol. 56, No. 12, 2011, pp. 1429–1439.

[24] Koshelev, V. I., and V. V. Plisko, "Energy Characteristics of Four-Element Arrays of Combined Antennas," *Izv. Vyssh. Uchebn. Zaved., Fiz.*, Vol. 56, No. 8/2, 2013, pp. 134–138.

[25] Koshelev, V. I., et al., "Ultrawideband Radiators of High-Power Pulses," *IEEE Pulsed Power Plasma Science Conf.*, Las Vegas, June 17–22, 2001, Vol. 2, pp. 1661–1664.

[26] Lebedev, V. I., *Microwave Technology and Devices, Vol. 1*, Moscow: Vysshaya Shkola, 1970 (in Russian).

[27] Feldshtein, A. L., and V. P. Smirnov, *Handbook on Waveguide Elements*, Moscow: Sov. Radio, 1967 (in Russian).

[28] Romanchenko, I. V., et al., "Repetitive Sub-Gigawatt RF Source Based on Gyromagnetic Nonlinear Transmission Line," *Rev. Sci. Instrum.*, Vol. 83, No. 7, 2012, p. 074705.

[29] Efremov, A. M., et al., "Generation and Radiation of Ultra-Wideband Electromagnetic Pulses With High Stability and Effective Potential," *Laser Part. Beams*, Vol. 32, No. 3, 2014, pp. 413–418.

[30] Kragalott, M., W. R. Pickles and M. S. Kluskens, "Design of 5:1 Bandwidth Stripline Notch Array From FDTD Analysis," *IEEE Trans. Antennas Propagat.*, Vol. 48, No. 11, 2000, pp. 1733–1741.

[31] Holter, H., and H. Steyskal, "On the Size Requirement for Finite Phased-Array Models," *IEEE Trans. Antennas Propagat.*, Vol. 50, No. 6, 2002, pp. 836–840.

[32] Holter, H., and H. Steyskal, "Some Experiences From FDTD Analysis of Infinite and Finite Multi-Octave Phased Arrays," *IEEE Trans. Antennas Propagat.*, Vol. 50, No. 12, 2002, pp. 1725–1731.

[33] Lee, J. J., S. Livingston, and A. Neto, "Recent Development of Wide Band Arrays," *Proc. European Conference on Antennas and Propagation*, Nice, Nov. 6–10, 2006 (ESA SP-626, October 2006).

[34] Bell, J. M., M. F. Iskander, and J. J. Lee, "Ultrawideband Hybrid EBG / Ferrite Ground Plane For Low-Profile Array Antennas," *IEEE Trans. Antennas Propagat.*, Vol. 55, No. 1, 2007, pp. 4–12.

[35] Koshelev, V. I., and V. V. Plisko, "Optimization of the Structure of Ultrawideband Combined Antenna Arrays," *Proc. 2nd All-Russian Microwave Conf.*, Moscow, Nov. 26–28, 2014, pp. 19–24 (in Russian).

[36] Godard, A., et al., "A Transient UWB Antenna Array Used With Complex Impedance Surface," *Inter. J. Antennas Propagat.*, Vol. 2010, Article ID 243145, 8 pages, doi: 10.1155/2010/243145.

[37] Mehrdadian, A., and K. Forooraghi, "Design of a UWB Combined Antenna and an Array of Miniaturized Elements With and Without Lens," *Progress In Electromagnetic Res. C.*, Vol. 39, 2013, pp. 37–48.

[38] Cadilhon, B., et al., "Ultra Wideband Antennas for High Pulsed Power Applications," *Ultra Wideband Communications: Novel Trends—Antennas and Propagation*, 2011, Dr. Mohammad Matin (Ed.), ISBN: 978-953-307-452-8, InTech, DOI: 10.5772/20305. Available from: http://www.intechopen.com/books/ultra-wideband-communications-novel-trends-antennas-and-propagation/ultra-wideband-antennas-for-high-pulsed-power-applications

[39] Mikheev, O. V., et al., "Approximate Calculation Methods for Pulse Radiation of a TEM-Horn Array," *IEEE Trans. Electromagn. Compat.*, Vol. 43, No. 1, 2001, pp. 67–74.

[40] McGrath, D. T., and C. E. Baum, "Scanning and Impedance Properties of TEM Horn Arrays for Transient Radiation," *IEEE Trans. Antennas Propagat.*, Vol. 47, No. 3, 1999, pp. 469–473.

[41] Koshelev, V. I., and V. V. Plisko, "Investigation of the Patterns of Linear Arrays under Wave-Beam Scanning," *Izv. Vyssh. Uchebn. Zaved., Fiz.*, Vol. 55, No. 9/2, 2012, pp. 33–36.

[42] Koshelev, V. I., and V. V. Plisko, "Interaction of Ultrawideband Radiation in Linear Array With Wave Beam Steering," *Proc. 14th Inter. Symposium on High Current Electronics*, Tomsk, Sep. 14–15, 2006, pp. 413–416.

[43] Andreev, Yu. A., V. I. Koshelev, and V. V. Plisko., "Combined Antenna and Linear Arrays for Radiation Low-Power Picosecond Pulse," *J. Commun. Technol. Electron.*, Vol. 56, No. 7, 2011, pp. 812–823.

[44] Kardo-Sysoev, A. F., et al., "Ultra Wide Band Solid State Pulsed Antenna Array," In *Ultra-Wideband, Short-Pulse Electromagnetics 5*, pp. 343–349, P. D. Smith and S. R. Cloude (eds.), New York: Plenum Press, 2002.

[45] Schmitz, J., et al., "Ultra-Wideband 4×4 Phased Array Containing Exponentially Tapered Slot Antennas and a True-Time Delay Phase Shifter at UHF," In *Ultra-Wideband, Short-Pulse Electromagnetics 9*, pp. 241–248, F. Sabath, et al. (eds.), New York: Springer, 2010.

[46] Maloratsky, L. G., and L. R. Yavich, Development and Design of Microwave Stripline Elements, Moscow: Sov. Radio, 1972 (in Russian).

[47] Andreev, Yu. A., V. I. Koshelev., and V. V. Plisko, "Characteristics of Receiving-Transmitting TEM Antennas," *Proc. 5th Scientific and Technical Conf. Radar and Radio Communications*, Moscow, Nov. 21–25, 2011, pp. 77–82 (in Russian).

[48] Andreev, Yu. A., et al., "High-Power Sources of Ultrawideband Radiation Pulses," *Proc. 15th Inter. Symposium on High Current Electronics*, Tomsk, Sep. 21–26, 2008, pp. 447–450.

[49] Balzovsky, E. V., et al., "Dual Polarized Receiving Steering Antenna Array for Measurement of Ultrawideband Pulse Polarization Structure," *Rev. Sci. Instrum.*, Vol. 87, No. 3, 2016, p. 034703.

CHAPTER 9

High-Power Ultrawideband Radiation Sources

Introduction

High-power UWB radiation sources are designed mainly to solve the problems of high-resolution radar and to investigate the susceptibility of electronic systems to the influence of strong electromagnetic fields. Classification and a brief overview of the research areas of high-power UWB radiation sources in different laboratories were given in the first chapter. More detailed information can be found in the monographs [1–3] and reviews [4–6].

This chapter presents the results of studies of high-power UWB radiation sources, developed mainly at the Institute of High Current Electronics (Tomsk), that depend for their operation on the excitation of single KAs and antenna arrays by bipolar high-voltage pulses. These studies have been carried out in two stages that are distinguished by the designs and characteristics of the KAs and bipolar pulse formers (BPFs) used in the UWB sources.

At the first stage (1993–2000), a KA combining an electric monopole and a loop analog of magnetic dipole was used together with a BPF equipped with one or two gas-filled spark gaps developed using earlier proposed circuit designs [7–9]. As a result, high-power UWB radiation sources were created based on a single KA (1994) and on a two-element array (1995) excited by bipolar voltage pulses of duration 4 ns at a pulse repetition frequency of 50 Hz. Subsequently, high-power UWB sources were created based on a single KA [10–12] and on a four-element array [13, 14] excited by 3-ns bipolar voltage pulses that were capable to produce radiation pulses with an effective potential of 100 kV and 500 kV, respectively, at a pulse repetition frequency of 100 Hz.

The disadvantages of the first version of a combined antenna were the presence of cross-polarized radiation, low peak power and peak field efficiencies [15] and low electric strength at short bipolar voltage pulses due to the small electrode gaps. The BPFs with one and two spark gaps used in first UWB sources had the obstacle to reducing the bipolar pulse duration to less than 3 ns.

In UWB sources being developed at the second stage, a KA with extended bandwidth [16] and a BPF based on a new circuit [17] are used. New versions of the KA, which is a combination of a TEM horn and (active and passive) magnetic dipoles, and of the BPF are free, to a large extent, of the above disadvantages. This has made it possible to use bipolar voltage pulses of duration as short as 200 ps and produce UWB pulses with multimegavolt effective potentials. The relevant studies are discussed below.

The efforts were focused on the development of high-power sources of linearly polarized UWB radiation. To extend the application area of high-power UWB sources, sources radiating orthogonally polarized field pulses have also been developed. Along with this, the possibilities to control the source characteristics, such as the radiation pattern and spectrum, at high peak power levels have been investigated.

9.1   The Limiting Effective Radiation Potential of a UWB Source

As previously mentioned, the main parameter used to assess the performance of a UWB source is the effective radiation potential defined as the product of the peak electric field strength by the distance in the far-field region, rE_p. Consider the problem of maximizing the amplitude of the field produced by an arbitrary antenna at a given observation point in the far-field region a given time following [18, 19] and compare the well-known UWB radiation sources using antennas of different types.

Assume that the maximum linear dimension of an antenna is such that it fits entirely within an imaginary sphere of radius a. The antenna is excited by a pulse of limited bandwidth and the input energy W is completely radiated by the antenna. It is well known [20] that the field of the antenna outside the sphere (for a harmonic radiation mode) can be represented as multipole expansions whose coefficients are determined by the distributions of electric and magnetic current densities within the sphere. It is convenient to use the following representations for the only non-zero radial components of the electric $[A_r^e(r,\theta,\varphi)]$ and the magnetic vector potential $[A_r^m(r,\theta,\varphi)]$ [20]:

$$A_r^e(r,\theta,\varphi) = \sum_{n=0}^{\infty}\sum_{m=0}^{n} a_{mn}h_n(kr)P_n^m(\cos\theta)\cos(m\varphi + \alpha_{mn}) \tag{9.1}$$

$$A_r^m(r,\theta,\varphi) = \sum_{n=0}^{\infty}\sum_{m=0}^{n} b_{mn}h_n(kr)P_n^m(\cos\theta)\cos(m\varphi + \beta_{mn}) \tag{9.2}$$

where r,θ,φ are the spherical coordinates of the observation point, $h_n(kr)$ are the spherical Hankel functions of the second kind expressed in Debye's terms, k is the wave number, and $P_n^m(\cos\theta)$ are the associated Legendre functions; the coefficients a_{mn}, b_{mn} are functions of frequency independent of the location of the observation point; α_{mn}, β_{mn} are constants determining the radiation polarization; the time dependence of the fields is described by $\exp(i\omega t)$.

The electromagnetic field components transverse to the radial direction that are needed for the subsequent analysis can be expressed in terms of the potentials $A_r^e(r,\theta,\varphi)$ and $A_r^m(r,\theta,\varphi)$ as

$$\begin{aligned} E_\theta = \frac{1}{r}\sum_{n=0}^{\infty}\sum_{m=0}^{n} \Bigg\{ &-a_{mn}iZ_0h_n'(kr)\frac{dP_n^m(\cos\theta)}{d\theta}\cos(m\varphi + \alpha_{mn}) \\ &+ b_{mn}\frac{m}{\sin\theta}h_n(kr)P_n^m(\cos\theta)\sin(m\varphi + \beta_{mn}) \Bigg\} \end{aligned} \tag{9.3}$$

$$E_{\varphi} = \frac{1}{r} \sum_{n=0}^{\infty} \sum_{m=0}^{n} \left\{ a_{mn} i Z_0 \frac{m}{\sin\theta} h'_n(kr) P_n^m(\cos\theta) \sin(m\varphi + \alpha_{mn}) \right.$$
$$\left. + b_{mn} h_n(kr) \frac{dP_n^m(\cos\theta)}{d\theta} \cos(m\varphi + \beta_{mn}) \right\} \tag{9.4}$$

$$H_{\theta} = \frac{1}{r} \sum_{n=0}^{\infty} \sum_{m=0}^{n} \left\{ -b_{mn} \frac{i}{Z_0} h'_n(kr) \frac{dP_n^m(\cos\theta)}{d\theta} \cos(m\varphi + \beta_{mn}) \right.$$
$$\left. - a_{mn} \frac{m}{\sin\theta} h_n(kr) P_n^m(\cos\theta) \sin(m\varphi + \alpha_{mn}) \right\} \tag{9.5}$$

$$H_{\varphi} = \frac{1}{r} \sum_{n=0}^{\infty} \sum_{m=0}^{n} \left\{ b_{mn} \frac{i}{Z_0} \frac{m}{\sin\theta} h'_n(kr) P_n^m(\cos\theta) \sin(m\varphi + \beta_{mn}) \right.$$
$$\left. - a_{mn} h_n(kr) \frac{dP_n^m(\cos\theta)}{d\theta} \cos(m\varphi + \alpha_{mn}) \right\} \tag{9.6}$$

where Z_0 is the characteristic impedance of the free space surrounding the antenna and $h'_n(kr)$ is the derivative of the nth Hankel function with respect to the total argument.

Choosing a spherical coordinate system so that the direction $\theta = 0$ would coincide with the optimization direction of the radiation field and taking into account the equations

$$\frac{dP_n^m(\cos\theta)}{d\theta}\bigg|_{\theta=0} = \frac{mP_n^m(\cos\theta)}{d\theta}\bigg|_{\theta=0} = \left\{ \begin{array}{cc} 0 & m \neq 1 \\ -\dfrac{n(n+1)}{2} & m = 1 \end{array} \right\} \tag{9.7}$$

we arrive at the conclusion that the maximum field strength in the direction $\theta = 0$ is determined only by the coefficients a_{1n} and b_{1n} entering into the expressions for the field components E_{θ}, E_{φ}, H_{θ}, H_{φ}. Hence, the coefficients a_{mn} and b_{mn} with $m \neq 1$ determine the energy transferred in the directions $\theta \neq 0$. Therefore, subsequently, in the procedure of maximizing the electric field amplitude at a fixed energy at the antenna input, we will assume $a_{mn} = b_{mn} = 0$ ($m \neq 1$) in the expressions for the field components E_{θ}, E_{φ}, H_{θ}, H_{φ}. We can also assume without loss of generality that the field radiated in the direction $\theta = 0$ is polarized along the x-axis associated with the spherical coordinate system, so that we have $\alpha_{1n} = \pi$ and $\beta_{1n} = \pi/2$.

With the above constraints and assumptions, $E_{\theta}(r,\theta,\varphi,\omega)$ in the plane $\varphi = 0$ is described by the expression

$$E_{\theta}(r,\theta,0,\omega) = \frac{1}{r} \sum_{n=0}^{\infty} \left\{ a_{1n} i Z_0 h'_n(kr) \frac{dP_n^1(\cos\theta)}{d\theta} + b_{1n} h_n(kr) \frac{1}{\sin\theta} P_n^1(\cos\theta) \right\} \tag{9.8}$$

Physically, the individual terms of the expansion (9.8) can be interpreted as the natural waves of a free-space waveguide. These waves satisfy the radiation condition at infinity and are characterized by the existence of a critical frequency, allowing one to restrict the number of significant terms of the expansion. Thus, the problem of elimination of physically unrealizable current distributions leading to a superdirective radiation mode is naturally solved. It is important that the number of significant terms in the expansion is directly related to the volume occupied by the radiation sources. Namely it is necessary to sum up N expansion terms that provide an acceptable value of the antenna quality factor. The number N can be chosen in various ways. For narrowband antennas, it is generally assumed that $N = [\omega_0 a/c]$, where the symbols [...] imply the integer part of the corresponding number; ω_0 is the center cyclic frequency of the radiation spectrum, and c is the velocity of light in free space. If an antenna radiates in a wide frequency band, N should increase with ω but in such a way as to prevent the occurrence of a superdirectivity mode. This is the recipe used in the study [18]. However, analysis of the results presented by Giri et al and Yankelevich and Pokryvailo [21, 22] has shown that it is more correctly to choose N in accordance with the criterion $N = [\omega_0 a/c + 2\pi]$. This provides more accurate estimates of the maximum electric field, especially for small-sized antenna, such as the KA, and for UWB pulses whose wavelengths near the lower bound of the frequency band can be substantially greater than the dimension a of the antenna.

Suppose that an antenna operating in a nonstationary excitation mode radiates a pulse in a limited frequency band characterized by the ratio $2\Delta\omega/\omega_0$. Using the notation $\Omega = \{\omega: -\omega_0 - \Delta\omega < \omega < -\omega_0 + \Delta\omega; \ \omega_0 - \Delta\omega < \omega < \omega_0 + \Delta\omega\}$, we can describe the field produced by the antenna in a region $r > a$ by the equation

$$E_\theta(r,\theta,0,\omega) = \frac{1}{2\pi} \int_\Omega E_\theta(r,\theta,0,\omega) \exp(i\omega t)\, d\omega \tag{9.9}$$

in which integration is performed over both positive and negative frequencies.

The total radiated energy is determined as

$$W = \frac{1}{2\pi} \int_\Omega \oint_S \left[\mathbf{E}(r,\theta,0,\omega),\ \mathbf{H}^*(r,\theta,0,\omega) \right] \mathbf{n}\, ds\, d\omega \tag{9.10}$$

where S is the surface area of a sphere of radius a and $\mathbf{n}$ is the external normal to the surface. Using the expressions for the field components and the orthogonality relations for the trigonometric functions and associated Legendre functions $P_n^1(\cos\theta)$ and integrating over the surface S, we obtain

$$W = \int_\Omega \left\{ \sum_{n=1}^{N(\omega)} \frac{n^2(n+1)^2}{2n+1} \left[i|a_{1n}|^2 Z_0 h_n'(ka) h_n^*(ka) - \frac{1}{Z_0} |b_{1n}|^2 h_n(ka) h_n'^*(ka) \right] \right\} d\omega \tag{9.11}$$

Introducing new coefficients

$$A_n = i^{n+1} Z_0 n(n+1) a_{1n}, \quad B_n = i^{n+1} n(n+1) b_{1n} \tag{9.12}$$

we see that (9.8), for $\theta = 0$, $kr \to \infty$, and (9.11) become symmetrical with respect to the coefficients A_n and B_n, and, therefore, $|E_\theta(r,0,0,\omega)|$ will be a maximum on condition that $A_n = B_n$. If this condition is fulfilled and $kr \to \infty$, (9.8) becomes

$$E_\theta(r,\theta,0,\omega) \approx \frac{\exp(-ikr)}{r} \sum_{n=1}^{N(\omega)} \frac{A_n}{n(n+1)} L_n(\theta) \tag{9.13}$$

where $L_n(\theta) = -(dP_n^1(\cos\theta)/d\theta) - (P_n^1(\cos\theta)/\sin\theta)$ and $L_n(0) = n(n+1)$

Accordingly, in time domain we have

$$E_\theta(r,\theta,0,\tau) = \frac{1}{2\pi r} \int_\Omega \sum_{n=1}^{N(\omega)} \frac{A_n}{n(n+1)} L_n(\theta) \exp(i\omega\tau)\, d\omega \tag{9.14}$$

where $\tau = t - r/c$.

Equation (9.11) for the radiated energy, with the Wronskian relation applied to the functions $h_n(ka)$ and $h_n^*(ka)$, becomes

$$W = \frac{2}{Z_0} \int_\Omega \sum_{n=1}^{N(\omega)} \frac{|A_n|^2}{2n+1}\, d\omega \tag{9.15}$$

We next optimize the amplitude of the field radiated in the direction $\theta = 0$ at a time $\tau = 0$. The functional for the electric field amplitude in the far-field region in the direction $\theta = 0$ at $\tau = 0$ is convenient to determine by writing (9.14) in terms of scalar product as [18]

$$E_\theta(r,0,0,0) = \left([A]^t, [F]\right) \tag{9.16}$$

where $[A]$ is the column vector composed of A_n, $[F]$ is the column vector with $F_n = 1/2\pi r$, and the symbol t denotes the operation of transposition. The scalar product is defined as

$$\left([A]^t, [B]\right) = \int_\Omega [A]^t [B]^*\, d\omega \tag{9.17}$$

The expression for the energy (9.15) can be represented as [18]

$$W = \left([A]^t, [H][A]\right) \tag{9.18}$$

where $[H]$ is a square diagonal matrix with elements $H_{mn} = 2\delta_{mn}/Z_0(2n+1)$ and δ_{mn} is the Kronecker symbol.

The field magnitude determined by (9.3) can be maximized for a given energy W if the following equation holds:

$$\nabla\left(E_\theta(r,0,0,0) - \lambda W\right) = 0 \tag{9.19}$$

Its solution yields the expression for the coefficients

$$A_n = \frac{Z_0(2n + 1)}{8\pi\lambda r} \tag{9.20}$$

where λ is the Lagrangian multiplier selected so that (9.15) would yield a specified value of W.

The algorithm for calculating the limiting effective radiation potential (in this case, the product of r by E_θ) is as follows: on substitution of (9.20) in (9.15), the Lagrangian multiplier λ is found. Thus, the values of the optimal coefficients A_n are completely determined:

$$A_n = \sqrt{Z_0 W} \frac{(2n + 1)}{\sqrt{2 \int\limits_\Omega \left[\sum\limits_{n=1}^{N(\omega)} (2n + 1) \right] d\omega}} \tag{9.21}$$

Substitution of them in (9.14) yields the final calculation [19]

$$rE_\theta(r,0,0,0) = \frac{\sqrt{Z_0 W}}{2\sqrt{2\pi}} \sqrt{\int\limits_\Omega \left[\sum\limits_{n=1}^{N(\omega)} (2n + 1) \right] d\omega} \tag{9.22}$$

We first compare the limiting effective radiation potentials estimated by (9.22) with $N = [\omega_0 a/c + 2\pi]$ [19] and $N = [\omega_0 a/c]$ [18]. The effective potential ratio $rE_{2\pi}/rE_0$ versus antenna electric length is given in Figure 9.1 for radiators with a 10% (curve 1) and a 150% frequency band (curve 2).

It can be seen that this quantity can be over 10 for electrically small antennas and tends to 1 for large antennas. Thus, the use of N chosen in [19] yields more accurate estimates of the limiting effective radiation potential for any antenna, especially for small antennas excited by UWB pulses.

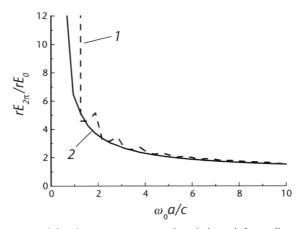

Figure 9.1 Effective potential ratio versus antenna electric length for radiators with a 10% (curve 1) and a 150% frequency band (curve 2).

It is of interest to use the chosen criterion to compare high-power UWB radiation sources with well-known types of antenna, such as the IRA [21], TEM [22], and KA [23]. Based on the data reported in the mentioned papers, the energy and the radiation spectrum were estimated for the IRA and the energy and the input voltage pulse spectrum for the TEM and KA. In the latter case, the antenna energy efficiency was assumed to be 100%. The bandwidth was estimated for the -10-dB level, and $\omega_0 = 2\pi f_0$ was taken as an average in this band: $f_0 = (f_L + f_H)/2$, where f_L and f_H are the lower and the upper bound frequency, respectively. The spectrum width for the test antennas ranged between 160% and 190%. The characteristics of the UWB sources for which the limiting effective radiation potential was estimated are given in Table 9.1.

For comparative analysis, $rE_\theta = rE_{2\pi}$ was estimated for UWB sources with different radiators and the efficiency factor was obtained as the ratio of the measured effective radiation potential to its limiting value: $k = rE_{exp}/rE_\theta$. The factor k for the IRA, TEM, and KA equals, respectively, 0.5, 0.35, and 0.26 (represented by points in Figure 9.2). For a source with a 16-element KA array excited by a 2-ns bipolar pulse [23], k was estimated to be 0.58 and for a source with a 64-element KA array excited by a 1-ns bipolar pulse [24] to be 0.65. These results are also shown by points

Table 9.1   Characteristics of UWB Radiation Sources with Single Antennas

	KA [23]	IRA [21]	TEM [22]
τ_p, ns	2	—	1
W, J	1.17	0.015	0.175
f_L, MHz	100	50	50
f_H, MHz	934	2900	2110
f_0, MHz	517	1475	1080
$\Delta f/f_0$, %	161	193	190

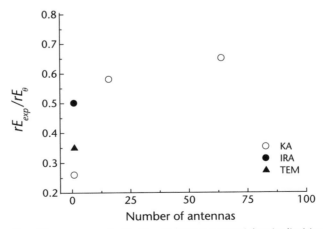

Figure 9.2   The ratio of the measured effective radiation potential to its limiting value versus the number of combined antennas in the array. The points represent measurements for single antennas of different types (KA, IRA, and TEM) and for a 16-element and a 64-element KA array. (Reprinted with permission Cambridge University Press.)

in Figure 9.2. In the calculations, it was assumed, like for single antennas, that there was no energy loss in the antenna–feeder system. From the above data, it follows that the efficiency of a radiator based on a multielement KA array increases with the number of elements in the array, despite the energy loss in the feeder system.

Estimates also show that for KAs and KA-based arrays, the ratio rE_θ/V, where V is the volume of the radiator, is greater than that for radiators of other types.

9.2   A Bipolar High-Voltage Pulse Generator

9.2.1   A Monopolar Voltage Pulse Generator

A generator of bipolar voltage pulses consists of a monopolar pulse generator and a BPF. The monopolar pulse generator, a coaxial line with a built-in Tesla transformer, is intended to charge the BPF. A series of generators of this type, collectively called SINUS [1, 25], was developed and built at the Institute of High Current Electronics (Tomsk). In most of the UWB radiation sources that are discussed below, the SINUS-160 monopolar pulse generator [17, 23] was used. Its design is shown schematically in Figure 9.3.

The pulse-forming line (1) with a built-in Tesla transformer (TT) is placed in a stainless-steel cylindrical case filled with transformer oil. The open ferromagnetic core (5) of the transformer is made of 0.08-mm-thick segments of electrical strip steel. The transformer cores serve as the conductors of the pulse-forming line. The primary winding (6) comprises one turn made of 0.3-mm-thick copper strip. The secondary winding (7) is applied on a smooth cone made of 0.3-mm-thick electrical pressboard and contains about 800 turns of 0.18-mm-diameter wire. For measuring the operating parameters of the generator, two voltage dividers are provided: a capacitive divider of the pulse-forming line charging voltage (D_1) and a capacitive divider of the transmission line voltage (D_2). The main parameters of the generator [17] are given in Table. 9.2.

For a high-voltage switch in the generator, an uncontrolled spark gap with hemispherical electrodes (2) is used. The working gas is nitrogen pressurized to 11 atm. When the generator operates at a pulse repetition frequency of 100 Hz, the gas is forced to flow through the electrode gap across the discharge channel using

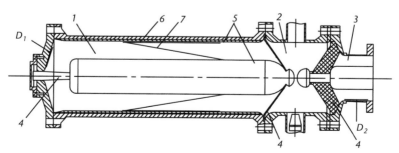

Figure 9.3 Schematic of the SINUS-160 high-voltage generator: 1—pulse-forming line, 2—uncontrolled gas-filled gap, 3—transmission line, 4—insulator, 5—ferromagnetic core, 6—primary winding, 7—secondary winding, D_1—capacitive divider of the pulse-forming line charge voltage, and D_2—capacitive voltage divider in the transmission line. (Reprinted with permission Cambridge University Press.)

Table 9.2 Parameters of the SINUS-160 Generator (Reprinted with permission Cambridge University Press.)

Electric length (TWTT)	4.5 ns
Characteristic impedance	40 Ω
Line capacitance	66 pF
Maximum charge voltage U_2^{max}	350 kV
Energy storage capacity	4 J
Charge voltage at 100 Hz	310 kV
Peak voltage across a 50-Ω load	150 kV
Time of charging to U_2^{max}	5 μs
Tesla transformer efficiency	50%
Frequency detuning between the transformer circuits	1.3

a blowdown system comprising a fan, a collimator, and a gas-cooling device. The transmission line (3) serves to transfer the generated high-voltage pulse to the BPF. The generator offers three operating modes: single-pulse operation, repetitive pulse operation at a pulse repetition frequency of 100 Hz, and operation with repetition frequency adjustable from 1 to 100 Hz.

The schematic circuit diagram of the generator is shown in Figure 9.4. The transmission line incorporates the following additional elements: a grounding inductor (L_0) to bring the potential of the transmission line center conductor to zero by the time of generation of the next voltage pulse, and a limiting resistor (R_0) to absorb the

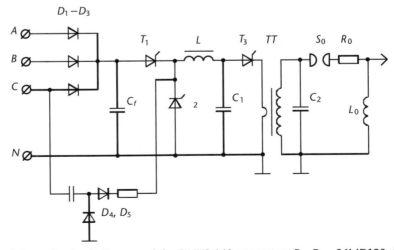

Figure 9.4 Schematic circuit diagram of the SINUS-160 generator: D_1–D_3—26MD120 diodes, D_4, D_5—D112-25-14 diodes, T_1, T_2—IRKU91/12 thyristors, and T_3—two parallel-connected TBI 153-1000 thyristors. (With permission from Pleiades Publishing, Ltd.)

excess energy after the operation of the chopping gap of the BPF. When the UWB pulse source operates for a long time at a pulse repetition frequency of 100 Hz, a significant average power is dissipated in resistor R_0; therefore, a cooling radiator is installed on the outer case of the lossy line.

The generator is powered from three-phase mains. The mains voltage rectified by D_1–D_3 diodes is applied to filter C_f, and simultaneously, capacitor C_1 is pre-charged via voltage multiplier D_4, D_5.

Consider one operation cycle of the generator connected according to the circuit shown in Figure 9.4. Initially, the filter capacitor (C_f) is charged to the rectified mains voltage (310–340 V). The primary capacitor of the Tesla transformer (C_1) is charged via the voltage multiplier to +600 V. The operation cycle begins as control pulse is applied to the main thyristor switch (T_3). On operation of switch T_3, capacitor C_1 is connected to the primary winding of the transformer, and the charging of the pulse-forming line capacitor (C_2) begins. As the coupling coefficient between the circuits $k \approx 1$, the voltage across capacitor C_2 varies according to the law

$$U_2 \approx U_1 \frac{N_2}{N_1} \frac{\alpha}{\alpha + 1}\left[1 - \cos\left(\frac{\pi t}{t_{\max}}\right)\right] \qquad (9.23)$$

where N_1, N_2 are the numbers of turns in the primary and the secondary circuit of the Tesla transformer, respectively; $\alpha = 1.3$ is the factor of natural frequency detuning between the circuits, and $t_{\max}$ is the time at which $U_2 \approx U_2^{\max}$.

Spark gap S_0 is always adjusted so that its breakdown occurs at a time $t_0 < t_{\max}$ to have $U_2 < U_2^{\max}$. This is necessary because if the spark gap fails to operate, large-amplitude voltage oscillations arise in the secondary circuit and persist for a long time, resulting in breakdown of the pulse-forming line insulation. On operation of spark gap S_0, a voltage pulse of duration 4.5 ns and amplitude 180 kV is formed that is used to charge the BPF lines. After a pause of 100 μs required for recovery of switch T_3, a control signal is applied to switch T_2. The switch pumps back the energy remaining in C_1 after the operation of the switch through choke L, and thus, the sign of the voltage across capacitor C_1 is reversed. When switch T_1 operates, capacitor C_1 is connected, via choke L, to the filter capacitor C_f, and then, capacitor C_1 is charged to the initial voltage level. The final voltage level across capacitor C_1 can be controlled by varying the operating time of switch T_1 relative to that of switch T_2. The filter capacitor C_f is charged through diode rectifier D_1–D_3. Choke L is made based on a ferromagnetic core. This made it possible, with the relatively small size of the choke, to attain significant inductance (1.8 mH) and thus reduce the circuit current amplitudes in switches T_1 and T_2 and the losses in energy pumpback and in charging capacitor C_1. The delay and repetition frequency of the generator pulses are remotely controlled from a computer via a control unit that generates triggering pulses for switches T_1, T_2, and T_3.

9.2.2 A Bipolar Pulse Former with an Open Line

A pulse former with an open line was proposed to produce bipolar high-voltage pulses of duration $\tau_p = 1$ns in a UWB radiation source with a single KA [17]. The characteristic impedance of the coaxial feeder connecting the BPF and the antenna, ρ_f, is 50 Ω. The design features of the BPF were retained in subsequent UWB sources with bipolar pulses of duration 1 ns [26], 2 ns [23], and 3 ns [27] used to excite both single antennas [23, 26, 27], through a 50-Ω feeder, and 16-element arrays [23, 26], through a 50/3.125-Ω wave transformer and a power divider.

Consider the features of operation of the proposed pulse former by the example of the BPF generating bipolar voltage pulses of duration $\tau_p = 2$ ns [23]. The schematic circuit diagram of the pulse former is shown in Figure 9.5. The circuit contains four coaxial lines (FL_0–FL_3), a peaking gap (S_1), a chopping gap (S_2), an isolating inductor of inductance $L_1 = 250$ nH, and a load (R_L). Lines FL_0–FL_2 have electric length (one-way travel time) $\tau = 0.35$ ns and characteristic impedance $\rho = 25$ Ω. Transmission line FL_3 with an electric length of 3.7 ns and characteristic impedance ρ_f is matched to the load, so that $\rho_f = R_L = 2\rho = 50$ Ω. Pulse-forming line FL_0 is charged from a monopolar pulse generator, via charging inductor L_1, to a voltage $-U_0$. On operation of peaking gap S_1, a negative voltage pulse of amplitude $-U_0/2$ propagates through line FL_1 that is loaded at the end by in-series connected lines FL_2 and FL_3 whose total impedance equals 3ρ. The voltage wave entering line FL_3 produces a bipolar pulse-negative wave of amplitude $-U_0/2$ across the load. In a time 2τ at which the reflected wave arrives at the beginning of line FL_1 and the wave traveling through FL_2 reaches its end, chopping gap S_2 operates. Now, a positive voltage relieving wave propagates through line FL_1 and a doubled positive wave reflected from the open end propagates through line FL_2. These two voltage waves arrive, respectively, at the output of FL_1 and at the input of FL_2, are summed up, and produce a bipolar pulse positive half-wave of amplitude $U_0/2$. The transient phenomena cease within a time equal to 4τ. Thus, with perfectly operating spark gaps S_1 and S_2, a bipolar voltage pulse of amplitude $\pm U_0/2$ and duration 4τ arises across the load.

The operation of the pulse former circuit was numerically simulated using the well-known PSPICE code. The switching time, that is, the time during which the spark gap resistance decreases from 100 kΩ to 0.01 Ω was set equal to 1 ns for S_1 and S_2. Resistor R_1 of resistance 1000 Ω is required only to realize the calculations. The inductance of charging inductor L_1 was chosen so that the charge voltage across line FL_0 would be close to a maximum at a minimum charging time. The charging voltage across line FL_0 reaches a maximum of 555 kV within 5 ns. When spark gap S_1 operates at about the charge voltage maximum and S_2 operates with a relative

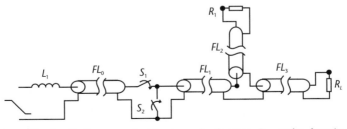

Figure 9.5 Schematic circuit diagram of a bipolar pulse former: *FL*—pulse-forming lines, *S*—spark gaps, *L*—charging inductor, and *R*—resistors. (With permission from Pleiades Publishing, Ltd.)

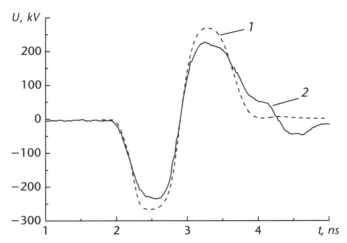

Figure 9.6 The calculated (curve 1) and the measured voltage pulse waveform (curve 2) at the pulse former output. (With permission from Pleiades Publishing, Ltd.)

delay of 0.7 ns, a bipolar voltage pulse of amplitude 270 kV and duration 1.8 ns is formed across the load (Figure 9.6, curve 1).

The pulse former design is presented schematically in Figure 9.7. Caprolon bushing insulators separate the working chamber into two sections, one filled with gas and the other with oil. A charging inductor (L_1), four coaxial lines (FL_0–FL_3), a peaking gap (S_1), and a chopping gap (S_2) are located in the section filled with nitrogen pressurized to 70–90 atm. The inner conductors of lines FL_0–FL_2 have diameters 33, 33, and 16 mm, respectively. Line FL_2 is insulated with PTFE. The ends of the inner conductors of lines FL_0 and FL_1 are the electrodes of peaking gap S_1; a disk of thickness 2 mm and an insert in the outer conductor of lines FL_0 and FL_1 are the electrodes of chopping gap S_2. The spark gap electrodes used in this and in the other pulse formers are made of copper to reduce erosion and, hence, increase the lifetime of the BPF.

Transmission line FL_3 connects the pulse former output to a resistive load or a radiating system (not shown in Figure 9.7). The left and right parts of line FL_3 are insulated with gas and oil, respectively. The voltage pulse at the pulse former output was recorded using a TDS 7404 oscilloscope with the help of a coupled line voltage divider [28] mounted in the oil-filled part of line FL_3.

At a nitrogen pressure of 90 atm in the pulse former chamber and at a 1.4-mm and a 1.2-mm gap spacing in spark gaps S_1 and S_2, respectively, the output voltage pulse shown in Figure 9.6 (curve 2) has two approximately equal amplitudes of 230 kV. As the chopping gap operates at a high rate of rise of voltage at the electrodes ($\sim 5 \cdot 10^{14}$ V/s), a multichannel switching mode with subnanosecond time and high stability is realized. The RMSD of the bipolar voltage pulse amplitude is less than $\sigma = 0.03$ for the RMSD of the generator pulse amplitude $\sigma = 0.01$ during the first hour of operation. Within 1–2 hours of operation of the spark gaps at a repetition frequency of 100 Hz, the pulse amplitude became unstable and the pulse waveform is distorted due to erosion of the spark gap electrodes. The pulse amplitude stability and waveform are almost completely restored by reducing the pressure in the pulse

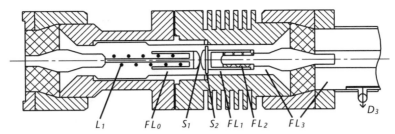

Figure 9.7 Schematic of a former of 2-ns bipolar pulses: *FL*—pulse-forming lines, *S*—spark gaps, *D*—voltage dividers, and *L*—charging inductor. (With permission from Pleiades Publishing, Ltd.)

former by 5–10 atm. As a result, the total time of operation at a frequency of 100 Hz without replacement of the spark gap electrodes reaches 5 hours with $\sigma = 0.05$.

Note that the formation of 1-ns bipolar voltage pulses has some features. In the experiment [17], a bipolar voltage pulse was used in which the amplitude of the second half-wave was much greater than that of the first half-wave. To reduce the difference between the amplitudes by reducing the amplitude of the second half-wave, an inductor of inductance $L_2 \approx 200$ nH was connected to the end of line FL_2 (Figure 9.5) [26].

To reduce the bipolar pulse duration from 1 ns to 0.5 ns [29] and to 200 ps [30], an intermediate line with a gas-filled spark gap was connected between the mono-polar pulse generator and the BPF to reduce the BPF charging time and increase the stability of the output pulse. A peaking section was also used to reduce the characteristic impedance of the output feeder, ρ_f, from 50 to 12.5 Ω for bipolar pulses of duration 1 ns [31] and 3 ns [32]. In this case, the characteristic impedance of the BPF lines ($\rho = 6.25$ Ω) accordingly decreased fourfold. Note that the efficiency of conversion of the energy stored in the pulse-forming line of the SINUS-160 generator into the bipolar pulse energy increases with pulse duration and reaches 30% for $\tau_p = 3$ ns [32]. The results show that every bipolar pulse duration and every amount of energy stored in the BPF call for a special monopolar pulse generator with an optimal energy stored in the pulse-forming line.

Consider the operation of a pulse former with a peaking section by the example of the picosecond bipolar pulse generator [30]. The generator consists of the SINUS-160 monopolar pulse generator, an intermediate peaking stage, and a BPF. In the schematic circuit diagram of the bipolar voltage pulse generator shown in Figure 9.8, the monopolar pulse generator is represented by pulse-forming line FL_0 and

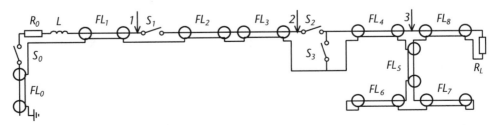

Figure 9.8 Schematic circuit diagram of a 200-ps bipolar voltage pulse generator: *FL*—pulse-forming lines, *S*—spark gaps, *R*—resistors, and *L*—charging inductor. (With permission from Pleiades Publishing, Ltd.)

spark gap S_0. This 40-Ω line was charged to a voltage of -360 kV from the secondary winding of a Tesla transformer at a pulse repetition frequency of 100 Hz.

The intermediate peaking stage consists of line FL_1, limiting resistor R_0, charging inductor L, and spark gap S_1. The BPF incorporates lines FL_2–FL_8, peaking gap S_2, chopping gap S_3, and load R_L. A similar generator circuit was used [30] to produce bipolar voltage pulses of duration 0.5 ns. The key point of this BPF circuit is that line FL_5 is connected between the ground conductors of lines FL_4 and FL_8, and its output is loaded by two series-connected high-impedance lines FL_6 and FL_7. The parameters of lines FL_6 and FL_7 were chosen by calculations taking into account their design features. The advantage of this bipolar pulse forming circuit is that the solid dielectric insulating the open line, which was generally used in BPFs, was replaced by a gas, and this reduced the line losses and simplified the BPF design.

The operation of the generator circuit was simulated using a computer. The spark gaps were assumed to perform nearly perfect switching. Pulse-forming line FL_0 is switched by spark gap S_0 to intermediate line FL_1 via limiting resistor R_0 and charging inductor L. The charge voltage across line FL_1 reaches a maximum of 500 kV within 3.1 ns (Figure 9.9, curve 1). The points for which the waveforms were calculated are indicated in Figure 9.8. Spark gap S_1 operates as the charging voltage reaches a maximum and connects line FL_1, via high-resistance line FL_2, to pulse-forming line FL_3 that is charged to 600 kV within 220 ps (curve 2). Once spark gap S_2 operates at a maximum charge voltage across line FL_3 and then spark gap S_3 operates with a relative delay time equal to two-way travel time for line FL_4, a bipolar pulse of amplitude ±350 kV and duration 200 ps (curve 3) is formed in transmission line FL_8 with characteristic impedance ρ_f whose end is loaded by a matched load $R_L = \rho_f = 50\ \Omega$. After the passage of the bipolar pulse through transmission line FL_8, voltage oscillations are observed across the line that are induced by reflections of the voltage wave from the short-circuited ends of lines FL_6 and FL_7. The oscillation amplitude decreases with increasing the characteristic impedances and electric lengths of lines FL_6 and FL_7, making up no more than 10% of the bipolar voltage pulse amplitude.

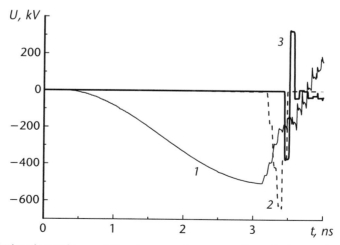

Figure 9.9 Calculated waveforms of the charge voltage across lines FL_1 (curve 1) and FL_3 (curve 2) and of the bipolar voltage pulse passing in line FL_8 (curve 3). (With permission Pleiades Publishing, Ltd.)

The design of the intermediate peaking stage and BPF is shown schematically in Figure 9.10. It consists of three gas-filled sections separated by polycarbonate insulators (1 and 3). In the first section filled with nitrogen pressurized to 90 atm, a disk line (FL_1), a peaking gap (S_1), and a capacitive voltage divider (D_1) are mounted. The charging voltage pulse is applied to line FL_1 from the monopolar pulse generator via a charging inductor (L). In the second section filled with nitrogen or hydrogen pressurized to 90–100 atm, lines FL_2–FL_7, a portion of transmission line FL_8, peaking gap S_2, and chopping gap S_3 are located. The ends of the inner conductors of lines FL_3 and FL_4 are the electrodes of spark gap S_2, and a 1-mm-thick disk (2) and the inner conductor of line FL_4 are the electrodes of spark gap S_3. The electrode gap spacing was varied. To reduce the amplitude of the voltage oscillations induced by the passed bipolar pulse, 10 35VCh17 ferrite rings of dimensions $39 \times 17 \times 6$ mm (7) are mounted in lines FL_6 and FL_7. The third section filled with SF_6 gas pressurized to 3 atm incorporates the second portion of FL_8 and a coupled line voltage divider (D_2) intended to detect the output bipolar voltage pulse.

The adjustment of the bipolar voltage pulse generator was reduced to sequentially setting up the gap spacing and pressure in spark gaps S_1–S_3. The operation delay times for spark gaps S_1 and S_2 were chosen so that their breakdown occurred without misses at a nearly maximum charging voltage. Then, by adjusting the gap spacing in chopping gap S_3, a symmetrical waveform of the output bipolar pulse was attained. At an optimum gas pressure and an optimum gap spacing, spark gap S_3 was broken down with a delay of 2.7 ns, which is close to the calculated delay time.

Figure 9.11 presents the voltage waveform at the output of voltage divider D_2 recorded for the second section filled with nitrogen pressurized to 90 atm. The electrode spacing in spark gaps S_2 and S_3 was 0.4 and 0.2 mm, respectively. The leading edge of the bipolar voltage pulse is extended due to the prepulse cause by the passage of the voltage pulse charging pulse-forming line FL_3 through the transfer capacitance of spark gap S_2 (Figure 9.8). The pulse duration, estimated at a 0.1 amplitude level by a linear approximation of the leading edge to the intersection with the zero line, was 230 ps. The maximum pulse amplitude differs threefold from the calculated value. This is due to the simplifying assumptions used in the simulation and to the operation of the spark gaps at a voltage below the maximum charging voltage. For the second section filled with hydrogen at 100 atm, the electrode spacing in spark

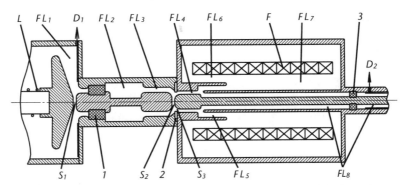

Figure 9.10 Schematic of the intermediate peaking stage and 200-ps bipolar voltage pulse former: 1, 3—bushing insulators, 2—electrode, S—spark gaps, FL—pulse-forming lines, F—ferrite rings, D—voltage dividers, and L—charging inductor. (With permission Pleiades Publishing, Ltd.)

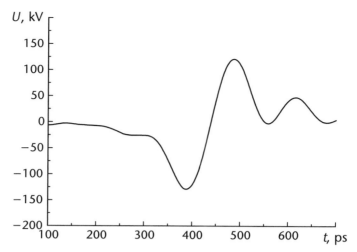

Figure 9.11 Waveform of the output bipolar voltage pulse. (With permission from Pleiades Publishing, Ltd.)

gap S_2 was increased to 0.8 mm (the greatest possible value). In this case, bipolar voltage pulses of amplitude -50 kV and $+35$ kV and duration 260 ps were obtained. The decrease in amplitude is due to the lower electric strength of hydrogen compared to nitrogen. In the subsequent investigations, hydrogen was not used.

The difficulties in the development of BPFs increase dramatically in attempting to simultaneously reduce the bipolar pulse duration and increase its peak power. This is due to the increasing probability of electrical breakdowns that reduce the BPF limiting parameters. Consider a BPF producing voltage pulses of duration 1 ns and peak power 3.2 GW [31, 24]. In the first BPF version [31], the pulse-forming line consisted of two series-connected disk lines, and the lines were switched by a ring-shaped spark gap of small diameter (15 mm). The bipolar pulse, whose amplitude reached 200 kV at a repetition frequency of 100 Hz, was highly unstable due to the electrical breakdowns occurred in the BPF. In the second BPF version [24], to increase the electric strength, a coaxial pulse-forming line and a ring spark gap of diameter 68 mm were used. The results of the investigations performed with the second BPF version are discussed below.

The bipolar voltage pulse generator consists of a monopolar pulse generator, an intermediate peaking stage, and a BPF. For a monopolar pulse generator, the SINUS-200 high-voltage pulse generator was used. In the schematic circuit diagram of the bipolar voltage pulse generator shown in Figure 9.12, the monopolar pulse generator is represented by an output pulse-forming line of characteristic imped-ance 28.3 Ω and electric length 3.9 ns (FL_0) and a spark gap (S_0). The line could be charged from the secondary winding of a Tesla transformer to a maximum voltage of 485 kV within 4 μs at a pulse repetition frequency of 100 Hz. The intermediate peaking stage consists of a limiting resistor (R_0), a leakage inductor (L), a trans-mission line (FL_1), an intermediate line (FL_2), and a spark gap (S_1). Transmission line FL_1 of variable characteristic impedance (from 65 to 88 Ω) connects spark gap S_0 with line FL_2. The bipolar pulse former is connected in an open-line circuit that incorporates lines FL_3–FL_7, peaking gap S_2, chopping gap S_3, and a load of rated resistance 12.5 Ω (R_L). Leakage inductor L serves to remove residual charge

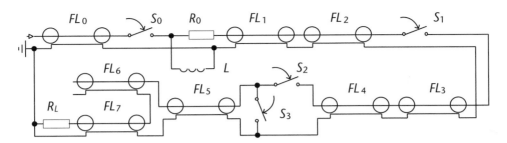

Figure 9.12 Schematic circuit diagram of a 1-ns bipolar voltage pulse generator: *FL*—pulse-forming lines, *S*—spark gaps, *R*—resistors, and *L*—inductors. (Reprinted with permission Cambridge University Press.)

from the electrodes of spark gaps S_0 and S_1 by the time of the next voltage pulse generation. Using resistor R_0 makes it possible to moderate voltage oscillations in the FL_0–S_0–R_0–FL_1–FL_4–S_1–S_2 circuit after the formation of a bipolar pulse and reduce erosion of the spark gap electrodes.

Figure 9.13 presents schematically the design of the intermediate peaking stage and BPF. It consists of two gas-filled sections and one oil-filled section. In the first section filled with nitrogen at 85 atm, lines FL_1 and FL_2, spark gap S_1, and capacitive voltage divider D_1 are located. Line FL_1 is not shown in Figure 9.13. The electrodes of spark gap S_1 are set with a gap spacing of 2.8 mm. The second section filled with nitrogen at 87 atm comprises lines FL_4 and FL_4, capacitive voltage divider D_2, and spark gaps S_2 and S_3. Lines FL_3 and FL_6 and the left portion of line FL_7 are insulated with caprolon. The electrode spacing in spark gaps S_2 and S_3 is 1.7 and 1 mm, respectively. The characteristic impedance of each of lines FL_2, FL_4, and FL_5 is 6.25 Ω. The right part of transmission line FL_7 is an oil-filled coaxial line of characteristic impedance 12.5 Ω.

The charging voltage pulses produced by the monopolar pulse generator arrived at line FL_2. On sequential operation of spark gaps S_1–S_3, the output voltage was transferred, via transmission line FL_7, to load R_L. The adjustment of the bipolar voltage pulse generator was reduced, as in previous cases, to sequentially setting up the gap spacing and pressure in spark gaps S_1–S_3. The operation delay times for spark gaps S_1 and S_2 were chosen so that their breakdown occurred without misses at a nearly maximum charging voltage of lines FL_2 and FL_4 at a pulse repetition frequency of 100 Hz. Then, by adjusting the gap spacing in chopping gap S_3, a

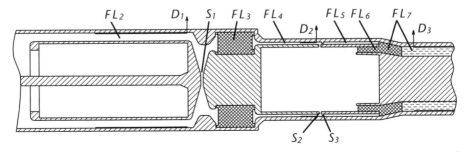

Figure 9.13 Schematic of the intermediate peaking stage and 1-ns bipolar voltage pulse former: *FL*—pulse-forming lines, *S*—spark gaps, and *D*—voltage dividers. (Reprinted with permission from Cambridge University Press.)

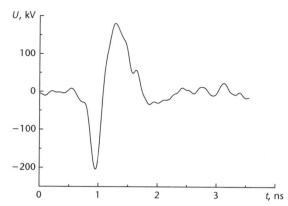

Figure 9.14 Waveform of the output voltage pulse. (Reprinted with permission from Cambridge University Press.)

symmetrical waveform of the output bipolar pulse was attained. The breakdown delay time for spark gaps S_1 and S_2 was 7.3 and 1.7 ns, respectively. The output bipolar voltage pulse recorded at the output of voltage divider D_3 is shown in Figure 9.14. Its amplitudes are −205 kV and +180 kV and the duration at 10% amplitude is 1 ns. The voltage amplitude jitter is no more than 4%. The pulse energy dissipated in the matched load R_L = 12.5 Ω is 1.2 J, making up 9% of the energy stored in pulse-forming line FL_0.

Owing to the high rise rate of voltage at the electrodes of spark gaps S_2 and S_3, multichannel switching was achieved for the 6.25-Ω lines with ring spark gaps of diameter 68 mm. The ratio $D/\tau_p c$ did not exceed the criterial level $1/\left(\pi\sqrt{\varepsilon_r}\right)$ for the excitation of higher modes (8.5). The rise rate of the negative voltage at the BPF output was a factor of 1.4 greater than that attained for the BPF used in the study [31].

9.3   Single-Antenna Radiation Sources

High-power UWB radiation sources with a single combined antenna [17, 23, 26, 27, 29, 30] have been designed based on the standard scheme comprising the SINUS-160 monopolar pulse generator, a bipolar pulse former, and an antenna. The duration of bipolar voltage pulses was varied in the range 0.2–3 ns and the amplitude in the range 100–230 kV. To prevent electrical breakdown, the antennas were placed in dielectric containers filled with SF_6 gas pressurized to 2.2–5 atm, depending on the pulse duration and amplitude. Typically, the dielectric containers were fabricated of polyethylene, and the gas pressure in them was not above 2.5 atm. Only for the source with an antenna excited by a 0.5-ns bipolar voltage pulse of amplitude 200 kV [29], the container was made of caprolon and the SF_6 gas pressure reached 5 atm. The dielectric containers had a thickness $\sim\lambda_0/30$, where $\lambda_0 = \tau_p c$, and had little effect on the radiation characteristics.

The antenna dimensions were varied in accordance with the similarity law to be about half the spatial length of the bipolar pulse. However, the antenna lead and the bushing insulator were designed by recognizing that the antenna would operate at high-voltage amplitudes. In some cases, this increased the voltage standing wave ratio (VSWR) to $K_V \leq 3$ in the voltage pulse frequency band. The design of the combined antenna was varied in the course of accumulation of service experience. To optimize the antenna geometry, the 4NEC2 code was used, and the VSWR of high-voltage antennas was reduced to $K_V \leq 2$ (see Section 7.5).

Let us briefly summarize the main results of investigations performed for the single-antenna UWB sources. The energy, peak power, and peak field efficiencies of the high-voltage antennas were, respectively, $k_w = 0.85$–0.93, $k_p = 0.6$–1.2, and $k_E = 0.9$–2. The boresight directivity achieved $D_0 = 4$–5.7. The methods used for estimating the above parameters are described elsewhere [33] (Section 2.4.). The half-peak-power pattern width in the H- and E-planes varied in the ranges 75–90° and 75–110°, respectively. With the radiation sources continuously operated for up to 5 h at a pulse repetition frequency of 100 Hz, UWB radiation pulses with a peak power of 120–1200 MW and an effective potential of 100–440 kV were obtained.

Consider in more detail only one UWB radiation source with a single antenna excited by a 3-ns bipolar voltage pulse [27]. The choice is dictated by the available data on the stability of the BPF output pulse in relation to the monopolar charging voltage pulse, which is important for the study discussed elsewhere. The appearance of the UWB radiation source is shown in Figure 9.15. As mentioned earlier, the source consists of three main components: a monopolar pulse generator (1), a bipolar pulse former (2), and a transmitting antenna enclosed in a dielectric container (3). The container is filled with SF_6 gas at 2.4 atm.

In the schematic circuit diagram of the bipolar voltage pulse generator (Figure 9.16), the SINUS-160 monopolar pulse generator is represented by the output pulse-forming line (FL_0) and the spark gap (S_0). The arrows indicate the points at which the voltage pulses were recorded using voltage dividers D_0–D_2. The BPF incorporates lines FL_1–FL_4, a peaking gap (S_1) and a chopping gap (S_2). A bipolar voltage pulse is formed in line FL_4 whose end is connected to load R_L, as spark gap S_1 operates at a nearly maximum charging voltage of line FL_1 and then spark gap S_2 operates with a relative delay equal to the two-way travel time of the voltage pulse through line FL_2. On operation of spark gap S_0, the charging voltage pulse passes

Figure 9.15 Appearance of a UWB radiation source with a single antenna excited by a 3-ns bipolar voltage pulse: 1—SINUS-160 monopolar pulse generator, 2—bipolar pulse former, and 3—antenna enclosed in a dielectric container. (With permission from David Publishing Company.)

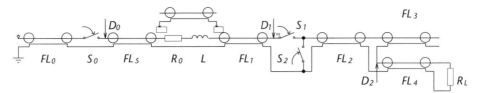

Figure 9.16 Schematic circuit diagram of a 3-ns bipolar pulse generator: *FL*—pulse-forming lines, *S*—spark gaps, *R*—resistors, *L*—inductor, and *D*—voltage dividers. (With permission from David Publishing Company.)

via transmission line FL_5 and arrives at line FL_1. Three charging modes were used: (1) via a limiting resistor of resistance $R_0 = 6\ \Omega$ and an inductor of inductance $L = 850$ nH ($L + R_0$ mode), (2) via line FL of characteristic impedance 66 Ω and electric length 0.14 ns and a resistor R_0 ($FL + R_0$ mode), and (3) via line FL (FL mode).

The design of the bipolar voltage pulse former is shown schematically in Figure 9.17. Three coaxial lines FL_1–FL_3, peaking gap S_1, and chopping gap S_2 are placed in a case filled with nitrogen pressurized to 90 atm. The charging voltage pulse generated by the SINUS-160 generator arrives at line FL_1 via inductor L (top of Figure 9.17) or via line FL (bottom of Figure 9.17). Transmission line FL_4 connects the output of the bipolar pulse former with resistive load R_L or an antenna (not shown in Figure 9.17).

An oscilloscope and voltage dividers D_0–D_2 were used to record the waveforms of the incident voltage wave generated by the SINUS-160, the voltage charging line FL, and the output voltage of line FL_4, respectively. Figure 9.18 shows the voltage waveforms taken from voltage divider D_1 for three modes of charging line FL_1.

The maximum charging voltage of line FL_1, determined by the hold-off voltage of spark gap S_1, was 310–340 kV, and the charging time varied between 3.1 and 9.7 ns. The amplitudes of the bipolar voltage pulse varied from ±150 to ±170 kV; its duration was 3 ns. The amplitude jitter of the incident voltage wave was 1.5%. The operation time jitter Δt of spark gap S_1 was estimated by the following simple method. The oscilloscope was started by the negative leading edge of the bipolar voltage pulse. The operation time jitter of spark gap S_1 depends on the rise rate of voltage at the electrodes and equals 80–100 ps for the charging of line FL_1 via (high-impedance) line FL. The measurement data for three modes of charging line FL_1 are given in Table 9.3.

The data of Table 9.3 show that the charging time and jitter are minimal for the FL mode of charging. However, the absence of a limiting resistor results in an increased current on operation of chopping gap S_2 and, consequently, in a shorter

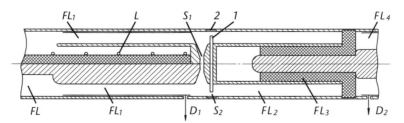

Figure 9.17 Schematic diagram of a 3-ns bipolar pulse generator: *FL*—pulse-forming lines, *S*—spark gaps, *D*—voltage dividers, *L*—inductor, and 1, 2—spark gap electrodes. (With permission from David Publishing Company.)

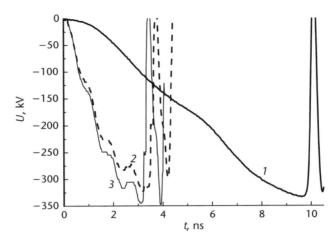

Figure 9.18   Waveforms of the voltage charging line FL_1 in different modes: $L + R_0$ (1), $FL + R_0$ (2), and FL (3). (With permission from David Publishing Company.)

Table 9.3   Parameters of Three Modes of Charging Line FL_1 (With permission from David Publishing Company)

Charging mode	Charging time, ns	Voltage rise rate, kV/ns	Jitter, ps
$L + R_0$	9.73	34	260–290
$FL + R_0$	3.44	92	120–140
FL	3.14	110	80–100

operational life of the BPF. In view of this, for the UWB radiation sources, the $FL + R_0$ charging mode was chosen. For this mode, the operation time jitter Δt of spark gap S_1 versus number of pulses is shown in Figure 9.19. It can be seen that the jitter increases slightly with the number of pulses.

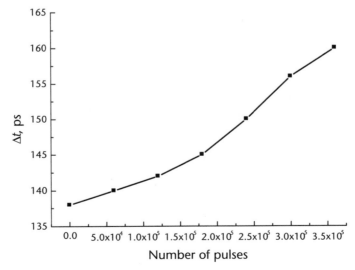

Figure 9.19   Operation time jitter versus for spark gap S_1. (With permission from David Publishing Company.)

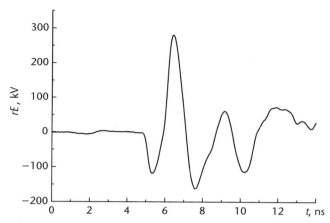

Figure 9.20 Waveform of a pulse radiated by an antenna excited by a 3-ns bipolar voltage pulse. (With permission from David Publishing Company.)

The UWB radiation source was tested for stability within an hour of continuous operation at a pulse repetition frequency of 100 Hz. The pressure in the spark gaps was 80 atm and was not varied for this hour. In the experiment, the amplitude of the electromagnetic pulse radiated in the direction of radiation maximum (rE_p), the bipolar voltage pulse amplitude, and the RMSD of the maximum amplitudes were measured. Averaging was performed over 100 pulses. Radiation pulses with an effective potential of 280 kV were produced with an RMSD of the measured amplitudes of up to 0.04 and a peak field efficiency $k_E = 1.7$. The waveform of the far-field radiation pulse (rE) is shown in Figure 9.20. The signal oscillations following the main pulse (the first three time lobes) are due to its reflections from surrounding objects. For these and subsequent measurements of the parameters of high-power UWB radiation pulses, TEM receiving antennas [34] were used.

9.4 Radiation Sources with Synchronously Excited Multielement Arrays

9.4.1 The Radiation Source with a Four-element Array

The radiation source with a four-element array has the simplest design among UWB radiation sources. In this case, a BPF with the output impedance $\rho_f = 12.5$ Ω matched to the feeder system impedance (50/4 Ω) is used, and, hence, there is no wave impedance transformer. The bipolar voltage pulse passes through a four-channel power divider and 50-Ω cable feeders and arrives at the array elements. Consider a source of this type [32] excited by a 3-ns voltage pulse.

The radiation source (Figure 9.21) consists of the SINUS-160 monopolar pulse generator (1), a BPF (2), a power divider (not shown in the figure), and a four-element antenna array (3) excited from the BPF via the power divider. The BPF pulses are applied to the array elements via cord-insulated PK 50-17-51 cable. To increase the electric strength of the cables, they were filled with SF_6 gas pressurized to 5 atm. The radiation characteristics were measured in an anechoic chamber.

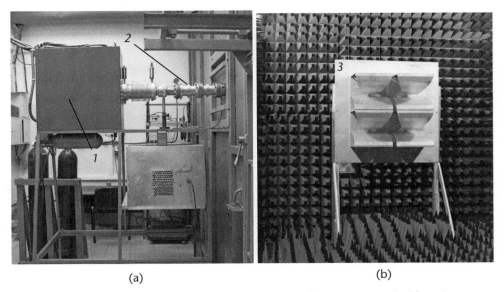

Figure 9.21 Appearance of the radiation source with a 4-element array excited by a 3-ns bipolar voltage pulse: 1—monopolar pulse generator, 2—bipolar pulse former, and 3—antenna array. (With permission from Pleiades Publishing, Ltd.)

With the electrode gap spacing in spark gaps S_1 and S_2 of the BPF equal, respectively, to 1.5 and 0.5 mm and the gas pressure equal to 65 atm, the line connected upstream of spark gap S_1 was charged to 180 kV within 7 ns. The output bipolar voltage pulse shown in Figure 9.22 (curve 1) has the negative (U_-) and the positive half-wave amplitude (U_+) equal to −83 and +90 kV, respectively. The bipolar pulse duration, determined at the $0.1U_+$ and $0.1U_-$ levels by linearly approximating the pulse trailing edge to its intersection with the zero line, was 3 ns. The energy delivered to the load in a pulse is 0.94 J, which makes up 30% of the energy stored in the pulse-forming line of the SINUS-160 generator. The operation time jitter of

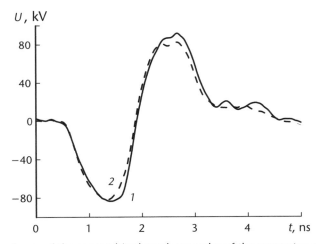

Figure 9.22 Waveforms of the output bipolar voltage pulse of the generator operated for 1 h (curve 1) and 5 hours (curve 2). (With permission from Pleiades Publishing, Ltd.)

spark gap S_1, Δt, was determined, like for a single-antenna UWB radiation source, by the RMSD of the bipolar pulse leading edge from the leading edge of the incident charging voltage pulse. As can be seen from the histogram shown in Figure 9.23, where N is the number of pulses, it was about 200 ps.

The radiating system is a four-element (2×2) array (Figure 9.21). The array elements are fastened on a dielectric plate and integrated in two vertical sections, each containing two elements. The neighboring elements in the vertical sections are galvanically connected to each other. The sections are spaced by $d_b = 50$ cm. This configuration is optimal for a KA array. The array aperture is 95×90 cm. For an array element a KA of dimensions $45 \times 45 \times 47$ cm (Figure 7.61) was used similar in design to that described elsewhere [27]. The design of the lead connecting PK 50-17-51 cables to the array antennas was changed.

For the antenna array, peak power patterns were measured in the far-field region ($r > 3.5$ m), whose boundary location was estimated by (2.72) and determined experimentally from the measurements of the relation $rE_p(r)$. The distance r was measured from the radiation center (partial phase center) that corresponded to the geometric center of the array radiating system. The pattern width at the half-peak-power level was estimated to be $35°$ (Figure 9.24). The boresight directivity of the array, D_0, was 18. The peak field efficiency of the array, k_E, measured at low voltages, was 8.3.

The UWB radiation source was tested for stability and operational life at a pulse repetition frequency of 100 Hz. Every hour of continuous operation was followed by two hours of timeout for the monopolar pulse generator would cool down. Initially, the nitrogen pressure in the spark gaps was set equal to 62 atm. Subsequently, it was reduced to compensate the increase in gap spacing caused by electrode erosion. The variation of the pressure in the course of testing is shown in Figure 9.25 (curve 3). Reducing the pressure in the BPF by 2–3 atm after each hour ensured stable operation of the spark gaps. The positive and negative half-wave amplitudes of the bipolar voltage pulse, U_+ and U_- (Figure 9.25, curves 1 and 2), and the amplitude

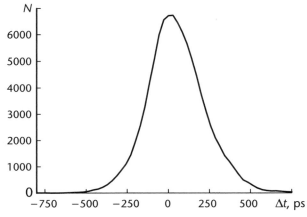

Figure 9.23 Histogram of the operation time jitter of peaking gap S_1 determined relative to the leading edge of the incident charging voltage wave. (With permission from Pleiades Publishing, Ltd.)

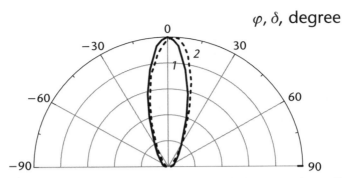

Figure 9.24   The *H*-plane (curve 1) and the *E*-plane peak power pattern (curve 2) measured for a 2 × 2 array excited by a 3-ns bipolar voltage pulse. (With permission from Pleiades Publishing, Ltd.)

of the electromagnetic pulse, rE_p, and its RMSD were measured simultaneously as functions of the number of pulses (Figure 9.26). All measurements were averaged over 100 pulses.

The RMSDs of the bipolar pulse amplitudes U_+ and U_- relative to their averages over five hours were 0.03 and 0.016, respectively. With the RMSD of the amplitude of the charging voltage incident wave over 100 pulses $\sigma = 0.01$, the RMSD of the amplitudes of the positive and negative half-waves of the bipolar pulse ranged between 0.035 and 0.05. The variation of the voltage pulse amplitude for 5 h of operation at a pulse repetition frequency of 100 Hz was not over 11%. Within the operation time, the duration of the bipolar pulse decreased to 2.7 ns (Figure 9.22, curve 2). This was due to that chopping gap S_2 operated at less than optimal delay

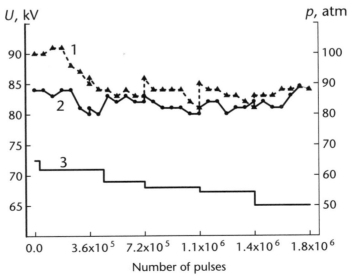

Figure 9.25   The positive (U_+, curve 1) and negative half-wave amplitudes (U_-, curve 2) of the bipolar voltage pulse and the pressure in the spark gaps (curve 3) versus number of pulses. (With permission from Pleiades Publishing, Ltd.)

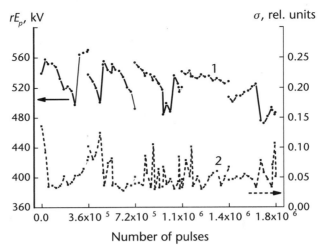

Figure 9.26 Effective radiation potential (curve 1) and its RMSD (curve 2) versus number of pulses. (With permission Pleiades Publishing, Ltd.)

times. The RMSD of the operation time of spark gap S_1 in relation to the leading edge of the charging voltage incident wave varied during the operation period in the range 180–250 ps.

As follows from Figure 9.26 (curve 2), the time it took for the radiation source to attain stable operation at the RMSD of the electromagnetic pulse amplitude (rE_p) $\sigma < 0.05$ was 5–10 min from the beginning of each hour. The average rE_p for 5 h of operation was 530 kV with an average RMSD $\sigma = 0.06$. The effective radiation potential varied within 17 %. The peak field efficiency of the UWB radiation source, k_E, was 6.2. The difference in k_E obtained in the low-voltage and high-voltage measurements is accounted for by that in the low-voltage measurements, the losses in the four-channel power divider and in the cable feeders connected between the power divider and the antenna leads were not taken into account.

Figure 9.27 presents the waveforms of electromagnetic pulses radiated by the antenna array in the first (curve 1) and fifth hours of operation (curve 2). It can be

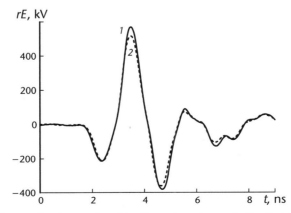

Figure 9.27 Waveforms of pulses radiated by the antenna array in the first (curve 1) and the fifth hour of operation (curve 2). (With permission from Pleiades Publishing, Ltd.)

seen that the radiation pulse waveform remains almost unchanged for nearly two million pulses.

9.4.2 Radiation Sources with 16-element Arrays

The UWB radiation pulse sources with a 16-element (4 × 4) array, connected according to the same circuit, are designed for excitation by bipolar voltage pulses of duration 2 ns [23], 1 ns [26], 0.5 ns [29], and 200 ps [30]. With these sources, radiation pulses of effective potential 0.4–1.7 MV were obtained at a pulse repetition frequency of 100 Hz. We consider only the sources that are significantly different in design and produce radiation pulses different in duration by an order of magnitude.

The radiation source [23] (Figure 9.28) comprises the SINUS-160 monopolar pulse generator (1), a BPF producing 2-ns voltage pulses (2), a wave transformer (3), and a rectangular 16-element array (4). The array elements are connected to the transformer output through a 16-channel power divider by PK 50-17-17 cable feeders insulated with solid polyethylene. The BPF design is shown in Figure 9.7.

The antenna array consists of four vertical sections of dimensions 30 × 120 cm, each containing four galvanically connected antennas. A distinctive feature of the section is that it is made not of individual antennas, but as an integral unit with a common metal back plate. The sections mounted on a dielectric plate are spaced by d_h = 36 cm. The array aperture is 138 × 120 cm.

The radiation characteristics were measured at a distance r = 9 m, which is close to the location of the far-field region boundary (r = 11 m) estimated by (2.72) and by the measured relation $rE_p(r)$ plotted in Figure 9.29. The E-plane and H-plane peak power patterns of the array are given in Figure 9.30. The patterns are symmetric, and their half-peak-power width is about 20°.

The UWB radiation source was tested for 1 h at a 100-Hz pulse repetition frequency. Note that the total number of pulses produced by the generator without replacement of the spark gap electrodes in the BPF by the beginning of testing was $2.16 \cdot 10^6$ (six-hour operation). Figure 9.31 presents the pulses at the input of an array element and the pulses radiated by the array. As can be seen from the figures,

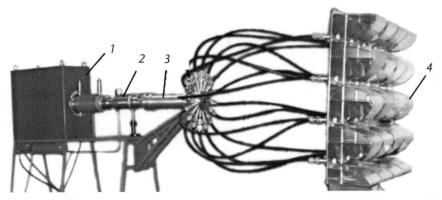

Figure 9.28 Appearance of a radiation source with a 16-element array excited by a 2-ns bipolar voltage pulse: 1—monopolar pulse generator, 2—bipolar pulse former, 3—wave transformer, and 4—antenna array. (With permission from Pleiades Publishing, Ltd.)

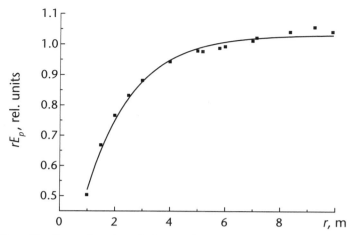

Figure 9.29 Quantity rE_p as a function of distance for the 16-element array excited by a bipolar pulse of duration 2 ns. (With permission from Pleiades Publishing, Ltd.)

erosion of the spark gap electrodes resulted in a decrease in voltage rise rate with the voltage pulse amplitude retained, reducing the amplitude of the radiated pulse. An additional factor responsible for the decrease in field amplitude might be partial electrical breakdowns of the cable feeders. Figure 9.31(b) shows that the effective radiation potential reached 1.67 MV at a pulse repetition frequency of 100 Hz. The peak field efficiency of the source, k_E, was 7.3. As the pulse repetition frequency was reduced to 1 Hz, rE_p reached 1.8 MV. The continuous operation time of the radiation source UWB at 100 Hz, limited by the breakdown of the coaxial cables, was no more than 1 h.

Subsequently, a UWB radiation source [35] was created which consists of a nine-element (3 × 3) array of KAs excited by a 2-ns bipolar voltage pulse and a gas-insulated feeder system. The source used the SINUS-160-30 monopolar pulse

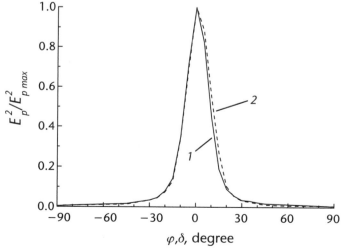

Figure 9.30 The H-plane (curve 1) and the E-plane peak power pattern (curve 2) of a 16-element array excited by a 2-ns bipolar voltage pulse. (With permission from Pleiades Publishing, Ltd.

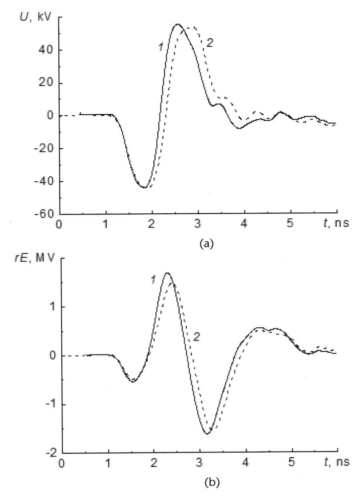

Figure 9.31 Voltage pulses at the input of an array element (a) and the electromagnetic pulses radiated by the 16-element array in the beginning (curve 1) and at the end of testing (curve 2). (With permission from Pleiades Publishing, Ltd.)

generator in which the characteristic impedance of the high-voltage pulse-forming line was reduced, compared with that used in the SINUS-160 generator, from 40 to 30 Ω. The output impedance of the bipolar pulse generator was also reduced from 50 to 12.5 Ω. The feeder system, comprising a wave impedance transformer (12.5/5.56 Ω), a nine-channel power divider and cord-insulated PK 50-17-51 cables filled with SF_6 gas pressurized to 5 atm. In experiments, UWB radiation pulses with an effective potential of 1 MV were produced at a 100-kV amplitude of the bipolar voltage pulse and a 100-Hz pulse repetition frequency. The peak field efficiency of the source, k_E, was equal to 10, which is about 1.4 times higher than that of the previous source with a 16-element array. The continuous operation time of the source, similar to the previously discussed UWB radiation source [32], was limited by erosion of the BPF electrodes.

Of considerable interest are the results of investigations of a UWB radiation source with an antenna array excited by a 200-ps bipolar voltage pulse [30]. The

Figure 9.32 Appearance of a UWB radiation source with a 16-element array excited by a 200-ps bipolar voltage pulse: 1—monopolar pulse generator, 2—bipolar pulse former, 3—wave transformer, 4—power divider, and 5—antenna array. (With permission from Pleiades Publishing, Ltd.)

appearance of the radiation source with a 16-element array is shown in Figure 9.32. The source comprises the SINUS-160 monopolar pulse generator (1), a bipolar pulse former (2), a wave transformer (3), a power divider (4), and an antenna array (5). The array elements are connected to the transformer output through a 16-channel power divider with cord-insulated PK 50-7-58 cable feeders. The feeder system was filled with SF_6 gas pressurized to 5 atm, which provided its electric strength. The design of the BPF used for this source is shown in Figure 9.10.

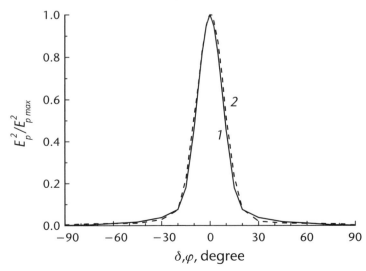

Figure 9.33 The *H*-plane (curve 1) and *E*-plane (2) peak power patterns of the 16-element array excited by a 200-ps bipolar voltage pulse. (With permission from Pleiades Publishing, Ltd.)

The array consists of separately fabricated KAs of dimensions $4 \times 4 \times 4.3$ cm fastened on a dielectric plate (Figure 8.16). The array elements are integrated in vertical sections of four elements each. The neighboring elements in the vertical section are galvanically connected. The spacing between the sections, d_h, is 4.4 cm. The array aperture is 17.2×16 cm.

The radiation characteristics were measured in the far-field region at distances of more than 6 m. The array peak power patterns are shown in Figure 9.33. The pattern width at half maximum is 18° in the horizontal H-plane (curve 1, solid line) and 20° in the vertical E-plane (curve 2, dashed line). According to the low-voltage measurements with a 200-ps bipolar pulse, the array directivity D_0 is 50. The peak field efficiency of the radiator, k_E, was equal to 3.

Figure 9.34 shows the waveforms of the pulses radiated by the array for two modes of operation of the BPF. Peaking gap S_2 (Figures 9.8 and 9.10) operates at a maximum charging voltage in the first mode and within the charging voltage rise time in the second mode. For the first mode, the effective radiation potential was 370 kV (curve 1) at the bipolar voltage pulse amplitude equal to 130 kV and duration equal to 230 ps (Figure 9.11), and for the second mode, it was 270 kV (curve 2) at the bipolar voltage pulse amplitude equal to 100 kV and duration equal to 215 ps. The peak field efficiency in these measurements reached $k_E = 2.8$. The duration of the radiation pulses at half maximum was 80 ps for both modes. The measurements were performed using a 6-GHz bandwidth Tektronix TDS 6604 oscilloscope.

The source was tested for stability and operational life at a pulse repetition frequency of 100 Hz. Every hour of continuous operation was followed by a two-hour break needed for the monopolar pulse generator to cool down. Initially, the nitrogen pressure in the BPF spark gaps was 95 atm. To compensate for the increase in gap spacing due to electrode erosion, the gas pressure was reduced. The pressure values for both operating modes of the UWB source are given in Figure 9.35. In

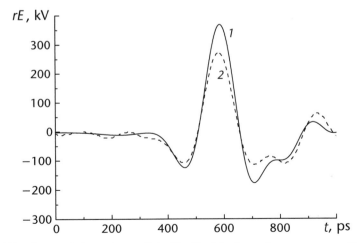

Figure 9.34 Waveforms of the pulses radiated by the 16-element array source with the BPF operated in the first (curve 1) and in the second mode (curve 2). (With permission from Pleiades Publishing, Ltd.)

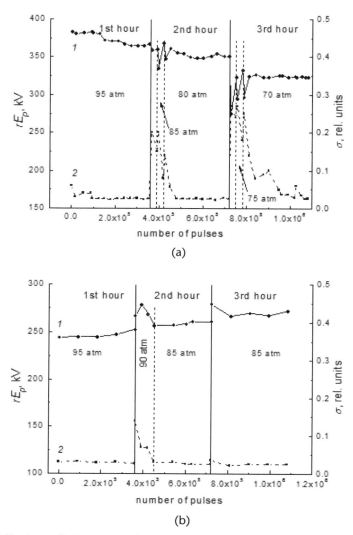

Figure 9.35 Effective radiation potential (curve 1) and its RMSD (curve 2) versus number of pulses for the first (a) and the second operating mode (b). (With permission from Pleiades Publishing, Ltd.)

the experiment, the electromagnetic pulse amplitude (rE_p) and its RMSD (σ) were measured. Averaging was performed over 100 pulses.

The results for the first mode are presented in Figure 9.35(a). To attain stable operation $(\sigma \leq 0.03)$, a time of up to 20 min is required. During the second hour of operation, the instability of the radiation increases for the first 20 min. During the third hour, the radiation instability increases significantly. The effective radiation potential decreases within three hours of operation of the UWB source from 380 to 320 kV. This is because the spacing of spark gap S_2 increases due to electrode erosion and it operates later that the charging voltage reaches a maximum.

To increase the stable operation time, the amplitude of the exciting bipolar pulse was reduced by adjusting spark gap S_2 to operate within the rise time of the charging voltage. The results for the second mode are presented in Figure 9.35(b). For this mode, the effective radiation potential increased from 245 to 270 kV within three hours of operation of the UWB source. This was because the gap spacing increased due to electrode erosion, and spark gap S_2 operated at a charging voltage closer to the maximum. The radiation was stable to within $\sigma \le 0.03$ with the exception of the period with a pressure of 90 atm during the second hour of operation.

More correct measurements of the characteristics of the UWB radiation source with an antenna array were performed in an anechoic chamber using a 12.5-GHz bandwidth Tektronix MSO 70000 oscilloscope. With a bipolar voltage pulse of duration 230 ps and amplitude 120 kV, radiation pulses of duration 70 ps at half maximum and effective potential 450 kV were produced at a pulse repetition frequency of 100 Hz. The peak field efficiency $k_E = 3.7$ was attained.

It seemed interesting to compare the above results with those obtained for the known UWB sources producing radiation pulses of duration ~100 ps at half maximum by using a parameter defined as the ratio of the effective radiation potential to the area of the radiator aperture. The estimates have shown that the UWB source [30] under consideration outperforms in this parameter by 1–2 orders of magnitude the radiation sources using antennas with a parabolic reflector [21, 36] and arrays of transmitting TEM antennas [37, 38] excited by monopolar voltage pulses. This is due to the compact KAs excited by bipolar voltage pulses used in the radiation source [30]. These estimates are consistent with the results of the comparative analysis of the radiators [19], given at the beginning of this chapter.

9.4.3   A Radiation Source with a 64-element Array

Structurally, the source under consideration [31, 24] (Figure 9.36) consists of the SINUS-200 monopolar pulse generator, a 1-ns bipolar pulse former, a power divider

Figure 9.36 Appearance of the radiation source with a 64-element array excited by a 1-ns bipolar voltage pulse: 1—monopolar pulse generator, 2—bipolar pulse former, 3—power divider, and 4—antenna array. (Reprinted with permission Cambridge University Press.)

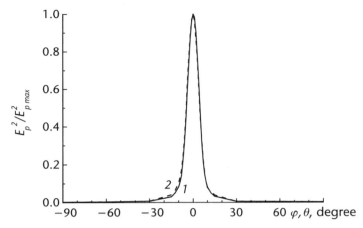

Figure 9.37 The *H*-plane (curve 1) and the E-plane peak power pattern (curve 2) of the 64-element array excited by a 1-ns bipolar voltage pulse. (With permission from Springer.)

with an impedance transformer, and a 64-element array. The schematic circuit diagram and design of the BPF [24] are presented, respectively, in Figures 9.12 and 9.13.

The 8×8 array consists of 64 separately fabricated KAs of dimensions $15 \times 15 \times 16.5$ cm fastened on a metal plate (it was discussed in detail above, see Figure 8.22). The array aperture is 141×141 cm. Figure 9.37 shows the radiation patterns of the 64-element array [39]. The pattern width at half peak power is 10° in both planes. Measurements were performed at distances over 10 m, but closer to the location of the far-field region boundary ($r = 26.7$ m) estimated by (2.72). The relation $rE_p(r)$ is plotted in Figure 9.38. It can be seen that the relation yet did not come to saturation at a distance of 13 m from the array.

In the first experiments [31], a BPF based on radial lines was used, and the feeder system was insulated with transformer oil. The array was excited by a bipolar voltage pulse of amplitude 200 kV. The radiated pulse was detected using a TEM antenna, installed at a distance of 10.7 m, and a Tektronix TDS 6604 oscilloscope. Figure 9.39

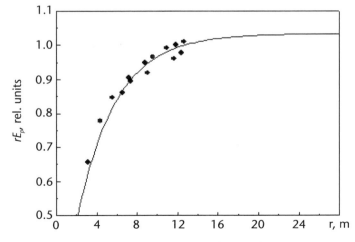

Figure 9.38 Quantity rE_p as a function of distance for the 64-element array excited by a bipolar pulse of duration 1 ns. (Reprinted with permission Cambridge University Press.)

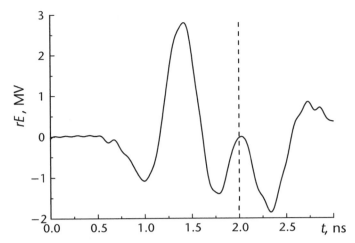

Figure 9.39 Waveform of the pulse radiated by the source with a 64-element array.

shows the waveform of the radiated pulse. The curve on the right of the dashed line represents the waveform of the pulse reflected from the surrounding metal objects. The effective radiation potential was 2.8 ± 0.2 MV at a pulse repetition frequency of 100 Hz. The radiation instability was caused largely by electrical breakdowns in the BPF. The peak field efficiency of the radiation source, k_E, was equal to 14.

Subsequently, a new BPF design was developed [24] in which the conical forming lines were replaced by cylindrical ones. In addition, BM-1 vacuum oil was used instead of transformer oil for insulation in the feeder system (Figure 8.19). This made it possible to reduce power loss in the power divider. The nitrogen pressure in the spark gaps of the intermediate peaking stage and of the BPF was the same (84–85 atm). The waveform of the bipolar voltage pulse at the input of the 64-element array is shown in Figure 9.14. The radiated pulse was detected using a TEM antenna, installed at a distance of 10.5 m, and a 30-GHz bandwidth LeCroy Wave-Master 830Zi oscilloscope. Figure 9.40 shows the waveform of the radiated pulse.

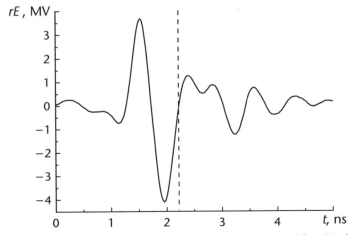

Figure 9.40 Waveform of the pulse radiated by the upgraded source with a 64-element array. (Reprinted with permission Cambridge University Press.)

The curve on the right of the dashed line represents the reflections from the surrounding metal objects imposed on the pulse. In these measurements, the product of the peak field strength by the distance reached 4.1 MV.

As mentioned earlier, the distance to the boundary of the far-field region for a 64-element array excited by a 1-ns bipolar pulse, estimated by (2.72), is $r = 26.7$ m. As the measurements of rE_p for the 64-element antenna array under consideration were carried at a distance of 10.5 m, the relation $rE_p(r)$ was extrapolated up to the boundary of the far-field region using the data previously obtained for a 4×4 array excited by a 2-ns bipolar voltage pulse (Figure 9.29) and the data available for the array under consideration (Figure 9.38). The results show that the effective radiation potential determined in the far-field region is 5.8% greater than rE_p ($r = 10.5$ m) and equals 4.3 MV with the radiation source peak field efficiency $k_E = 21$. Note that k_E of the IRA-based UWB source [35] is not over 6.

The UWB radiation source under consideration was investigated for stability and continuous operation time at a pulse repetition frequency of 100 Hz (Figure 9.41). Every 10 min of continuous operation was followed by an hour break needed for the BPF to cool down. Initially, the nitrogen pressure in the spark gaps was 84–86 atm, and after ten minutes of operation, it was increased by 2 atm. As a result, rE_p decreased by 20%. The amplitudes of the bipolar voltage pulse also decreased, but its duration remained unchanged. Cooling of the BPF did not lead to full restoration of the original value of $rE_p = 4.1$ MV. The radiation instability was substantial at the beginning of 10-minute pulse packets and the RMSD decreased to $\sigma = 0.03$. To ensure long-term continuous operation of the radiation source with small variations in rE_p, it is necessary to perform forced cooling of the bipolar pulse former and control the pressure in the spark gaps.

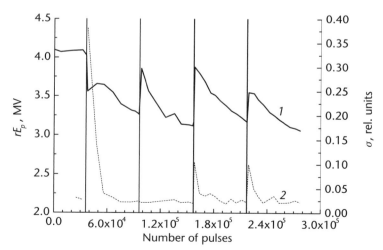

Figure 9.41 Effective radiation potential (curve 1) and its RMSD (curve 2) versus number of pulses for the UWB radiation source excited by a 1-ns bipolar voltage pulse. (Reprinted with permission Cambridge University Press.)

9.5   Production of Orthogonally Polarized Radiation Pulses

Multielement antenna arrays excited by bipolar high-voltage pulses via cable feeders open prospects for the production of high-power electromagnetic pulses with orthogonal polarizations. For this purpose, an array is divided into two subarrays with elements arranged at 90° relative to each other. The cable feeders of the two subarrays are different in electric length by the duration of the radiated pulse or more. In this case, the radiated electromagnetic pulses are separated in free space.

Low-voltage investigations of arrays of this type excited by 1-ns bipolar pulses are discussed in Section 8.3. Their geometries are shown in Figure 8.29(a–c) for the arrays labeled, respectively, as AA1–AA3. Array AA1 is an array version with synchronously excited elements (4 × 4) whose electric field vectors have the same direction. The radiation sources with arrays of this type are described in detail earlier. Arrays of types AA2 and AA3 were used to produce high-power electromagnetic pulses with orthogonal polarizations. Each subarray consisted of eight elements arranged in a straight line or diagonally.

The appearance of the UWB radiation source [26] with an AA2-type array is shown in Figure 9.42. The array consists of antennas of dimensions 15 × 15 × 16 cm optimized for excitation by 1-ns bipolar voltage pulses. The array elements are fastened on a dielectric plate with the element spacing $d_v = d_h = 18$ cm. A 200-kV bipolar pulse (Figure 9.43) was applied from the BPF to the array elements at a frequency of 100 Hz through a wave transformer with a 16-channel power divider and PK 50-17-17 cable feeders. For an AA1-type array, the electric lengths of the cables are the same, and for arrays of type AA2 and AA3, they differ by 2 ns. Typical waveforms of the pulses radiated by an AA2-type array with vertical and horizontal polarizations, spaced in time by 2 ns, are given in Figure 9.44.

Table 9.4 presents the amplitudes of bipolar pulses at the inputs of the array elements (U) and the effective radiation potentials (rE_p) for different array types.

Figure 9.42 Appearance of a source producing orthogonally polarized radiation pulses: 1—monopolar pulse generator, 2—bipolar pulse former, 3—wave transformer, and 4—16-element array consisting of two subarrays. (With permission from Pleiades Publishing, Ltd.)

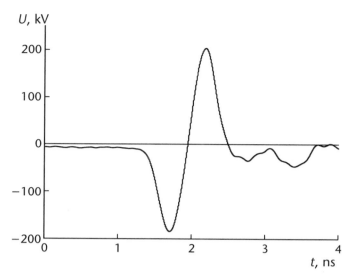

Figure 9.43 Waveform of a 1-ns bipolar voltage pulse at the output of a 50-Ω BPF. (With permission from Pleiades Publishing, Ltd.)

The increase in bipolar pulse amplitudes for the AA2-type array is due to changes in the operating conditions of the BPF. In this case, the pulse duration decreased by ~10%.

Note that sources of orthogonally polarized UWB radiation pulses spaced in time can be used both in radar, for instance, to detect objects on the background of randomly inhomogeneous surface [40, 41], and in research on their action on electronic systems. In the latter case, the conditions for testing electronic systems may be more stringent.

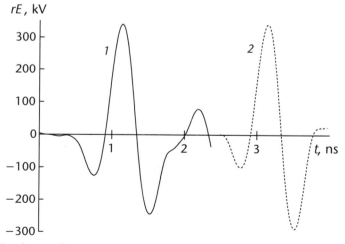

Figure 9.44 Waveforms of a vertically polarized (curve 1) and a horizontally polarized electromagnetic pulse (curve 2) radiated by an AA2-type array. (With permission from Pleiades Publishing, Ltd.)

Table 9.4 Parameters of voltage and radiation pulses. (With permission from Pleiades Publishing, Ltd.)

Array type	U, kV	Vertical polarization, rE_p, kV	Horizontal polarization, rE_p, kV
AA1	+44/−33	+780/−600	—
AA2	+55/−40	+340/−240	+340/−280
AA3	+44/−33	+320/−260	+290/−220

9.6   A Four-Channel Source Radiating in a Controlled Direction

The UWB radiation sources with a single bipolar voltage pulse former are limited in peak power by the hold-off voltage of the BPF. In this connection, a new UWB radiation source circuit was proposed [42, 27] and realized [43, 44], which comprises a monopolar pulse generator, a multichannel bipolar pulse former, and an array with the number of antennas equal to the number of independent BPFs. The circuit does not contain a wave transformer with a power divider, and the exciting voltage pulse is supplied to each array element from an individual BPF. The radiation stability of this type of source is determined by the stability of operation of the (uncontrolled) pulse formers.

Thus, possibilities arise for further increasing the total peak power of the multichannel pulse former and for controlling the radiation direction using pulse formers producing time-shifted voltage pulses of the same duration. Using multichannel pulse formers producing bipolar voltage pulses of different duration and the corresponding combined antennas opens additional prospects. The use of radiation sources of this type will make it possible to synthesize radiation pulses with an arbitrary waveform and, hence, an arbitrary spectrum in the far-field region by summing up electromagnetic pulses with different frequency bands [45, 46].

The radiation source [43, 44] consists of the SINUS-160 monopolar pulse generator, four BPFs, and a 4-element array whose elements are excited immediately from the BPFs. The voltage pulses produced by the BPFs are applied to the array elements via coaxial lines and cord-insulated PK 50-17-51 cables that are filled with SF_6 gas pressurized to 5 atm. The radiation characteristics were measured in an anechoic chamber.

The bipolar pulse generator consists of the SINUS-160 high-voltage monopolar pulse generator, an intermediate peaking stage, and a unit of four BPFs. In the schematic circuit diagram of the bipolar voltage pulse generator shown in Figure 9.45, the SINUS-160 generator is represented by an output pulse-forming line (FL_0) and a spark gap (S_0). Line FL_0 was charged from the secondary winding of a Tesla transformer to a voltage of −360 kV with a pulse repetition frequency of 100 Hz and switched by spark gap S_0, through a high-impedance line (FL_1) and a limiting resistor (R_0), to an intermediate line (FL_2).

Spark gap S_1 operates at a near-maximum voltage and switches line FL_2, via distribution line FL_3, to lines FL_{41}–FL_{44} of respective bipolar pulse formers F_1–F_4.

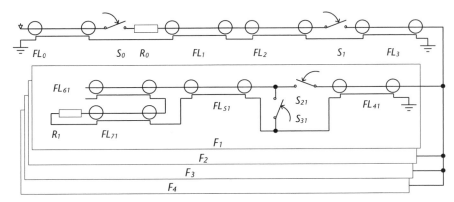

Figure 9.45 Schematic circuit diagram of a voltage pulse generator with a four-channel bipolar pulse former: *FL*—lines, *S*—spark gaps, R_1–R_4—loads, R_0—limiting resistor, and F_1–F_4—pulse formers. (With permission from Pleiades Publishing, Ltd.)

All pulse formers are connected in circuits with open lines FL_{61}–FL_{64}. Once peaking gaps S_{21}–S_{24} operate at near-maximum voltages and chopping gaps S_{31}–S_{34} operate with a relative delay equal to the two-way travel time for lines FL_{51}–FL_{54}, bipolar voltage pulses of duration 3 ns are formed in transmission lines FL_{71}–FL_{74} loaded at the ends by 50-Ω loads R_1–R_4.

Structurally, the intermediate peaking stage with the BPF unit (Figure 9.46) consists of six gas-filled sections separated by caprolon insulators (1 and 2). In the first section filled with nitrogen at 28–36 atm, lines FL_1–FL_3, spark gap S_1, and capacitive charging voltage divider D_1 are located. Four identical isolated sections bordered by insulators 1 and 2, which are filled with nitrogen at 25–35 atm, contain lines FL_{41}–FL_{44}, FL_{51}–FL_{54}, and FL_{61}–FL_{64}; spark gaps S_{21}–S_{24} and S_{31}–S_{34}, and capacitive charging voltage dividers D_{21}–D_{24}. Four identical transmission lines FL_{71}–FL_{74} with integrated coupled-line voltage dividers D_{31}–D_{34} and loads

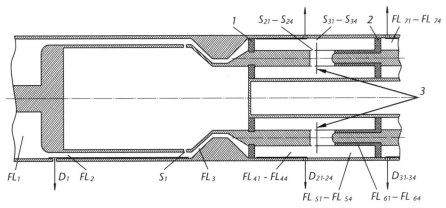

Figure 9.46 Schematic diagram of the intermediate peaking stage with a 4-channel BPF unit: *FL*—lines, *S*—spark gaps, *D*—voltage dividers, 1, 2—insulators, and 3—disk electrodes. (With permission from Pleiades Publishing, Ltd.)

R_1–R_4 are insulated with SF_6 gas at 3 atm. The pressure in the sections can be controlled independently.

The ends of the inner conductors of lines FL_2, FL_3 and FL_{41}–FL_{44}, FL_{51}–FL_{54} serve as the electrodes of ring peaking gaps S_1 and S_{21}–S_{24}, respectively. The electrode spacing in spark gaps S_1 and S_{21}–S_{24} is 1.4 and 0.6 mm, respectively. The electrodes of chopping gaps S_{31}–S_{34} are 2-mm-thick disks (3) and cylindrical inserts mounted with 0.5-mm spacing on the outer conductors of lines FL_{51}–FL_{54}.

The switching spark channels in spark gap S_1 shift over the electrodes (70 mm in diameter) from shot to shot. This may result in nonuniform charging of pulse-forming lines FL_{41}–FL_{44} and, hence, in unstable operation of pulse formers F_1–F_4. To make the charging more uniform, line FL_3 with the central conductor diameter decreasing from 70 to 24 mm is connected in the charging circuit. The line impedance varies from 6.25 to 12.5 Ω. The axes of the four pulse formers are arranged at diametrically opposite points of a circle of diameter 53 mm. The charging voltage pulse generated by the SINUS-160 passed through line FL_1, and the bipolar pulses formed on operation of spark gaps S_{21}–S_{24} and S_{31}–S_{34} passed through 50-Ω lines FL_{71}–FL_{74} to matched resistive loads or to transmitting antennas.

The output bipolar pulses of voltage dividers D_{31}–D_{34} were recorded with a LeCroy WaveMaster 830Zi oscilloscope and the charging voltages of lines FL_2 and FL_{41}–FL_{44} coming from capacitive voltage dividers D_1 and D_{21}–D_{24}, respectively, with a Tektronix TDS 6604 oscilloscope. Voltage dividers D_1 and D_{31}–D_{34} were calibrated, whereas D_{21}–D_{24} were used as-received only to evaluate the time it took to charge lines FL_{41}–FL_{44}.

Spark gap S_1 was broken down with a 6.9-ns delay at a voltage of 145 kV. By fitting the gap spacing in peaking gaps S_{21}–S_{24} and in chopping gaps S_{31}–S_{34} to a precision of 0.05 mm, symmetrical waveforms were attained for the bipolar pulses produced by pulse formers F_1–F_4. Varying the breakdown delay time of spark gaps S_{21}–S_{24} in the range from 0.5 to 0.9 ns by varying the pressure in the pulse formers in the range from 25 to 35 atm could synchronize the bipolar pulses with respect to voltage zero time or shift the voltage zero time of one pulse former relative to that of another by Δt of up to 300 ps.

The output bipolar pulses received from voltage dividers D_{31}–D_{34} are shown in Figure 9.47 for the following modes: all pulses in the pulse formers are synchronized with respect to voltage zero time (a) and the pulses in pulse formers F_2 and F_3 are delayed in zero voltage time by 300 ps relative to the pulses in pulse formers F_1 and F_4 (b). The pulses have amplitudes ±(50–60) kV and duration 3 ns at a repetition frequency of 100 Hz. The prepulses occurred because the voltage pulses charging pulse-forming lines FL_{41}–FL_{44} passed to the load through the transfer capacitances of the respective spark gaps S_{21}–S_{24}.

With the amplitude jitter of the charge voltage of the SINUS-160 generator $\sigma < 0.01$, the amplitude jitter of the bipolar voltage pulses is no more than $\sigma = 0.05$. With $\Delta t > 300$ ps, the output bipolar voltage pulses became more unstable and their waveforms were distorted. The stability of the output pulses of the pulse formers relative to each other was evaluated by the spread in measurements of the time interval t_1 between the time of the oscilloscope triggering by the trailing edge

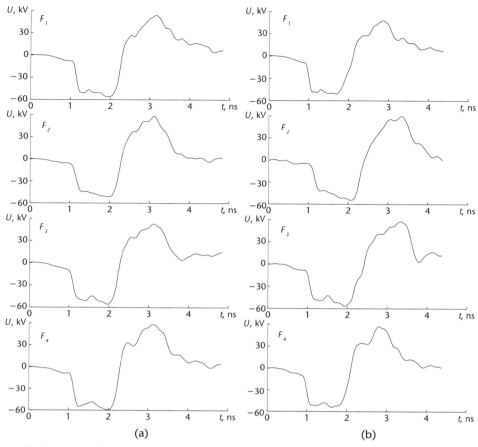

Figure 9.47 Output bipolar pulses of pulse formers F_1–F_4 received from voltage dividers D_{31}–D_{34}, respectively, for the following modes: all pulses in the pulse formers are synchronized with respect to voltage zero time (a) and the pulses in pulse formers F_2 and F_3 are delayed in voltage zero time by 300 ps relative to the pulses in pulse formers F_1 and F_4 (b) (Permission Pleiades Publishing, Ltd.)

of the pulse of one pulse former and the voltage zero time of another one (hold-off time). In the hold-off time histogram given in Figure 9.48, where N is the number of pulses, the RMSD is 50–70 ps. The energy delivered to the intermediate line and to the load in a pulse makes up, respectively, 53% and 14% of the energy stored in pulse-forming line FL_0, equal to 3.2 J. Note that in the radiation source with one pulse former [32], 30% of the stored energy was transferred to the load (antenna–feeder system). The decrease in energy efficiency of the 4-channel BPF is due to the losses in the pulse-peaking section.

The radiating system is similar to that used in the radiation source [32] (Figure 9.21), except for the absence of a four-channel power divider, as the array elements are connected directly to the bipolar pulse formers. For the detection of electromagnetic radiation, a TEM antenna and a Tektronix TDS 6604 oscilloscope were used. The waveform of the radiated pulse (rE) in the far-field region averaged over 100 shots is shown in Figure 9.49.

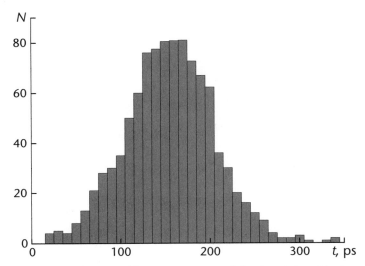

Figure 9.48 Histogram of the spread in measurements of the time interval between the time of the oscilloscope triggering by the trailing edge of the pulse of pulse former F_1 and the voltage zero time of pulse former F_2. (With permission from Pleiades Publishing, Ltd.)

The UWB radiation source was tested for operational life in the mode of one-hour operation and a two-hour break for cooling of the Tesla transformer. The total operation time of the source at a pulse repetition frequency of 100 Hz was more than 5 h without replacement of spark gap electrodes. At the end of the first hour of operation, the output pulses became unstable due to that spark gap S_1 was broken down when the charging voltage already had passed over its maximum. This took place because the gap spacing increased due to electrode erosion and the pressure in the spark gap increased due to the gas heating by the end of one-hour operation. Hooking up an additional ballast gas container of volume 3 l to the case of spark gap S_1 and reducing the pressure by about 2 atm after each hour of operation made it possible to stabilize the operation of the bipolar voltage pulse

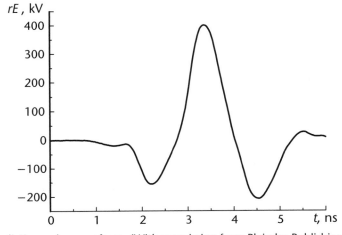

Figure 9.49 Radiation pulse waveform. (With permission from Pleiades Publishing, Ltd.)

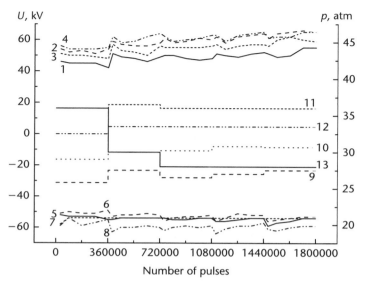

Figure 9.50 The negative (curves 1–4) and the positive bipolar pulse amplitude (curves 5–8) for F_1 (curves 1, 5), F_2 (curves 2, 6), F_3 (curves 3, 7), and F_4 (curves 4, 8), and the changes of pressure in pulse formers F_1 (curve 9), F_2 (curve 10), F_3 (curve 11), F_4 (curve 12) and in spark gap S_1 (curve 13) versus number of pulses. (With permission from Pleiades Publishing, Ltd.)

generator. Figure 9.50 presents the amplitudes of bipolar voltage pulses versus number of pulses for pulse formers F_1–F_4.

In the experiment, the amplitudes of the positive and negative half-waves of the bipolar voltage pulse (Figure 9.50, curves 1–8), and the electromagnetic pulse amplitude (rE_p) and its RMSD σ (Figure 9.51) were measured versus the number of pulses. Figure 9.50 also shows the pressure in the spark gap of the intermediate

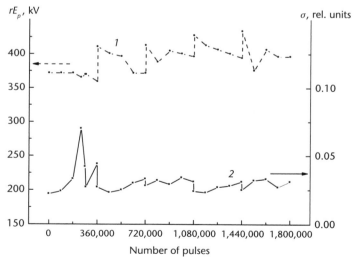

Figure 9.51 Effective radiation potential (curve 1) and its RMSD (curve 2) versus number of pulses. (With permission from Pleiades Publishing, Ltd.)

peaking stage (curve 13) and pulse formers (curves 9–12) versus the number of pulses. In all measurements, averaging was performed, as previously, over 100 pulses.

The average value of rE_p for five-hour operation was 400 kV, and the effective radiation potential decreased by no more than 20% with $\sigma = 0.03$. The peak field efficiency of the UWB radiation source, k_E, was 6.6, which is a little greater than $k_E = 6.2$ of the radiation source with a single pulse former and a 4-channel power divider [32].

Using four independent pulse formers made it possible to investigate the deviation of the main beam from the normal to the array. To do this, the voltage zero times of two pulse formers exciting the vertical row of antennas were shifted by 200 or 300 ps with respect to those of the other two pulse formers. Unfortunately, the size of the anechoic chamber allowed measuring only a radiation pattern fragment of width ≤20°. In addition, a code for calculating the array pattern was used that simulated the summation of radiation pulses in view of their waveforms, the patterns of the elements, and the delays of excitation of the elements in the array. Figure 9.52 presents the H-plane patterns of the array with simultaneously excited elements. The radiation pattern at a low voltage and the method of its measuring are presented elsewhere [32]. It can be seen that the calculated and the experimental pattern are in good agreement.

Figure 9.53(a) presents the H-plane patterns of the array with a 200-ps delay in excitation of two elements. The experimental (curve 1) and the calculated pattern maxima (curve 2) are in agreement, and the deviation angle is 5.5°. However, for the delay in excitation of two elements equal to 300 ps [Figure 9.53(b)], the locations of the maxima of the experimental (curve 1, 12°) and calculated (curve 2, 8°) patterns are different. Perhaps, this is due to the change in the duration of the first half-period of the voltage pulse [Figure 9.47(b)] that was not taken into account in the calculations.

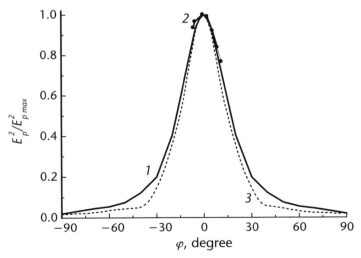

Figure 9.52 The H-plane peak power patterns of the 2×2 array measured at low (curve 1) and high voltage (curve 2), and the calculated pattern (curve 3). (With permission from Pleiades Publishing, Ltd.)

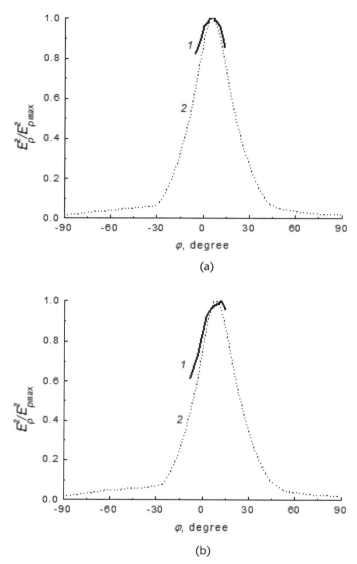

Figure 9.53 The experimental (curve 1) and calculated (curve 2) *H*-plane peak power patterns of the 2 × 2 array with a delay in excitation of two elements equal to 200 (a) and 300 ps (b). (With permission from Pleiades Publishing, Ltd.)

9.7 A Controlled-Spectrum Radiation Source

The appearance of the radiation source [47] is shown in Figure 9.54. The monopolar voltage pulse produced by the SINUS-200 generator (1) is supplied to a nonlinear transmission line (2), and then, it passes through a bandpass filter (3) and a feeder (4) and arrives at a transmitting antenna (5). To increase the electric strength of the source, the antenna is placed in a dielectric container filled with SF$_6$ gas at a pressure

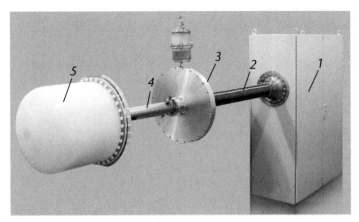

Figure 9.54 Appearance of a source of high-power UWB radiation pulses with a controlled spectrum: 1—SINUS-200 generator, 2—nonlinear transmission line, 3—bandpass filter, 4—feeder, and 5—transmitting antenna. (With permission from Pleiades Publishing, Ltd.)

of 2.4 atm, and the feeder, the bandpass filter, and the nonlinear transmission line are filled with vacuum oil.

The nonlinear transmission line is a uniform coaxial line partially filled with saturated ferrite [48]. The ferrite is saturated in the initial state by the axial field of the solenoid. Once the leading edge of the high-voltage pulse [Figure 9.55(a)] arrives at the line input, the resulted magnetic field in the ferrite starts increasing. This leads to an increase in the group velocity of some portions of the pulse leading edge. The nonlinearity of the group velocity gives rise to the formation of a shock front behind which magnetization precession is excited in the ferrite. As a result, a packet of high-frequency oscillations superimposed on the monopolar pulse is formed at the output of the nonlinear line of characteristic impedance 28 Ω [Figure 9.55(b)].

The power dissipated in the solenoid is about 1 kW, and so there is no need in its forced cooling when operated in a packet mode. The pulse-forming line of the SINUS-200 generator can produce voltage pulses of duration 8 ns at half maximum and 2–3 ns rise time to half amplitude. The voltage amplitude is determined by the pressure in the spark gap operating in the self-breakdown mode. When a 28-Ω load was connected to the generator, the voltage amplitude could be set in the range from 100 to 300 kV in the single-pulse mode. In a packet mode (up to 5 s), the generator is capable of operating at a pulse repetition frequency of up to 200 Hz. A prolonged operational period at a maximum pulse repetition frequency is provided with the voltage pulse amplitude limited by 250 kV. Thus, the limiting power supplied to the nonlinear line was 3.3 GW with energy of up to 26 J. The center frequency of the oscillations excited in the line is over 1.2 GHz at a maximum and 0.5 GHz at a minimum voltage. The duration of the oscillation packet at a level of −3 dB is 3–4 ns throughout the voltage range.

The bandpass filter was used to isolate the high-frequency component of the output voltage pulse. The low-frequency component of the pulse, having passed through the inductor of the bandpass filter, is absorbed in its matched resistor.

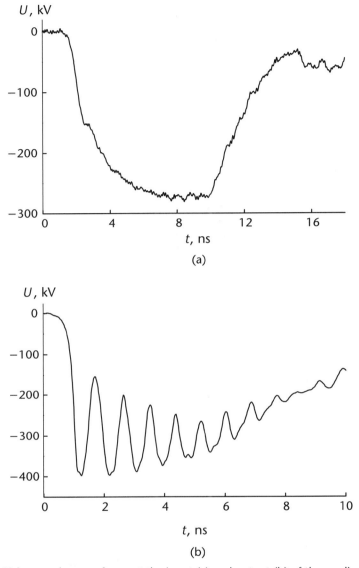

Figure 9.55 Voltage pulse waveforms at the input (a) and output (b) of the nonlinear transmission line at a 270 kV input pulse amplitude and a 50-kA/m bias field. (With permission from Pleiades Publishing, Ltd.)

The high-frequency component [Figure 9.56(a)] passes through the filter capacitor, forming a high-frequency pulse with a maximum energy of 1.6 J and a peak power of 2.8 GW in the 50-Ω antenna feeder (reference 4 in Figure 9.54). Measurements of the parameter S_{21} in the frequency range from 0.5 to 1.3 GHz performed using an Agilent 8719ET Network Analyser show [Figure 9.56(b)] that the value of S_{21} ranges between 0.7 and 0.8, with the exception of a small dip at about 0.9 GHz.

A previously used KA of dimensions $32 \times 30 \times 30$ cm excited by 2-ns high-voltage bipolar voltage pulses [24] was chosen for the base transmitting antenna. The antenna was optimized for VSWR and radiation pattern in a specified frequency

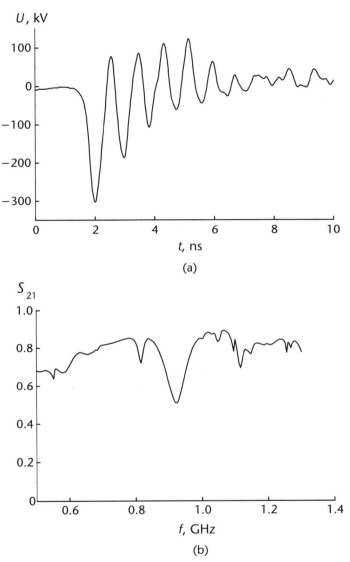

Figure 9.56 Voltage pulse waveform at the output of the matched bandpass filter at a 270-kV input pulse amplitude and a 50-kA/m bias field (a), and the bandpass filter parameter S_{21} versus frequency (b). (With permission from Pleiades Publishing, Ltd.)

range using the 4NEC2 code [49]. Figure 9.57 shows the measured (curve 1) and calculated VSWR (curve 2) versus frequency. The frequency dependence of the K_V for the antenna placed in a container is also shown in Figure 9.57. As can be seen from the figure, the VSWR of the antenna placed in a container (curve 3) is slightly different from that of the antenna located in free space.

The high-power radiation pulses transmitted by the KA were detected in the far-field region with a receiving TEM antenna and recorded with a Tektronix TDS 6604 oscilloscope. The frequency tuning of the radio pulses was performed by varying either the input voltage of the nonlinear transmission line, through a

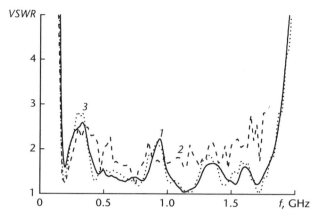

Figure 9.57 Frequency dependence of the VSWR of the combined antenna: 1—experiment, 2—calculation, and 3—experiment for the antenna placed in a dielectric container. (With permission from Pleiades Publishing, Ltd.)

control of the pressure in the spark gap of the SINUS-200 generator, or the bias field H_z.

The waveform and amplitude spectrum $S(f)$ of the radio pulse produced at the lowest input voltage and an optimal bias are shown in Figure 9.58. The radio pulse

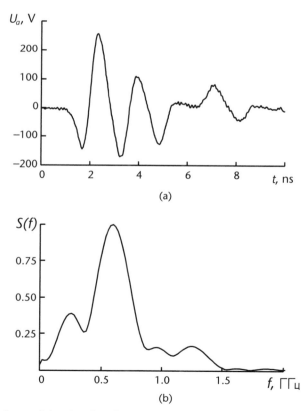

Figure 9.58 Waveform of the signal at the output of the receiving antenna (a) and its amplitude spectrum (b) for a 135-kV input voltage amplitude and a 40-kA/m bias field. (With permission from Pleiades Publishing, Ltd.)

full duration is about 6 ns and the width of its spectrum $S(f)$ at a level of −3 dB is 0.23 GHz with center frequency $f_0 = 0.6$ GHz. Here, the center frequency refers to the frequency corresponding to the radiation spectrum maximum. The relative spectral width at −10 dB level is equal to 0.4, which meets the criterion of UWB radiation [50]. The pulse energy at the input of the nonlinear line is 5.9 J. The energy at the input of the UWB antenna is 0.2 J and the energy of the radiated pulse, estimated from the voltage pulse spectrum at the antenna input and $K_V(f)$ of the antenna (8.7), is 0.15 J. In this case, the energy efficiency of the radiation source was 2.6%.

Figure 9.59 presents the effective radiation potential in the far-field region as a function of bias field H_z. Figure 9.59(b) illustrates the tuning of the radio pulse center frequency by varying the bias field at a minimum voltage at the nonlinear line input. The frequency tuning band at −3 dB is from 0.47 to 0.69 GHz and makes up 19%.

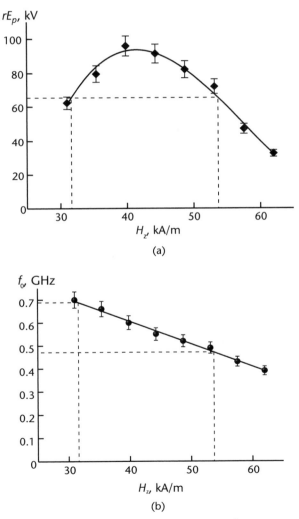

Figure 9.59 The effective radiation potential (a) and the center frequency tuning band at −3 dB (b) versus bias field at a 135-kV amplitude of the nonlinear line input voltage. (With permission from Pleiades Publishing, Ltd.)

Increasing the amplitude of the incident pulse to a maximum value of 300 kV increases the radio pulse full duration to 8 ns [Figure 9.60(a)]. The center frequency increases to 1.15 GHz and the radiation spectrum width at −3 dB to 0.27 GHz [Figure 9.60(b)]. The energy of the voltage pulse at the antenna input is equal to 1.6 J. The energy of the radiated pulse, estimated from the voltage pulse spectrum and $K_V(f)$ of the antenna, is 1.15 J. The energy efficiency of the radiation source is 4.5%, and it can be increased by properly optimizing the source characteristics.

The effective radiation potential at an optimal bias and a maximum voltage reaches a maximum of 310 kV [Figure 9.61(a)]. With the maximum voltage, the center frequency tuning band at −3 dB is from 1.06 to 1.25 GHz and makes up 15% [Figure 9.61(b)].

Figure 9.62 illustrates the center frequency tuning of radio pulses by varying the input voltage amplitude U_0 at an optimum magnetic field (solid line). Dashed lines indicate the limits to which the center frequency can be varied by varying the bias field. It can be seen that the center frequency increases linearly with input voltage. Thus, joint control of voltage and bias field allows tuning the center frequency of the radio pulses produced by the UWB radiation source under consideration from

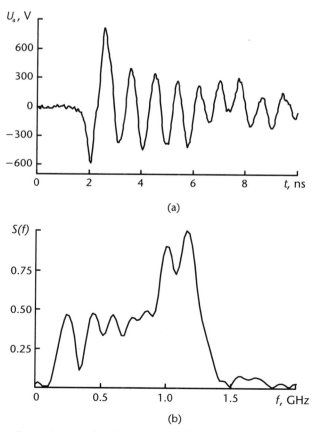

(a)

(b)

Figure 9.60 The waveform of a signal at the output of the receiving antenna (a) and its amplitude spectrum (b) at a 300-kV input voltage amplitude and a 53 kA/m bias field. (With permission from Pleiades Publishing, Ltd.)

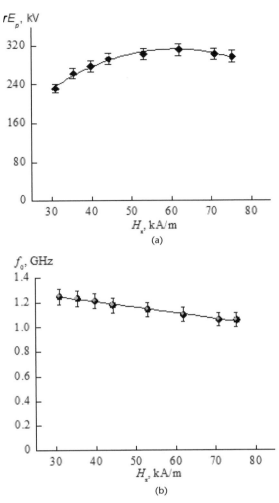

Figure 9.61   The effective radiation potential (a) and the center frequency tuning band at −3 dB (b) versus bias field at a 300-kV amplitude of the nonlinear line input voltage. (With permission from Pleiades Publishing, Ltd.)

0.5 to 1.3 GHz. The radiation source operated at a pulse repetition frequency of 50 Hz has demonstrated stability of the radiation parameters at a 250-kV amplitude of the voltage pulse at the input of the nonlinear line.

Note that a UWB radiation source depending for its operation on the excitation of a nonlinear transmission line by a high-voltage monopolar pulse produced by the SINUS-200 generator that operates with a center frequency of high-frequency oscillations f_0 = 1.2 GHz and a bandwidth of 0.4 GHz at −10 dB is described in detail elsewhere [51]. To produce linearly polarized electromagnetic radiation pulses, a profiled horn antenna of large diameter was used. At the antenna input, a converter was connected to convert the TEM wave of the coaxial line into the TE_{11} mode of a circular waveguide that was subsequently converted into a Gaussian wave beam. The effective radiation potential reached 560 kV at a 250-kV amplitude of the voltage pulse at the input of the nonlinear transmission line. The radiation source operated in a packet mode at a pulse repetition frequency of 200 Hz with 5-s packet duration.

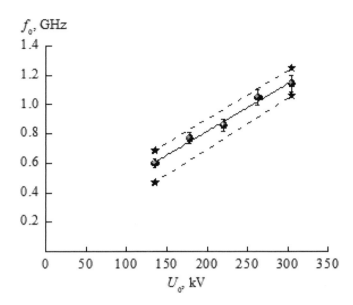

Figure 9.62 Radiation center frequency as a function of the voltage amplitude at the input of the nonlinear line. (With permission from Pleiades Publishing, Ltd.)

Johnson and coworkers [52] report on the creation of a high-power microwave source based on four-element array of TEM antennas, each excited from an independent nonlinear transmission line. The lines are charged from a common voltage pulse generator. The use of gyromagnetic controlled delay line allowed the authors to realize wave beam scanning within ±16.7°. In the experiments, the microwave source produced radiation pulses of effective potential 90 kV at a repetition frequency of 100 Hz.

Conclusion

Investigations have been performed and high-power UWB radiation sources have been developed in which KA arrays are excited by bipolar voltage pulses produced by a single generator. Two circuit designs have been realized for these UWB sources. In the first circuit, the antenna array is excited from a single BPF via a power divider. In the second one, the number of BPFs is equal to the number of KAs, and this circuit offers considerable opportunity to control the UWB radiation characteristics. A feature of this circuit is the use of uncontrolled two-electrode spark gaps in the BPFs.

Based on the first circuit, a series of high-power UWB radiation sources has been created in which a multielement (2×2, 3×3, 4×4, 8×8) array of KAs is excited by a bipolar voltage pulse of duration 0.2–3 ns. With these sources, radiation pulses with effective potential $rE_p = 0.4$–4 MV have been produced at a pulse repetition frequency of 100 Hz. The sources show high stability of the radiation parameters at a total number of pulses of up to 2 million without replacement of the electrodes in the BPF spark gaps.

The above theoretical estimates and experimental results show that the radiators based on KA arrays excited by bipolar voltage pulses have a maximum specific efficiency, defined as the ratio of the effective radiation potential to the array volume or area.

It has been demonstrated that high-power pulse sources with orthogonal polarizations are feasible. Orthogonally polarized radiation pulses can be produced due to the excitation of two perpendicularly oriented KA subarrays by bipolar voltage pulses shifted in time. Using cable feeders of different length makes the radiation source capable of producing a pulse packet with the number of pulses equal to the number of antennas in the array for every generator voltage pulse. The repetition frequency of radiation pulses in a packet is expected to be some orders of magnitude higher than that of the exciting voltage pulses.

It has been shown that the use of KAs offers the possibility to create high-power UWB radiation sources with nonlinear transmission lines excited by high-voltage pulses with the spectral center frequency tunable over wide limits.

Problems

9.1 What is the limiting effective radiation potential? Formulate the requirements for the radiator (single antenna or array) that must be met to achieve it.

9.2 Describe various BPF circuits and identify their advantages and disadvantages.

9.3 By which means the reproducibility of the bipolar voltage pulse at the output of a BPF with an open line is achieved?

9.4 What limits the operational life of the spark gaps in a BPF and what is done to increase it?

9.5 Depict the block diagram of a UWB source radiating a packet of electromagnetic pulses with a pulse repetition frequency of 1 GHz in one shot.

9.6 What parameters determine the center frequency tuning of the radiation produced by a source with a nonlinear transmission line, and how the tuning affects the radiated power?

References

[1]   Mesyats, G. A., *Pulsed Power*, New York: Kluver Academic/Plenum Publishers, 2004.

[2]   Giri, D. V., *High-Power Electromagnetic Radiators: Nonlethal Weapons and Other Applications*, Cambridge: Harvard university press, 2004.

[3]   Benford, J., J. A. Swegle, and E. Schamiloglu, *High Power Microwaves. Second edition.* New York: Taylor & Francis, 2007.

[4]   Agee, F. J., et al., "Ultra-Wideband Transmitter Research," *IEEE Trans. Plasma Sci.*, Vol. 26, No. 3, 1998, pp. 860–872.

[5]   Prather, W. D., et al., "Survey of Worldwide High-Power Wideband Capabilities," *IEEE Trans. Electromagn. Compat.*, Vol. 46, No. 3, 2004, pp. 335–344.

[6]   Koshelev, V. I., "Antenna Systems for Radiation of High-Power Ultrawideband Pulses," *Proc. 3rd All-Russian Scientific and Technical Conference on Radar and Radio Communications*, Moscow, Oct. 26–30, 2009, Vol. 1, pp. 33–37 (in Russian).

[7] Vvedensky, Yu. V., "Nanosecond Pulse Thyratron Generator with a Universal Output," *Izv. Vyssh. Uchebn. Zaved., Radiotekh.*, No. 2, 1959, pp. 249–251.

[8] Auslender, V. L., Il'in O. G., and Shenderovich A. M., "Formation of Current Pulses of Controlled Duration," *Prib. Tekh. Eksperim.*, No. 3, 1962, pp. 81–83.

[9] Auslender, V. L., Il'in O. G., and Shenderovich A. M., "Formation of Pulses across a Varied Load," *Prib. Tekh. Eksperim.*, No. 2, 1963, pp. 173–174.

[10] Koshelev, V. I., et al., "High-Power Ultrawideband Electromagnetic Pulse Radiation," *Proc. SPIE*, Vol. 3158, 1997, pp. 209–219.

[11] Andreev, Yu. A., et al., "A High-Power Ultrawideband Electromagnetic Pulse Generator," *Instrum. Exp. Tech.*, Vol. 40, No. 5, 1997, pp. 651–655.

[12] Andreev, Yu. A., et al., "High-Power Ultrawideband Electromagnetic Pulse Radiation," *Proc. 11 IEEE Inter. Pulsed Power Conf.*, Baltimore, June 29–July 2, 1997, Vol. 1, pp. 730–735.

[13] Andreev, Yu. A., et al., "Gigawatt-Power-Level Ultrawideband Radiation Generator," *Proc. 12 IEEE Inter. Pulsed Power Conf.*, Monterey, June 27–30, 1999, Vol. 2, pp. 1337–1340.

[14] Andreev, Yu. A., et al., "An Ultrawideband Gigawatt Pulse Generator," *Instrum. Exp. Tech.*, Vol. 43, No. 2, 2000, pp. 224–229.

[15] Andreev, Yu. A., Yu. I. Buyanov, and V. I. Koshelev, "Small-Sized Ultrawideband Antennas Radiating High-Power Electromagnetic Pulses," *Zh. Radioelektron.*, No. 4, 2006: http://jre.cplire.ru/mac/apr06/1/text.html

[16] Koshelev, V. I., et al., "Ultrawideband Radiators of High-Power Pulses," *IEEE Pulsed Power Plasma Science Conf.*, Las Vegas, June 17–22, 2001, Vol. 2, pp. 1661–1664.

[17] Andreev, Yu. A., et al., "High-Power Ultrawideband Radiation Source," *Laser Part. Beams*, Vol. 21, No. 2, 2003, pp. 211–217.

[18] Pozar, D. M., D. H. Schaubert, and R. E. McIntosh, "The Optimum Transient Radiation from an Arbitrary Antenna," *IEEE Trans. Antennas Propagat.*, Vol. 32, No. 6, 1984, pp. 633–640.

[19] Belichenko, V. P., et al., "Estimation of an Utmost Efficient Potential of Ultrawideband Radiating System," *Proc. 14th Inter. Symposium on High Current Electronics*, Tomsk, Sept. 10–15, 2006, pp. 391–394.

[20] Harrington, R. F., "Effect of Antenna Size on Gain, Bandwidth, and Efficiency," *J. Res. Nat. Bur. Stand.*, Vol. 64D, No. 1, 1960, pp. 1–12.

[21] Giri, D. V., et al., "Design, Fabrication, and Testing of a Paraboloidal Reflector Antenna and Pulser System for Impulse-Like Waveforms," *IEEE Trans. Plasma Sci.*, Vol. 25, No. 2, 1997, pp. 318–326.

[22] Yankelevich, Y., and A. Pokryvailo, "A Compact Former of High-Power Bipolar Subnanosecond Pulses," *IEEE Trans. Plasma Sci.*, Vol. 33, No. 4, 2005, pp. 1186–1191.

[23] Gubanov, V. P., et al., "Sources of High-Power Ultrawideband Radiation Pulses with a Single Antenna and a Multielement Array," *Instrum. Exp. Tech.*, Vol. 48, No. 3, 2005, pp. 312–320.

[24] Efremov, A. M., et al., "Generation and Radiation of Ultra-Wideband Electromagnetic Pulses with High Stability and Effective Potential," *Laser Part. Beams*, Vol. 32, No. 3, 2014, pp. 413–418.

[25] Mesyats, G. A., et al., "Repetitively Pulsed High-Current Accelerators with Transformer Charging of Forming Lines," *Laser Part. Beams*, Vol. 21, No. 2, 2003, pp. 197–209.

[26] Efremov, A. M., et al., "Generation and Radiation of High-Power Ultrawideband Nanosecond Pulses," *J. Commun. Technol. Electron.*, Vol. 52, No. 7, 2007, pp. 756–764.

[27] Koshelev, V. I., et al., "Study on Stability and Efficiency of High-Power Ultrawideband Radiation Source," *J. Energy Power Eng.*, Vol. 5, No. 6, 2012, pp. 771–776.

[28] Efremov, A. M., and B. M. Kovalchuk, "A Subnanosecond Voltage Divider on Coupled Lines," *Instrum. Exp. Tech.*, Vol. 47, No. 1, 2004, pp. 61–62.

[29] Efremov, A. M., et al., "High-Power Sources of Ultra-Wideband Radiation with Subnanosecond Pulse Lengths," *Instrum. Exp. Tech.*, Vol. 54, No. 1, 2011, pp. 70–76.

[30] Andreev, Yu. A., et al., "Generation and Emission of High-Power Ultrabroadband Picosecond Pulses," *J. Commun. Technol. Electron.*, Vol. 56, No. 12, 2011, pp. 1429–1439.

[31] Koshelev, V. I., et al., "High-Power Source of Ultrawideband Radiation Wave Beams with High Directivity," *Proc. 15th Inter. Symposium on High Current Electronics*, Tomsk, Sep. 21–26, 2008, pp. 383–386.

[32] Andreev, Yu. A., et al., "A High-Performance Source of High-Power Nanosecond Ultrawideband Radiation Pulses," *Instrum. Exp. Tech.*, Vol. 54, No. 6, 2011, pp. 794–802.

[33] Andreev, Yu. A., Yu. I. Buyanov, and V. I. Koshelev, "A Combined Antenna with Extended Bandwidth," *J. Commun. Technol. Electron.*, Vol. 50, No. 5, 2005, pp. 535–543.

[34] Andreev, Yu. A., V. I. Koshelev, and V. V. Plisko., "Characteristics of Receiving-Transmitting TEM Antennas," *Proc. 5th Scientific and Technical Conf. Radar and Radio Communications*, Moscow, Nov. 21–25, 2011, pp. 77–82 (in Russian).

[35] Gubanov, V. P., et al., "A Source of High-Power Pulses of Ultrawideband Radiation with a Nine-Element Array of Combined Antennas" *Instrum. Exp. Tech.*, Vol. 60, No. 2, 2017, pp. 213–218.

[36] Baum, C. E., et al., "JOLT: A Highly Directive, Very Intensive, Impulse-Like Radiator," *Proc. IEEE*, Vol. 92, No. 7, 2004, pp. 1096–1109.

[37] Fedorov, V. M., et al., "Active Antennas Array with Control and Stabilization of Regimes of Synchronizing for UWB Video-Pulses," *Proc. 14th Inter. Symposium on High Current Electronics*, Tomsk, Sep. 10–15, 2006, pp. 405–408.

[38] Fedorov, V. M., et al., "Ultra-Wideband Sub-Nanosecond High Power Radiators," *Proc. 15th Inter. Symposium on High Current Electronics*, Tomsk, Sep. 21–26, 2008, pp. 403–406.

[39] Koshelev, V. I., V. V. Plisko, and K. N. Sukhushin, "Array Antenna for Directed Radiation of High-Power Ultra-Wideband Pulses," In *Ultra-Wideband, Short-Pulse Electromagnetics 9*, pp. 259–267, F. Sabath, et al. (eds.), New York: Springer, 2010.

[40] Koshelev, V. I., et al., "Detection of Metal Objects on the Medium with a Random Inhomogeneous Surface in Ultrawideband Pulses Probing," *Proc. 5th Scientific and Technical Conf. Radar and Radio Communications*, Moscow, Nov. 21–25, 2011, pp. 87–92 (in Russian).

[41] Koshelev, V. I., A. A. Petkun, and V. M. Tarnovsky, "Influence of Properties of the Medium with a Random Inhomogeneous Surface and Geometry of a Receiving Antenna Array on Detection of Metal Objects," *Izv. Vyssh. Uchebn. Zaved., Fiz.*, Vol. 56, No. 8/2, 2013, pp. 159–163.

[42] Koshelev, V. I., et al., "Increasing Stability and Efficiency of High-Power Radiation Source," *Proc. 16th Inter. Symposium on High Current Electronics*, Tomsk, Sep. 19–24, 2010, pp. 415–418.

[43] Koshelev, V. I., et al., "High-Power Source of Ultrawideband Radiation with Wave Beam Steering," *Izv. Vyssh. Uchebn. Zaved., Fiz.*, Vol. 55, No. 10/3, 2012, pp. 217–220.

[44] Efremov, A. M., et al., "A Four-Channel Source of High-Power Pulses of Ultrawideband Radiation," *Instrum. Exp. Tech.*, 2013. Vol. 56, No. 3, 2013, pp. 301–308.

[45] Andreev, Yu. A., et al., "Multichannel Antenna System for Radiation of High-Power Ultrawideband Pulses," In *Ultra-Wideband, Short-Pulse Electromagnetics 4*, pp. 181–186, E. Heyman, B. Mandelbaum, and J. Shiloh (eds.), New York: Plenum Press, 1999.

[46] Koshelev, V. I., V. V. Plisko, and E. A. Sevostyanov., "Synthesis of Ultrawideband Radiation of the Array of Combined Antennas Excited by Bipolar Pulses of Different Length," *Izv. Vyssh. Uchebn. Zaved., Fiz.*, Vol. 58, No. 8/3, 2015, pp. 54–58.

[47] Andreev, Yu. A., et al., "Generation and Radiation of High-Power Ultrawideband Pulses with Controlled Spectrum," *J. Commun. Technol. Electron.*, Vol. 58, No. 4, 2013, pp. 297–306.

[48] Gubanov, V. P., et al., "Effective Transformation of the Energy of High-Voltage Pulses into High-Frequency Oscillations Using a Saturated-Ferrite-Loaded Transmission Line," *Tech. Phys. Lett.*, Vol. 35, No. 7, 2009, pp. 626–628.

[49] NEC Based Antenna Modeler and Optimizer: http://www.qsl.net/4nec2

[50] Federal Communication Commission USA (FCC) 02-48, ET Docket 98-153, First Report and Order, April 2002.

[51] Romanchenko, I. V., et al., "Repetitive Sub-Gigawatt RF Source Based on Gyromagnetic Nonlinear Transmission Line," *Rev. Sci. Instrum.*, Vol. 83, No. 7, 2012, p. 074705.

[52] Johnson, J. M., et al., "Characteristics of a Four Element Gyromagnetic Nonlinear Transmission Line Array High Power Microwave Source," *Rev. Sci. Instrum.*, Vol. 87, No. 5, 2016, p. 054704.

General Symbols

$\mathbf{A}^e$	electric vector potential
$\mathbf{A}^m$	magnetic vector potential
A_e	effective area of a receiving antenna
$\mathbf{B}$	magnetic induction
$b = f_H/f_L$	ratio bandwidth
C	capacitance; capacity of a communications channel
C'	capacitance per unit length
c	velocity of propagation of electromagnetic waves in a medium, in particular in vacuum
$\mathbf{D}$	electric induction
D	diameter; the maximum transverse dimension of an antenna array
D_0	boresight directivity of an antenna
d	spacing between neighboring array elements
$\mathbf{E}$	electric field strength
$\mathbf{E}(\mathbf{r},t)$	electric field strength at a point $\mathbf{r}$ at a time t
$\mathbf{E}(\mathbf{r},\omega)$	electric field strength at a point $\mathbf{r}$ at an angular frequency ω
E_p	peak electric field strength
rE_p	effective radiation potential
F	probability of false alarm
$F(\theta,\varphi)$	normalized antenna pattern function
$f(\theta,\varphi)$	amplitude antenna pattern function
f	frequency
f_H, f_L	the higher and the lower frequency of a radiation pulse spectrum at a level of -10 dB
f_0	central frequency of a radiation pulse spectrum
Δf	radiation pulse spectrum width

G	gain of a transmitting antenna
$g_n(t,\theta,\varphi)$	genetic function
$\mathbf{H}$	magnetic field strength
$\mathbf{H}_i$	incident magnetic field strength
$\mathbf{H}(\mathbf{r},t)$	magnetic field strength at a point $\mathbf{r}$ at a time t
$\mathbf{H}(\mathbf{r},\omega)$	magnetic field strength at a point $\mathbf{r}$ at an angular frequency ω
$\mathbf{H}_a(i\omega)$	antenna transfer function
$|\mathbf{H}_a(\omega)|$	antenna amplitude-frequency response
$h(t)$	impulse response
$\mathbf{h}(t)$	polarization matrix of impulse responses
h	height
I, I^e	electric current
I^m	magnetic current
I_e, I_o	in-phase and antiphase current components
$\mathbf{j}_e$	electric current density
$\mathbf{J}_e$,$\mathbf{J}$	electric surface current density
$\mathbf{k}$	wave vector
k	wave number for a lossless medium
k_w	energy efficiency of an antenna
k_p	peak power efficiency of an antenna
k_E	peak field strength efficiency of an antenna
K_V	voltage standing wave ratio
K_Σ	reradiation factor of a receiving antenna
K_u	voltage gain
L	length, inductance
L'	inductance per unit length
l_e	effective length of a receiving antenna
$\mathbf{m}$	magnetic dipole moment
N	number
N_0	noise power spectral density
n	refraction index of a medium
$\mathbf{p}$	electric dipole moment

$\mathbf{p}_a(\theta,\varphi)$	polarization characteristic of an antenna
P	power; probability of detection
P_Σ	radiation power
P_S, P_N	signal and noise power
p	gas pressure in a spark gap
q	charge; signal-to-noise ratio
Q	antenna quality factor
r	distance
Γ, R	reflection coefficient
R_Σ	radiation resistance
$\mathbf{S}$	Poynting vector
S	area
$s(t)$	radiation pulse waveform
$S(\omega)$	radiation pulse spectral function
$|S(\omega)|^2$	radiation pulse energy spectrum
T	pulse repetition period, oscillation period
T_S	signal duration
t	time
$U(t), U_g(t)$	voltage pulse of a generator
$U_a(t)$	voltage at the output of a receiving antenna
v	propagation velocity of a current wave
V	volume
W	energy
W_S	energy of a signal
$W(u)$	voltage distribution function
w^e	electric field energy density
w^m	magnetic field energy density
Z_0	impedance of a medium
$Z_a = R_a + iX_a$	impedance of an antenna
Z_L	impedance of a load
α	decay coefficient; angle between vectors
β	phase coefficient

γ	propagation constant of an electromagnetic wave
ε	electric permittivity of a medium, electromotive force
$\varepsilon_r = \varepsilon/\varepsilon_0$	relative electric permittivity
ε_0	electric permittivity of vacuum
η	fractional bandwidth; object shape reconstruction accuracy
λ	wavelength
λ_0	central wavelength of a pulse spectrum
μ	magnetic permeability of a medium
$\mu_r = \mu/\mu_0$	relative magnetic permeability
μ_0	magnetic permeability of vacuum
$\mathbf{\Pi}^e$	Hertz electric potential
$\mathbf{\Pi}^m$	Hertz magnetic potential
ρ, ρ_a, ρ_f	line, antenna, and feeder characteristic impedance
ρ^e	electric charge density
σ	radar cross section of an object; root mean square deviation
σ^e	electric conductivity of a medium
σ_N	noise dispersion
τ_p	pulse duration
φ^e	scalar potential
$\Phi_a(\omega)$	phase-frequency response of an antenna
$\psi(\theta,\varphi)$	phase pattern of an antenna
$\omega = 2\pi f$	angular frequency

List of Main Abbreviations

AA	active antenna, antenna array
AE	active element
AFR	amplitude-frequency response
AM	antenna module
BPF	bipolar pulse former
CS	complex spectrum
EMC	electromagnetic compatibility
EMF	electromotive force
EMP	electromagnetic pulse
FET	field effect transistor
FFT	fast Fourier transformation
FWHM	full width at half maximum
GF	genetic function
IED	improvised explosive device
IoT	internet of things
IPCP	interperiod correlation processing
IR	impulse response
IRA	impulse radiating antenna
ITS	intelligent transport systems
KA	combined antenna
LC	inductor-capacitor circuit
LFM	linear frequency-modulated
M2M	machine-to-machine
MIMO	multiple input multiple output
MISO	multiple input single output
PFR	phase-frequency response

POD	probability of detection
PS	polarization structure
RCS	radar cross section
RF	radio frequency
RFID	radio frequency identification
RMSD	root mean square deviation
RTLS	real-time location system
SIMO	single input multiple output
SNR	signal-to-noise ratio
SWA	standing-wave antenna
TE	transverse electric
TM	transverse magnetic
TEM	transverse electromagnetic
TSA	tapered slot antenna
TT	Tesla transformer
TWA	traveling-wave antenna
UWB	ultrawideband
VRA	vector receiving antenna
VSWR	voltage standing wave ratio

About the Authors

Vladimir I. Koshelev graduated with honors Tomsk Polytechnic Institute as a specialist in Physical Electronics in 1971. Since 1971, he has been with the Institute of High Current Electronics, SB, RAS (formerly the High Current Electronics Department of the Institute of Atmospheric Optics, SB, USSR AS). Now he is Head of the Microwave Electronics Laboratory of the institute, DSc. (Physics & Maths) (1991), Professor (2006). In addition to his duties at the institute, he worked as Professor at Tomsk State University (1998–2006) and Tomsk Polytechnic University (since 2007). His research interests include High Power Microwaves and Ultrawideband Electromagnetics. He authored and coauthored over 250 scientific publications. V. I. Koshelev is a member of the URSI Russian National Committee. He is a recipient of the Prize of the Siberian Branch of the USSR Academy of Sciences (1986) and the second-class medal of the Order of Merit for the Motherland (1999).

Yury I. Buyanov graduated from Tomsk State University as a specialist in radiophysics and electronics in 1966. Since 1968, he has been with the Radiophysics Department of Tomsk State University, being currently Associate Professor. Since 1995, he is a part-time employee as Senior Researcher of the Microwave Electronics Laboratory of the Institute of High Current Electronics, SB, RAS. He received his PhD. (Physics & Maths) degree in 1980. He authored 18 inventions and over 120 scientific publications. His research interests include active receiving antennas, ultrawideband antennas, and antennas with enhances functionalities.

Victor P. Belichenko received his MS., PhD., and DSc. degrees in Radiophysics from Tomsk State University, Russia, in 1970, 1980, and 2010, respectively. Since 1973, he has been with the Siberian Physical-Technical Institute of Tomsk State University. Since 1993, he has been with the Radiophysics Department of Tomsk State University, where he is currently Professor. In addition to his duties at the university, since 1995, he is Senior Researcher of the Microwave Electronics Laboratory of the Institute of High Current Electronics, SB, RAS. His research interests include mathematical methods in the excitation, propagation, and scattering of electromagnetic waves, ultrawideband electromagnetics, antennas, and near-field microwave microscopy. Professor Belichenko was one of the initiators and Technical Committee Chairman of the International Conference Series on Actual Problems in Radiophysics (APR).

Index

Understanding Electromagnetic Scattering Using the Moment Method: A Practical Approach, Randy Bancroft

Wavelet Applications in Engineering Electromagnetics, Tapan Sarkar, Magdalena Salazar Palma, and Michael C. Wicks

For further information on these and other Artech House titles, including previously considered out-of-print books now available through our In-Print-Forever® (IPF®) program, contact:

Artech House
685 Canton Street
Norwood, MA 02062
Phone: 781-769-9750
Fax: 781-769-6334
e-mail: artech@artechhouse.com

Artech House
16 Sussex Street
London SW1V HRW UK
Phone: +44 (0)20 7596-8750
Fax: +44 (0)20 7630 0166
e-mail: artech-uk@artechhouse.com

Find us on the World Wide Web at: www.artechhouse.com